Springer Tracts in Electrical and Electronics Engineering

Springer Tracts in Electrical and Electronics Engineering (STEEE) publishes the latest developments in Electrical and Electronics Engineering - quickly, informally and with high quality. The intent is to cover all the main branches of electrical and electronics engineering, both theoretical and applied, including:

- Signal, Speech and Image Processing
- Speech and Audio Processing
- Image Processing
- Human-Machine Interfaces
- Digital and Analog Signal Processing
- Microwaves, RF Engineering and Optical Communications
- Electronics and Microelectronics, Instrumentation
- Electronic Circuits and Systems
- Embedded Systems
- Electronics Design and Verification
- Cyber-Physical Systems
- Electrical Power Engineering
- Power Electronics
- Photovoltaics
- Energy Grids and Networks
- Electrical Machines
- Control, Robotics, Automation
- Robotic Engineering
- Mechatronics
- Control and Systems Theory
- Automation
- Communications Engineering, Networks
- Wireless and Mobile Communication
- Internet of Things
- Computer Networks

Within the scope of the series are monographs, professional books or graduate textbooks, edited volumes as well as outstanding PhD theses and books purposely devoted to support education in electrical and electronics engineering at graduate and post-graduate levels.

Review Process

The proposal for each volume is reviewed by the main editor and/or the advisory board. The books of this series are reviewed in a single blind peer review process.

Ethics Statement for this series can be found in the Springer standard guidelines here https://www.springer.com/us/authors-editors/journal-author/journal-author-helpdesk/before-you-start/before-you-start/1330#c14214

Muhammad Suleman Tahir •
Muhammad Bilal Tahir • Muhammad Sagir •
Abdullah Mohamed Asiri
Editors

Lithium-Sulfur Batteries: Key Parameters, Recent Advances, Challenges and Applications

Springer

Editors
Muhammad Suleman Tahir
Khwaja Fareed University of Engineering and Information technology
Rahim Yar Khan, Pakistan

Muhammad Bilal Tahir
Instiitute of Physics
Khwaja Fareed University of Engineering and Information technology
Rahim Yar Khan, Pakistan

Muhammad Sagir
Institute of Chemical and Environmental Engineering
Khwaja Fareed University of Engineering
Rahim Yar Khan, Pakistan

Abdullah Mohamed Asiri
Chemistry Department
King Abdulaziz University
Jeddah, Saudi Arabia

ISSN 2731-4200 ISSN 2731-4219 (electronic)
Springer Tracts in Electrical and Electronics Engineering
ISBN 978-981-99-2795-1 ISBN 978-981-99-2796-8 (eBook)
https://doi.org/10.1007/978-981-99-2796-8

This Springer imprint is published by the registered company Springer Nature Singapore Pte Ltd.
The registered company address is: 152 Beach Road, #21-01/04 Gateway East, Singapore 189721, Singapore

Preface

Muhammad Suleman Tahir, Muhammad Bilal Tahir, Muhammad Sagir, and Abdullah Mohamed Asiri

Lithium-sulfur batteries have emerged as a promising alternative to traditional lithium-ion batteries due to their higher energy density and lower cost. This technology has the potential to revolutionize the way we store and use energy in a variety of applications, from electric vehicles to portable electronics. This book aims to provide a comprehensive overview of the current state of lithium-sulfur battery technology, including the materials and methods used in their production, the challenges facing their widespread adoption, and the potential solutions to these challenges. We begin by discussing the basic principles of lithium-sulfur batteries and the materials used in their construction. We then delve into the various methods used to improve the performance and stability of these batteries, including the use of advanced materials, advanced electrolytes, and advanced designs. We also examine the challenges facing the widespread adoption of lithium-sulfur batteries, such as the limited cycle life and the high cost of materials. Finally, we explore potential solutions to these challenges and discuss the future of lithium-sulfur battery technology. This book is intended for researchers, engineers, and students in the field of energy storage, as well as anyone with an interest in the future of lithium-sulfur batteries. It is hoped that the information and insights provided in this book will help advance the development and deployment of this important technology.

Despite several advantages such as high theoretical capacity, low-cost, and stability, lithium-sulfur batteries face many challenges. Some of the challenges are low electrical conductivity, the possible reaction of sulfur with lithium to form a soluble lithium salt, the formation of the dendrimer, the large volume variation of cathode materials during the electrochemical reaction, and the shuttle behavior of high soluble intermediate polysulfides in the electrolyte. This book provides some possible solutions to these issues through novel architecture, using composite materials, doping to improve low conductivity, etc. This book emphasizes novel materials, architectural concepts, and methods to improve the performance of lithium-sulfur batteries.

Acknowledgments

The journey to complete this book on lithium-sulfur batteries has been a long and challenging one, and we would like to extend our sincere gratitude to all those who have supported us along the way. We would like to thank our colleagues and peers in the research community for their valuable contributions to the field of lithium-sulfur batteries. We have been fortunate to have had the opportunity to learn from some of the leading experts in the field, and we are grateful for their guidance and support. We would also like to acknowledge the support of our families, who have been incredibly understanding and supportive throughout the writing process. Finally, we would like to thank the publisher and its team for their invaluable support and dedication to bringing this book to fruition. Without the contributions and support of these individuals and organizations, this book would not have been possible. We are deeply grateful for their help and support.

I would like to express my deepest gratitude to all those who have supported me throughout the writing of this book, "Lithium-Sulfur Batteries." First and foremost, I would like to thank my mentor beloved father ALLAH DITTA TAHIR, for their guidance and support throughout the research and writing process. Their expertise and encouragement were invaluable in the development of this book. I would also like to thank my family, my colleagues, and friends, for their valuable feedback on the manuscript and for their support during the challenging times. I am also grateful to the staff of editor, reviewer, and publisher for their support and assistance in bringing this book to publication. I would also like to acknowledge the contributions of the many researchers and scientists in the field of lithium-sulfur batteries, whose work has been the foundation of this book. Finally, I would like to thank my family for their unwavering support and understanding during the time it took to complete this book. This book would not have been possible without the support and encouragement of all of these individuals, and I am truly grateful for their contributions.

Contents

About the Editors

Muhammad Suleman Tahir was born in 1971 in Pakistan. He graduated (BSc Chemical Engineering) from the Department of Chemical Engineering, University of Punjab, Pakistan, in 1996. He worked with polymer industry (PET, manufacturing), and he worked with polymer industry (PET, manufacturing) and chloroalkali industry from 1996 to 1999 as chemical engineer (shift engineer). He completed MS chemical engineering in 2000 and joined academia at NFC, Institute of Engineering and Research. He won the national scholarship for PhD and did his PhD and postdoc from Technical University Graz, Austria, 2008–2011, in the topic "Application of Corona Discharge in Waste Treatment (Advanced Oxidation)." He joined KFUEIT from his previous assignment as Dean of Engineering Faculty, University of Gujrat, and brings with him a vast experience of 23 years of public and private sector. While serving in industry and academia, he has developed institutions like NFC Faisalabad and University of Gujrat. Besides being a Postdoc and PhD in Chemical Engineering from the Graz University of Technology Austria, he is a renowned national and international speaker. He is an author of various books and book chapters approved by HEC and several national and international research articles. His enriched experience of developmental projects includes completion of 15 construction projects at the University of Gujrat worth PKR 2 billion, 3 community training projects worth PKR 35 million under umbrella of NAVTTC, and 5 research projects worth PKR 28 million from HEC. He is a strong advocate of academia and industry linkage with proven and tangible gains for academia from the Chamber of Commerce Sialkot and other stakeholders of society. Prof. Dr. Muhammad Suleman is an innovative team leader and motivator with strong team oriented management skills that promote team's personal development and productivity with KPIs.

Muhammad Bilal Tahir has PhD in Physics with specialization in energy and environmental applications, material science and nanotechnology, environmental engineering, photocatalysis, etc. He is currently working as Director of Institute of Physics and Director of Centre for Innovative Material Research, KFUEIT, RYK, and Additional Director ORIC, KFUEIT, RYK Pakistan. He has published more than 500 papers on the topics of enhanced oil recovery (EOR), renewable energy, alternative fuels, waste-to-energy, and catalytic and resource recovery. He has delivered over 100 oral/invited talks to various national and international forums. His work is cited over 20000 times in the peer-review press. Dr. Tahir is serving as guest editor in several special issues *Current Analytical Chemistry*, *Frontiers in Energy Research*, and numerous others. He has completed many national and international funded research projects. He has been actively involved in community services and consultation to various international organizations for providing sustainable solutions for energy and waste-related issues.

Muhammad Sagir has over 18 years of research experience in the field of oil and gas, enhanced oil recovery, surfactants, and photocatalysis. In this period, he has been working in Malaysia and Pakistan. He is currently working as an assistant professor at the Department of Chemical Engineering, University of Gujrat, and is involved in four different research projects as PI, Co-PI, and collaborator. He has also been part of a number of research projects completed in various capacities. He has supervised MPhil student thesis, and a number of MPhil students are currently under his supervision. He has published 70 journal papers in international peer-reviewed journals (ISI Indexed) with high impact factors and over 40 papers/abstracts in international conferences. He has a total impact factor of 200, and his research work has been cited over 1500 times in peer-reviewed press.

Abdullah Mohamed Asiri is the head of the Chemistry Department at King Abdulaziz University since October 2009, and he is the founder and director of the Center of Excellence for Advanced Materials Research (CEAMR) since 2010 till now. He is a professor of Organic Photochemistry.

He graduated from King Abdulaziz University (KAU) with a BSc in Chemistry in 1990 and a PhD from University of Wales, College of Cardiff, UK, in 1995. He was promoted to a professor in 2004. His research interest covers color chemistry, synthesis of novel photochromic, thermochromic systems, synthesis of novel coloring matters and dyeing of textiles, materials chemistry, nanochemistry and nanotechnology, polymers, and plastics.

He is the principal supervisor of more than 17 Msc and 2 PhD thesis. He is the main author of ten books in different chemistry disciplines. He has published more than 400 research papers in international journals. He is the editor-in-chief of *King Abdulaziz University Journal of Science*. A major achievement of Prof. Asiri is the discovery of tribochromic compounds, a class of compounds that change from slightly or colorless to deep colored when subjected to small pressure or when grind. This discovery was introduced to the scientific community as a new terminology published by IUPAC in 2000. This discovery was awarded a patent from European and UK patent offices.

Prof. Asiri involved in many committees at the KAU level and also on the national level; he took a major role in the Advanced Materials Committee working for KACST to identify the national plan for science and technology in 2007.

Prof. Asiri played a major role in advancing the chemistry education and research in KAU, and he has been awarded the best researchers from KAU for the past 4 years. He also awarded the Young Scientist Award from the Saudi Chemical Society in 2009 and also the First Prize for the distinction in science from the Saudi Chemical Society in 2012. He also received a recognition certificate from the American Chemical Society (Gulf Region Chapter) for the advancement of chemical science in the Kingdome. Also he received a Scopus Certificate for the most publishing scientist in Saudi Arabia in chemistry in 2008.

He is also a member of the editorial board of *Pigments and Resin Technology* (UK), *Organic Chemistry in Sight* (New Zealand), *Designed Monomers and Polymers*, and *Journal of Single Molecule Research*. He is the vice-president of Saudi Chemical Society (Western Province Branch).

Chapter 1
Introduction to Electrochemical Energy Storage Technologies

Junaid Ahmad, Zia Ur Rehman, Muhammad Bilal, and Faheem K. Butt

Abstract Energy storage and conversion technologies depending upon sustainable energy sources have gained much attention due to continuous increasing demand of energy for social and economic growth. Electrochemical energy storage (EES) technologies, especially secondary batteries and electrochemical capacitors (ECs), are considered as potential technologies which have been successfully utilized in electronic devices, immobilized storage gadgets, and pure and hybrid electrical vehicles effectively due to their features, like remarkable energy and power density, excellent round-trip ability, long life cycle, cheap, and environment friendly. Among the developed batteries, Li-ion batteries are widely used at a large scale. Among secondary batteries, Li-ion, lithium-sulfur, and sodium-ion batteries have gained much attention of researchers across the globe and could deliver large-scale electric energy in the future. This chapter describes a short introduction to energy storage mechanisms and different types of EES devices.

Keywords Energy storage · Electrochemical capacitors · Lithium-sulfur · Sustainable energy

Sustainable Energy Conversion and Storage

Owing to the rapid increase in demand of energy and social evolution, sustainable energy conversion and storage now has been shifted towards renewable sources of energy such as solar, tidal, hydropower, wind, and nuclear energy, etc. which has grasped much attention. In today's world, the consumption of global primary energy

J. Ahmad · F. K. Butt (✉)
Department of Physics, Division of Science and Technology, University of Education Lahore, Lahore, Pakistan
e-mail: faheemk.butt@ue.edu.pk

Z. U. Rehman · M. Bilal
School of Physics, College of Physical Science and Technology and School of Environmental Science and Engineering, Yangzhou University, Yangzhou, P. R. China

M. S. Tahir et al. (eds.), *Lithium-Sulfur Batteries: Key Parameters, Recent Advances, Challenges and Applications*, Springer Tracts in Electrical and Electronics Engineering, https://doi.org/10.1007/978-981-99-2796-8_1

has been increased, in which the usage of renewable sources has increased and the consumption of conventional energy modes like has coal decreased. Improved energy storage and conversion methodologies are needed to observe the consumption of sustainable energy, particularly the renewables (Dudley 2018; Xin et al. 2019).

Although the words, energy storage and conversion are used together but they are two different terms, energy storage and energy conversion have different meanings. Energy conversion, as well as referred as energy evolution, is the mechanism through which one kind of energy converts into another form such as in a heat engine, where heat energy is turned into mechanical energy. On the other hand, energy storage means that to store the energy upon the production so that it can be used later on. There is a commonly known natural phenomenon called photosynthesis in which plants collect solar energy directly from sunlight and save it in organic materials. It may be noted that both processes obey the law of conservation of energy which is defined as "Energy cannot be either created or destroyed; it can only be transformed from one state to another" (Dudley 2018; Xin et al. 2019).

Since energy is gathered from various ways such as radiation, heat, gravity, and electricity, it is necessary to introduce the various energy storage devices in which energy can be converted from one kind to another and can be easily stored. For example, battery is an energy storage device that transforms and accumulates electrical energy via chemical processes (Dudley 2018; Xin et al. 2019).

The existence of energy storage is not a new idea, but it is present since ancient times, but it was not identified clearly. By crop raising and accumulating wood, people gather energy from the environment and then stored it for later use. Various techniques have been developed to deliver the power to civilization by storing the energy in a thermal or electrical form.

Background of Electrochemical Energy Storage (EES)

EES is utilized at a large scale to develop the power. Nowadays, it is utilized in various portable electronic appliances. It can be employed in transport field as batteries and ECs to deliver the power. Worldwide demand of energy is anticipated to shoot up in upcoming years. Therefore, various advanced countries have merged EES technologies with the power grid so that they can link with different renewable modes of energy like solar and wind for the sake of electric grid power. In spite of the fact that there are a number of chances for the implication of EES, these devices need to deal with different problems, and researches should be done to develop the most appropriate materials for energy storage and its transmission. Primarily, for customer products and stationary power, bountiful, harmless materials should be utilized to lessen the costs and enhance EES device protection. EES systems should be regarded as an important concern for the long-term advancement of energy

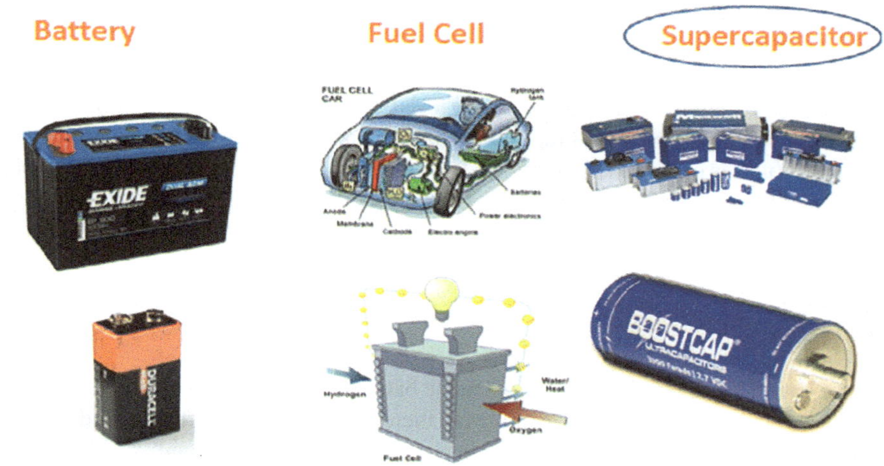

Fig. 1.1 Different types of EES devices (Siwal et al. 2020)

technology. Renewable modes of energy such as solar and wind can be utilized to produce electricity so that energy's requirements can be fulfilled easily.

Production of energy from these renewable modes involves adequate and reconcilable EES methods. Electrical energy must consistently be accessible for feasible urban end uses on an authentic basis. Therefore, there is need to develop improved EES systems to use solar or wind power at a large scale. There is a need of up-to-date EES apparatus to use them in hybrid electric vehicles, plug-in hybrids, and all electrovehicles. EES standard, solidity, and knowledge extension is necessary for advanced mobile equipment, including laptops and smartphones. Some of the EES technologies are illustrated in Fig. 1.1.

EES Devices

Batteries

Battery is an EES equipment and comprises of single or more electrochemical cells having output terminal to provide the electrical power to electrical devices. Batteries can be classified into the following categories:

- Primary batteries
- Secondary batteries
- Battery systems for grid-scale energy supply

Primary Batteries

Such types of batteries which are not able to charge again after their usage are known as primary batteries. These batteries are used for portable electronic devices and generally not recycled, are simple and convenient, and need small maintenance. They are further categorized on the basis of electrolyte material. They are available in different sizes such as AA, C, AAA, etc. Alkaline, lithium, and zinc-carbon (Zn-C) batteries are most common.

Zn-C Battery

These batteries are also known as Leclanche batteries. In 2004, these batteries played 39% role in markets of Europe, and then, there was a decline in their usage. These batteries exhibited a small development and watt-hour cost, and developed in different sizes, shapes, capacities, and voltages. Furthermore, they have small energy density, inferior leakage resistance, and voltage decline with discharging. In this type of batteries, carbon as a cathode is attached with the blend of MnO_2 having an acidic electrolyte, enfolded in a zinc cover which act as an anode. These types of batteries need a careful mechanism for dumping the disposal and recovery of metals.

Alkaline Battery

After the drawbacks of Zn-C batteries, alkaline batteries took the place in battery markets. These batteries crushed the Zn-C batteries and provided more efficiency. Alkaline batteries have excellent low temperature performance, extravagant performance, and adequate shelf life and are inexpensive. These batteries are similar to Zn-C batteries in construction: zinc and MnO_2 are utilized as an anode and cathode, but KOH acts as electrolytic material.

Lithium Primary Cells

Since 1990, lithium cells gained much attention as an excellent performance battery. These cells exhibited large voltage, smooth discharge, large shelf life, large range of working temperature, and satisfactory power density. Lithium batteries have lithium metal and combustible solvents, and combustible hydrogen gas can be developed after the contact of lithium with water. Lithium-air batteries are also the example of lithium cells and are under developmental process. These batteries have 5–10 times greater energy density as compared with lithium-ion batteries.

Secondary Batteries

These batteries can be charged again and therefore known as rechargeable cells. These batteries have various applications in our daily life such as car ignition and various handy electric gadgets (e.g., mobile phones, laptops), and they are also being developed for electric and hybrid vehicles. There are a lot of rechargeable batteries that are common such as lead-acid (Pb-A), lithium-sulfur (Li-S), lithium-ion (Li-ion), nickel-cadmium, nickel metal hydride, etc.

Pb-A Batteries

Pb-A batteries are the most popular and have the largest market contribution with respect to sales and MWh generation. They are mostly utilized in automotive industry and also power supply to the telecommunications and data networks. This kind of batteries has various advantages such as low production price, excellent performance under different temperatures, excellent rate performance, high voltage, and excellent charge retention. There are some disadvantages also such as short charge/discharge cycle, restricted energy density, acid stratification, acid which emanates if ruptured, and problem in de-escalating. There are various environmental concerns related to the production and utilization of lead, and recovery is also needed to mitigate the effects.

Li-Ion Batteries

High storage efficiency can be gained by the help of lithium batteries, and these can be used in sustainable transport as power source. This kind of batteries is a potential candidate for small-scale electronic gadgets and micro-grid frameworks. Li-ion batteries have various advantages such as closed cells which are maintenance-free, large cycle life, fast charging, large energy density, large range of working temperature, and enough flexibility for the design. A different type of salts can be used as electrolyte material in these batteries. Customarily, lithium hexafluorophosphate ($LiPF_6$) is used as electrolyte material. Because $LiPF_6$ salts are sensitive to moisture and degrade into dangerous chemicals, other salts are being monitored to overcome these issues. Solid-state electrolytes could be used but they are expensive; however, improvements are being made to make them economic. Li-ion batteries have different disadvantages such as large primary cost, remarkable charge/discharge volatility, repeated charging requirement, and inadequate cycle life (Fig. 1.2).

The type of material and its concentration in batteries is different according to the design, application, and kind of cathode. Various kinds of materials with their concentrations for the evolution of Li-ion batteries for battery electric vehicle (BEV), hybrid electric vehicle (HEV), and plug-in hybrid electric vehicle (PHEV) are illustrated in Table 1.1.

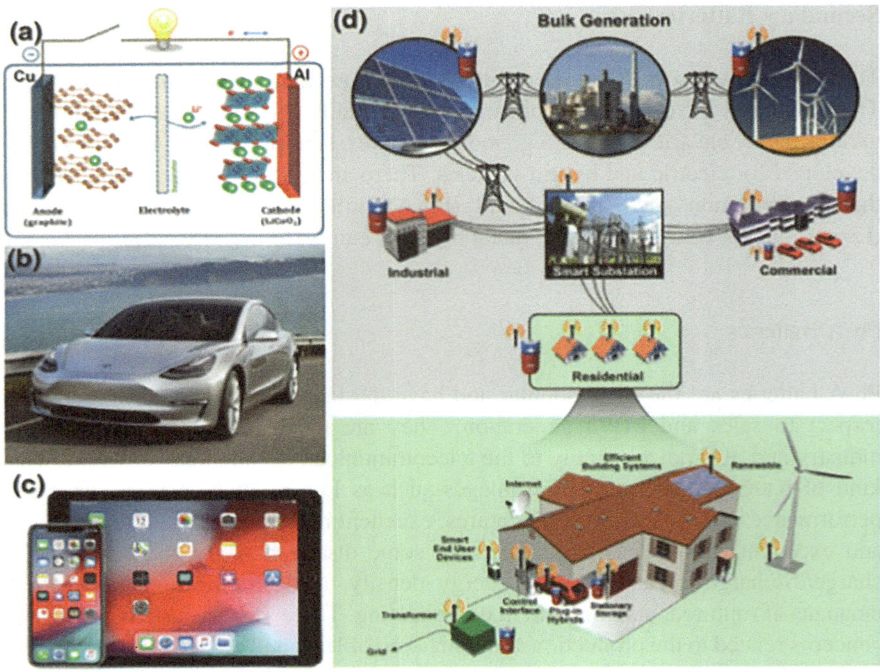

Fig. 1.2 (**a**) Schematic illustration of Li-ion battery, (**b–d**) Applications in the field of (**b**) automobiles, (**c**) consumable electronics, (**d**) grid-scale storage systems (Xin et al. 2019)

Table 1.1 Measured mass of some typical lithium-ion cell organic electrolyte components (HEV, PHEV, BEV) (Eriksson 2001)

	Mass (%)		
Component	HEV	PHEV	BEV
Lithium manganese oxide ($LiMn_2O_4$)	27	27	33
Binder	2.1	2.0	2.5
$LiPF_6$	1.5	1.6	1.
Glycol	2.3	1.2	1.0
Polyethylene terephthalate	2.2	1.6	1.2
Thermal insulation	0.43	0.33	0.34
Copper	13	15	11
Graphite/carbon	12	12	15
Polypropylene	2.0	2.2	1.7
Steel	2.8	1.8	1.4

Measured mass of some typical lithium-ion cell organic electrolyte components (HEV, PHEV, BEV)

Lithium-Sulfur (Li-S) Batteries

Li-S batteries are emerging as an appealing class for their large theoretical capacity and cheap price owing to plenty of sulfur. Despite having 3–5 times the theoretical energy density of Li-ion batteries, these batteries cannot be used on a large scale as expected. The implementation of Li-S batteries is restricted due to loss of capacity and less columbic efficiency owing to shuttling of polysulfide, less volumetric density, large internal resistance, fast capacity fading, and self-discharge. These drawbacks can be reduced by new designs of the cells; therefore, they are gaining much attention of the researchers.

Battery Systems for Grid-Scale Energy

Manufacturing unique battery systems with numerous important features is necessary for grid-scale energy storage. Flow batteries and sodium-sulfur batteries are two important types of grid-scale battery systems that are being developed.

Flow Batteries

Redox flow batteries (RFBs), another name for flow batteries, go through a chemical reaction with the electrolyte material that is kept in the external tanks inside the reaction chamber. In these systems, electroactive materials mixed in liquid electrolyte and energy are produced through the redox reactions in different cells. Reduction pulls out electrons and ions form electrolyte; oxidation recombines these electrons and ions with another electrolyte. Both of the cells are attached with the external tank. RFBs have different advantages such as flexible design ability, control over temperature, fast response time, no self-discharge, facile monitoring, and remarkable stability. It also has some drawbacks, such as requiring pump, power, and flow management, and having less power and energy density.

Sodium-Sulfur (Na-S) Batteries

Na-S batteries are considered as large-temperature batteries and use liquid Na and S which are advantageous for grid-scale as ESSs. These batteries have different advantages such as economic owing to abundance of materials, large cycle life, flexible working, and inconsiderate behavior to ambient circumstances. Despite the advantages, there are many disadvantages also associated with these batteries such as the requirement of high temperature for working and security issues related to the reactivity of materials.

Electrochemical Capacitors (ECs)

ECs are also known as supercapacitors that have remarkable cyclic stability, quick charging/discharging capability, and secure charge storage process, work for large range of temperature, and have excellent power density. They can be charged/discharged within a few seconds and exhibit higher power density as compared to other EES devices. ECs can be classified into two types on the basis of different energy storage processes. These two types are electrochemical double-layer capacitors (EDLCs) and pseudocapacitors (Butt et al. 2020; Sagadevan et al. 2021). The comparison among the EDLC and pseudocapacitors is shown in Fig. 1.3.

Initially, EDLCs store charges in dual electric layers developed in the neighborhood of the electrode and electrolyte interface. For this reason, the mechanism is exceptionally reversible, and the life span is necessarily boundless. Pseudocapacitors not just reserve energy in EDLCs through electric dual layer but also save energy due to quick redox reactions and admissible cross-linking of ion electrodes. The manufacturing stuff of electrode and electrolyte material manages the electrochemical condenser activity jointly. Most common materials that are used for electrode are conductive polymers, carbon-based materials, and metal transition oxides. Moreover, hydrophilic electrolytes, organic electrolytes, and ionic acid are commonly used electrolyte materials for ECs.

ECs provide same output as other EES devices. Figure 1.4 illustrates Ragone's map of ECs with power and energy densities. ECs are same like batteries and fuel cells and large energy densities indicated by capacities, though ECs show low energy densities owing to their ability of mechanical loading. On the other hand, fuel cells and batteries have high energy density, but they exhibit low density owing to their sluggish reactions. Electric condensers bridge the gap among the condensers and batteries or fuel cells. Electrochemical condensers will exhibit high energy density of battery by managing the capacity of high-power condenser (Sagadevan et al. 2021) (Fig. 1.4).

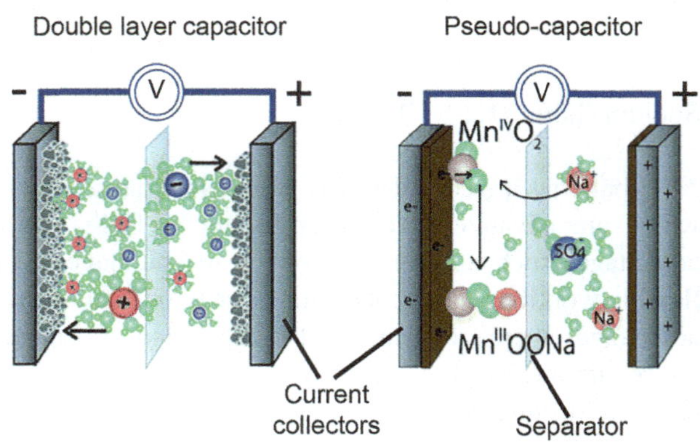

Fig. 1.3 Differentiation among EDLC and pseudocapacitors (Gulzar et al. 2016)

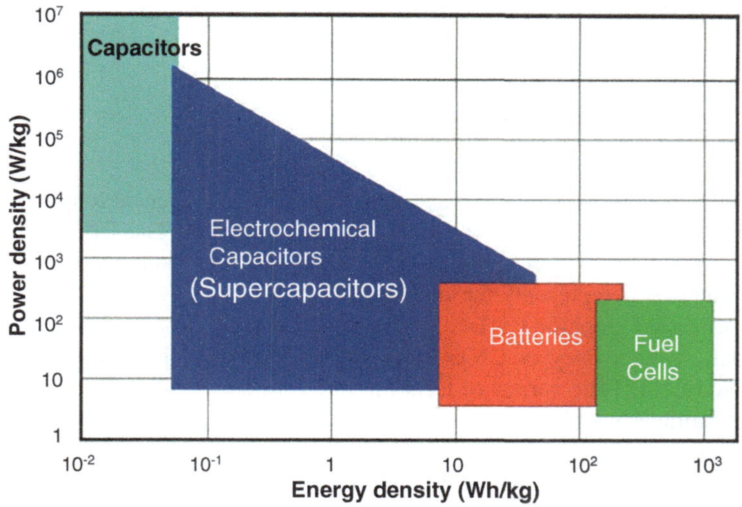

Fig. 1.4 Energy density versus power density of various EES devices (Sagadevan et al. 2021)

Fuel Cells

Fuel cells repeatedly transform chemical energy of fuel to electricity by exterior supply of fuel to straight oxidation substrate that produces electrical energy. Fuel cells are considered as primary processes that straightforwardly utilize fuels like H_2, and implicit processes that utilize fuels sequentially through catalytic and thermal mechanisms. A very familiar method is to produce methanol by using methane ($CH_4 \rightarrow CH_3OH$) with the help of syngas reaction to develop liquid methanol which can be carried comfortably. In case of implicit fuel cell, methanol is going through an agent such that $CH_3OH \rightarrow CO + 2H_2$, and $CO + H_2O \rightarrow CO_2 + H_2$, and H_2 develops electricity as it can be oxidized into water through a catalytic process. A fuel cell is like a battery and contains a cathode, an anode, and an electrolytic membrane. Fuel cells have different advantages such as efficient transformation in the power output cycle, solidity, moldable scaling, and minimum disintegration. The shortcomings of fuel cells are high upfront cost, high-cost conversion reactions to develop methanol, the requirement of fossil fuels, and minimum fuel framework for fuel cell vehicles.

EES Application's Key Components and Materials

Carbon Nanotubes (CNTs)

CNTs as Effective Electrode Materials

CNTs are nanocarbons with electrical attributes, good mechanical toughness, large SA, good chemical inertness, and good size distributions. The CNTs have been proposed as prospective elements for electrodes in Li-ion batteries (Leroux et al. 1999). Figure 1.5 represents the discharge/charge plots of an electrode made of highly structured graphitic CNTs.

In fact, the efficiency of CNTs displayed in Fig. 1.6 is just a special case for graphitized CNTs. The variety of CNTs in microstructure and surface activity with different storage mechanisms are described Li-storage capacities. There are some major features of CNTs that are necessary when we use as electrode materials. The texture of material is also very important in the implementation of Li-ion batteries.

Fig. 1.5 Galvanostatic discharge/charge plots of CNT specimen operated at a rate of C/5 in 1m $LiPF_6$ in EC/DMC solution (Dudley 2018)

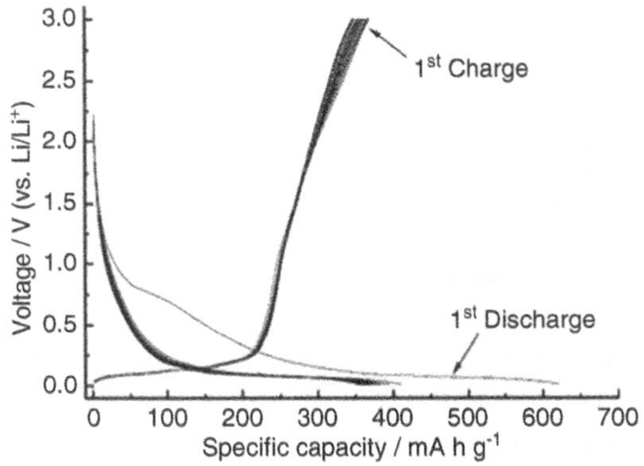

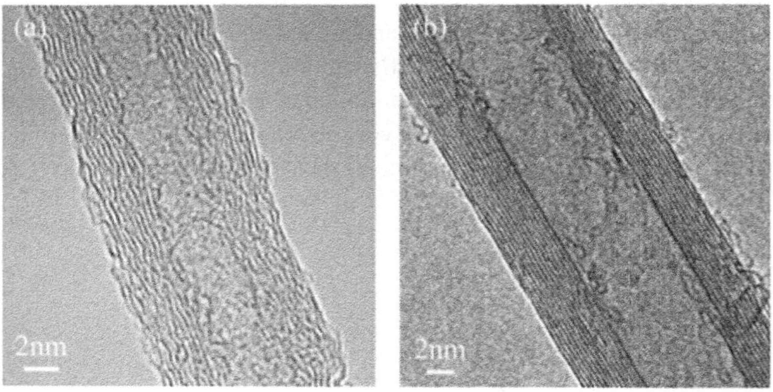

Fig. 1.6 High-resolution TEM pictures of CNTs with (**a**) defect-rich exterior and walls and (**b**) well-structured graphitic walls (Siwal et al. 2020)

Commercially available CNTs have a high defect density, in both the exterior and bulk (Tessonnier et al. 2009). Figure 1.6a, b shows the microstructure of a defective CNT with a graphitized wall CNT respectively. The concentration of defects enhances the Li-storage capability. Different procedures can be applied to enhance the defect's concentration and shorten the size of CNTs which have much for performance as electrode materials (Gao et al. 2000).

CNTs as Additives in Electrodes

1D configuration and dimensionally large CNTs simply form a system by interconnection. Due to the extraordinary properties of CNTs like increased the usage as the additives for electrode materials. This is helpful to improve their thermoelectric and mechanical sustainability to improve changes in volume throughout the charging and discharging mechanisms (Sotowa et al. 2008). Furthermore, insertion of the electrolyte also plays a good role in performance. CNTs as additives are currently used in Li-ion batteries on commercial level (Pandolfo and Hollenkamp 2006). Figure 1.7 depicts the appearance of such a composite anode.

The authors believe that an entire bottom-up technique, in which a hierarchically designed carbon is synthesized by joining CNTs with a susceptible organic matrix to develop the actual electrode, is the best option. This can only be done rationally if we enhance the structure, appearance, and interfacial characteristics of the CNTs while keeping the other reaction parameters in mind. It will be necessary to use both an incorporated synthesis method (demanding strong cooperation among polymer and inorganic carbon chemistry) and a manageable reaction engineering.

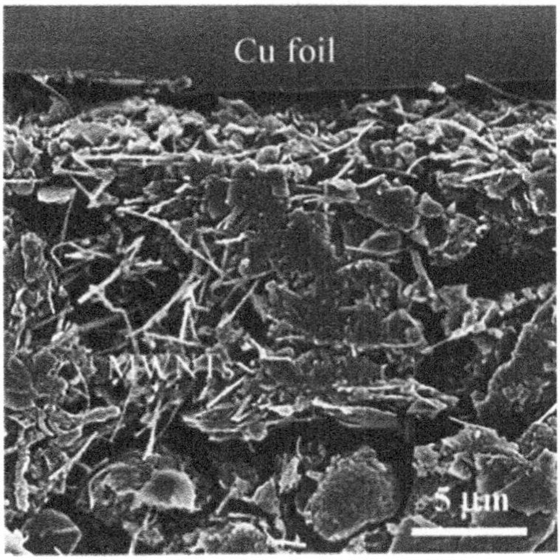

Fig. 1.7 SEM picture of a carbon anode in an industrial cell in which multiwalled CNTs were utilized as preservative. (Adopted from Sagadevan et al. 2021)

Carbon Nanofibers

CNFs are graphitized carbon nanostructures containing or lacking a tubular channel. Their structure is like a tiny graphene basic building structure which has a towerlike shape, and not has the rolled huge graphene framework. Any low-resolution microscopic pictures do not certainly display this disparity. These CNFs are generally synthesized by CVD method of hydrocarbons (Oberlin et al. 1976).

CNFs as Active Materials for Anodes in Li-Ion Batteries

CNFs also have a good efficiency as electrode materials like CNTs; it depends on their morphology of material (Yoon et al. 2004). The prices of CNF product decrease now, but electrodes made with CNFs remain more efficient, both in terms of cost and output outcomes, than those made with graphite. The one and only implementation of CNFs on commercial level in Li-ion batteries is as additive or fillers (Endo et al. 2001). Furthermore, porous CNFs are fabricated with the help of electrospinning of a two-component polymer solution by thermal technique, and these materials were used as Li-ion storage materials (Kim et al. 2007). Porous CNFs can also be used as porous mats directly in anodes for Li-ion battery without addition of any polymers and other such additives. The charging and discharging abilities were increased from 858 and 566 $mAhg^{-1}$ in the starting cycle of Li-ion battery (Ji and Zhang 2009). The key benefit to utilizing electron-spun CNTs as electrodes is that they are independent from binder or any other conductive additive material.

CNF-CNT Composites

The storage capacity and life cycle of electrochemical storage devices would be improved by a composite of these two carbon substances (CNF-CNT). To increase storage capability, Zhang et al. used well-graphitized CNTs as nano-containers with CNFs (Ji and Zhang 2009). The concept of this fabricated material is displayed in Fig. 1.8.

CNTs with an enrich graphitized configuration were used to develop this composite. As shown in Fig. 1.8, CNFs embedded in a CNT composite were created by (1) deposition of the active metal (Co) and (2) growth of CNFs via CCVD. This type of synthesis process is an outstanding demonstration of how to create a hierarchical structure of nanocarbons with this blend to obtain appealing electrochemical features such as large storage capability, high durability, and excellent electrical characteristics.

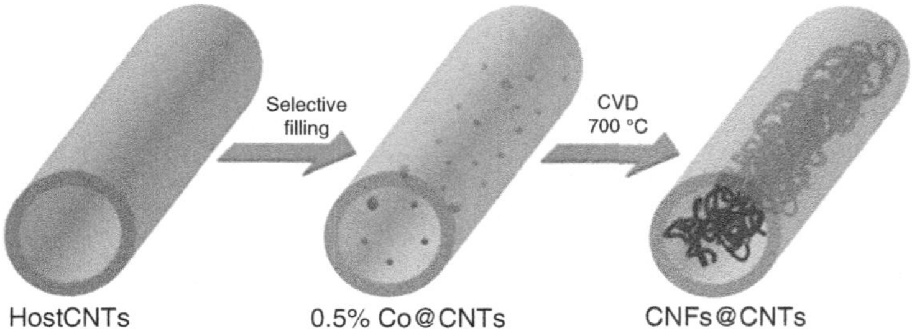

Fig. 1.8 Sectional diagram of the fabrication method of CNFs-CNTs. (Reprinted with permission from Ji and Zhang 2009)

Graphene

Graphene is an attractive material in the fields of nanoscience and nanomaterials. It is a 2D carbon sheet with the thickness of one atom which represents graphite structural unit. Graphene materials have a high SA and more prismatic spots that attach functional groups during the treatment process. Many studies on graphene and graphene-based nanocomposites as electrode materials for Li-ion batteries and other EES devices have been published (Liang and Zhi 2009). Anodes prepared using graphene nanosheet synthesis provided a specific capacity of 945 $mAhg^{-1}$ in the preliminary discharging and a reversible capacity of 650 $mAhg^{-1}$ in the early charging, as shown in Fig. 1.9 (Wang et al. 2009). So the intrinsic electrical features of graphene can play a crucial role for EES applications.

Ordered Mesoporous Carbon

The enhancement of ion pathways, without sacrificing chemical dispersion of lithium ions and electron transport inside the electrode materials, is a major issue for improving performance rate (Wang et al. 2001). Ordered mesoporous carbon (OMC) materials could be fabricated with suitable carbon precursors with template-free method which can increase the demand to use in the field of energy storage applications (Jun et al. 2000). Mesoporous materials have good irreversible storage capability and hysteresis in the charging and discharging processes. This is because these OMC materials have a greater surface area with many adsorption sites for electrode-electrolyte interface. These inherent properties are important in electrochemistry, which may lead to applications in lithium-ion batteries or other devices. For now, it appears that nanomaterials, particularly OMC substances with nanostructures, can provide a foundation for studying some basic aspects of EES devices, such as charge transfer constraint, the involvement of porous material, and electrode or electrolyte interface cooperation.

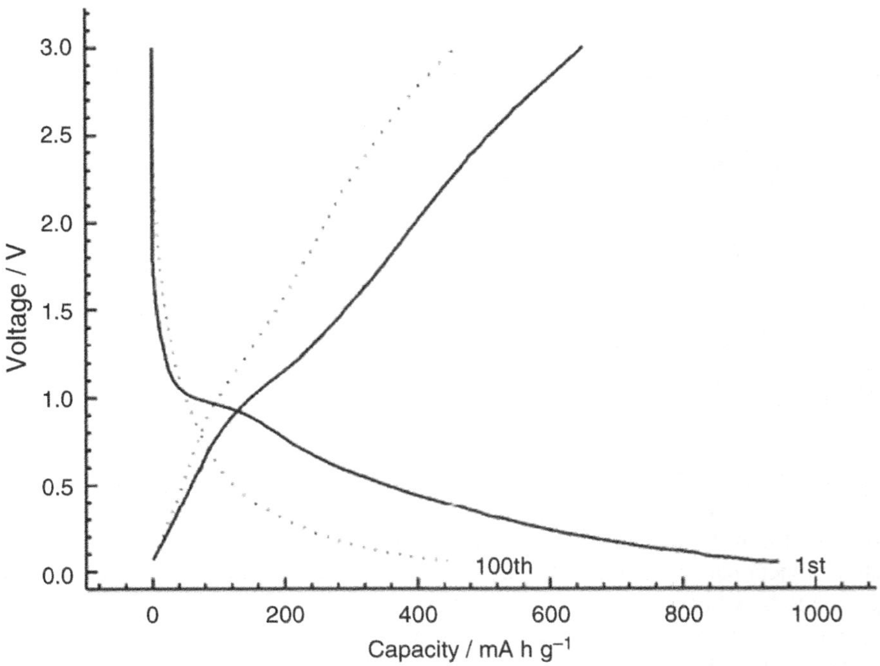

Fig. 1.9 Graphene nanosheet anode charge and discharge curves in Li-ion cells (Wang et al. 2009)

Conclusion

In this chapter, we briefly introduced the sustainable EES technologies that are considered as potential technologies to compete with the rapid demand of global energy. Furthermore, we briefly discussed the various types of EES technologies such as batteries, supercapacitors, and fuel cells. Major components and materials for these technologies were also described.

References

Butt FK et al (2020) Graphitic carbon nitride/metal oxides nanocomposites and their applications in engineering. In: Composite materials: applications in engineering, biomedicine and food science. Springer, Cham, pp 231–265

Dudley B (2018) BP statistical review of world energy. BP Statistical Review, London. Accessed Aug 2018. 6, p 00116

Endo M et al (2001) Vapor-grown carbon fibers (VGCFs): basic properties and their battery applications. Carbon 39(9):1287–1297

Eriksson T (2001) LiMn2O4 as a Li-ion battery cathode. From bulk to electrolyte interface divaportal.org

Gao B et al (2000) Enhanced saturation lithium composition in ball-milled single-walled carbon nanotubes. Chem Phys Lett 327(1–2):69–75
Gulzar U et al (2016) Next-generation textiles: from embedded supercapacitors to lithium ion batteries. J Mater Chem A 4(43):16771–16800
Ji L, Zhang XJN (2009) Fabrication of porous carbon nanofibers and their application as anode materials for rechargeable lithium-ion batteries. Nanotechnology 20(15):155705
Jun S et al (2000) Synthesis of new, nanoporous carbon with hexagonally ordered mesostructure. J Am Chem Soc 122(43):10712–10713
Kim C et al (2007) Synthesis and characterization of porous carbon nanofibers with hollow cores through the thermal treatment of electrospun copolymeric nanofiber webs. Small 3(1):91–95
Leroux F et al (1999) Electrochemical insertion of lithium in catalytic multi-walled carbon nanotubes. J Power Sources 81:317–322
Liang M, Zhi LJ (2009) Graphene-based electrode materials for rechargeable lithium batteries. J Mater Chem 19(33):5871–5878
Oberlin A, Endo M, Koyama TJC (1976) High resolution electron microscope observations of graphitized carbon fibers. Carbon 14(2):133–135
Pandolfo AG, Hollenkamp AF (2006) Carbon properties and their role in supercapacitors. J Power Sources 157(1):11–27
Sagadevan S et al (2021) Fundamental electrochemical energy storage systems. In: Advances in supercapacitor and supercapattery. Elsevier, Amsterdam, pp 27–43
Siwal SS et al (2020) Carbon-based polymer nanocomposite for high-performance energy storage applications. Polymers 12(3):505
Sotowa C et al (2008) The reinforcing effect of combined carbon nanotubes and acetylene blacks on the positive electrode of lithium-ion batteries. ChemSusChem 1(11):911–915
Tessonnier J-P et al (2009) Analysis of the structure and chemical properties of some commercial carbon nanostructures. Carbon 47(7):1779–1798
Wang Q et al (2001) Determination of chemical diffusion coefficient of lithium ion in graphitized mesocarbon microbeads with potential relaxation technique. J Electrochem Soc 148(7):A737
Wang G et al (2009) Graphene nanosheets for enhanced lithium storage in lithium ion batteries. Carbon 47(8):2049–2053
Xin S et al (2019) Introduction to electrochemical energy storage. In: Nanostructures and nanomaterials for batteries. Springer, Singapore, pp 1–28
Yoon S-H et al (2004) Novel carbon nanofibers of high graphitization as anodic materials for lithium ion secondary batteries. Carbon 42(1):21–32
Zhu AL, Wilkinson DP, Zhang X, Xing Y et al (2016) Zinc regeneration in rechargeable zinc-air fuel cells—a review. J Energy Storage 8:35–50

Chapter 2
Introduction to Lithium-Sulfur Battery

Afaq Hassan, Saima Nazir, Umair Azhar, Tausif Ahmad, and M. B. Tahir

Abstract The advancements in energy storage hold great significance in light of the escalating global energy demand and the rapid proliferation of rechargeable vehicles and portable automated devices. Oil-intensive transportation has bad effect on the quality of life, thus presenting the necessity of the clean energy sources like solar and wind energy. Sulfur-based batteries serve as the promising alternative for the traditional lithium-ion batteries, as the Li-ion batteries offer several limitations. It also examines the advantages of sulfur-based batteries over lithium-ion batteries and their potential to meet the energy requirements of larger rechargeable vehicles. Additionally, it addresses the challenges and solutions for the development of the sulfur-based batteries for the innovative large-scale energy storage system. Overall, this chapter serves as a comprehensive introduction to the field of lithium-sulfur batteries, providing a solid foundation for further exploration of their potential in energy storage applications.

A. Hassan · U. Azhar
Institute of Chemical and Environmental Engineering, Khwaja Fareed University of Engineering and Information Technology, Rahim Yar Khan, Punjab, Pakistan

S. Nazir
Nawaz Sharif medical College, University of Gujrat, Gujrat, Pakistan

University of Health Sciences, Lahore, Pakistan

T. Ahmad (✉)
Institute of Chemical and Environmental Engineering, Khwaja Fareed University of Engineering and Information Technology, Rahim Yar Khan, Punjab, Pakistan

Institute of Energy & Environmental Engineering, University of the Punjab, Lahore, Pakistan
e-mail: tausif.ieee@pu.edu.pk

M. B. Tahir
Institute of Physics, Khwaja Fareed University of Engineering and Information Technology, Rahim Yar Khan, Punjab, Pakistan
e-mail: m.bilaltahir@kfueit.edu.pk

M. S. Tahir et al. (eds.), *Lithium-Sulfur Batteries: Key Parameters, Recent Advances, Challenges and Applications*, Springer Tracts in Electrical and Electronics Engineering, https://doi.org/10.1007/978-981-99-2796-8_2

Keywords Lithium-sulfur batteries · Energy storage · Energy density · Cost · Carbon coatings · Large-scale energy storage systems

Introduction to Lithium-Sulfur Batteries

Advances in storage of the energy are significant because of the worldwide energy increased demand and the widespread development of rechargeable cars and handy automated devices (Qie et al. 2012; Ge et al. 2014; Chen et al. 2013a; Ji et al. 2009a, 2010, 2011a; Gu et al. 2012; Ji and Zhang 2009a, b, c; 2010; Zhang et al. 2004). Adverse effect on the environment and life quality has been recorded with the oil-intensive transportation means (Wang et al. 2010, 2011a; Xiong et al. 2012; Yin et al. 2003; Ellis et al. 2010). Solar and wind power are becoming more and more popular sources of clean energy. Wind and solar energy are no more feasible without the storage of the energy, which is difficult for the efficient and cost-effective treatments. Thus, a huge infrastructure is required for the electrical energy storage systems (ESS). One of the most attention-seeking ESS is the lithium-ion battery that is used in most of the automated systems.

However, existing lithium-ion battery technology is unable to handle extensive energy demands for larger rechargeable cars and automobiles. Lithium-ion cells are not recommended for army power supply system, which demand more capacity, cost-effectiveness, and improved safety (Tarascon and Armand 2001; Li et al. 2009; Ellis et al. 2010; Goodenough and Kim 2010; Etacheri et al. 2011).

As traditional LIBs founded on intercalation chemicals have reached their power density limits, numerous research studies have focused on developing novel large ESS (Bruce et al. 2012). Sulfur offers advantages such as high capacity (1675 $mAhg^{-1}$), cost-effectiveness, ubiquitous supply, and nontoxicity; the next generation of high theoretical specific power (Li-S) batteries (2600 $Whkg^{-1}$) has been widely used recently found (Yang et al. 2013a; Yin et al. 2013; Manthiram et al. 2014; Son et al. 2015; Pang et al. 2016) (Table 2.1).

Lithium-ion batteries (LIBs) are one of the fundamental parts in the transportable electrical gadgets for many years because of their good performance activity in comparison with the other batteries (Tarascon and Armand 2001; Dunn et al. 2011), though the growing demand for electrical automobiles and large level smart grid needs battery with excessive power density, less value, and lengthy cycle lifestyles (Bruce et al. 2012; Yang et al. 2013a; Manthiram et al. 2014). The less progress on

Table 2.1 Capacity of different cathode materials (Dixitar 2020)

Material of cathode	Nominal voltage (V)	Capacity ($mAhg^{-1}$)	
		Theoretical	Practical
$LiCoO_2$	3.6	274	110–140
$LiNi_{1/3}Co_{1/3}Mn_{1/3}O_2$	4.1	278	210
$LiMn_2O_4$	3.7	148	110–120
$LiFePO_4$	3.6	170	150
S	2.2	1672	–

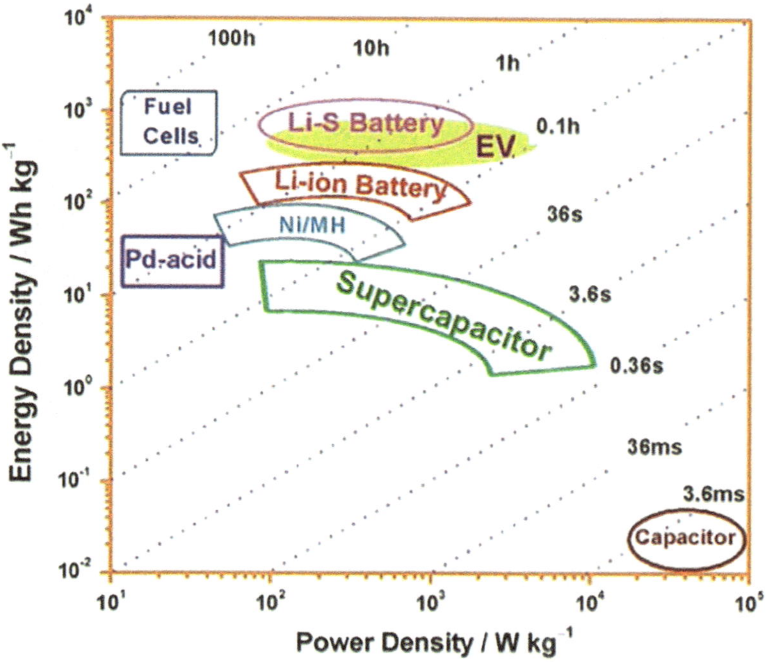

Fig. 2.1 Power and energy density comparison of different energy devices

cathodes (ability typically less than 200 mAh g^{-1}) is an obstacle for the rise in the energy and power compactness of the LIB battery has precipitated the examination of new electrochemical electricity storage structures (Fig. 2.1), inclusive of the Li-S batteries with an excessive theoretic ability and power density of 1675 mAh g^{-1} and 2600 Wh kg^{-1} (Liang et al. 2014; Manthiram et al. 2012). It has shown the five times better energy density as compared to cheaper lithium-ion batteries with $LiCoO_2$/ graphite device (387 Wh kg^{-1}) (Kolosnitsyn and Karaseva 2008; Wang et al. 2012, 2013a; Zhang et al. 2014; Ji et al. 2009b).

So as to deal with the abovementioned problems and recognize the usage of excessive power and energy compactness and large cycle life of lithium-sulfur batteries, researchers have looked into all of the important additives in Li-S batteries, and awesome development have presented in this field. The lithium-sulfur battery has been known as the power storage system and utilizes solar power for charging of the battery in daytime and offers power at night using the solar-charged Li-S batteries.

Lithium-sulfur batteries are highly observed because they have three to five times better energy density than LIBs (Choi et al. 2012). In a lithium-sulfur battery, the cathode is made up of a liquid organic electrolyte between carbon-sulfur compound and the anode metallic lithium (Mikhaylik et al. 2010). Figure 1.3 shows the lithium-sulfur battery configuration. Sulfur at the electrode is reduced to Li_2S as electricity during release by a complex process consisting of several intermediate polysulfides.

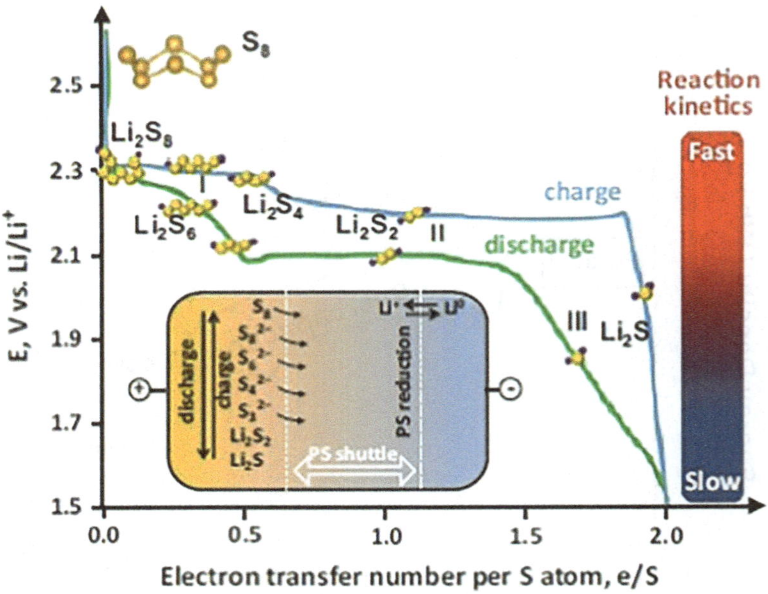

Fig. 2.2 Charge-discharge curve of the Li-S batteries (Wang et al. 2013a)

Figure 2.2 illustrates the typical Li-S cell discharge-charge curve. Activated sulfur is electrically reduced by a series of several intermediate polysulfides in the form of Li_2Sx (x = 2–8) on the electrode surface, a small chain of electrolyte, and polysulfide Li_2Sx (x = 4–8) which has very good solubility. Li_2Sx polysulfides (x = 2–4) has low solubility (Ji and Nazar 2010; Yang et al. 2013b).

Sulfur is a solid nonmetal in a yellow color with eight-atom cyclic atom known as S8. Sulfur has more than thirty (30) allotropic forms (Greenwood and Earnshaw 1997), but alpha-bromide sulfur (α-S8) is highly thermodynamically stable at room temperature (RT) (Templeton et al. 1976; Pastorino and Gamba 2000, 2001). The exponentional growth in alpha-bromide sulfur (α-S8) Li-S battery systems after charging (Walus et al. 2013; Kulisch et al. 2014; Villevieille and Nova'k 2013) or as the first direct electrode material has been reported in recent papers (found at the infiltration of sulfur based on CNT structure) carbon nanotube (Moon et al. 2013) (Fig. 2.3).

According to Liu et al. (2011a), with a theoretic volume of 1675 mAh g, sulfur is a very capable alternative for the cathode material with high energy storing ability. Instead of a topotactic process (Ji et al. 2009b), (Li-S) battery technology uses converting chemicals (2.1):

$$S_8 + 16Li \leftrightarrow 8Li_2S \left(E^0 = 2.2V \text{ vs } Li^+/Li^0\right. \tag{2.1}$$

Li-S battery is used for the electricity storage followed by the electrochemical method (Eq. 2.1). As a result, active material weight in the lithium-sulfur battery

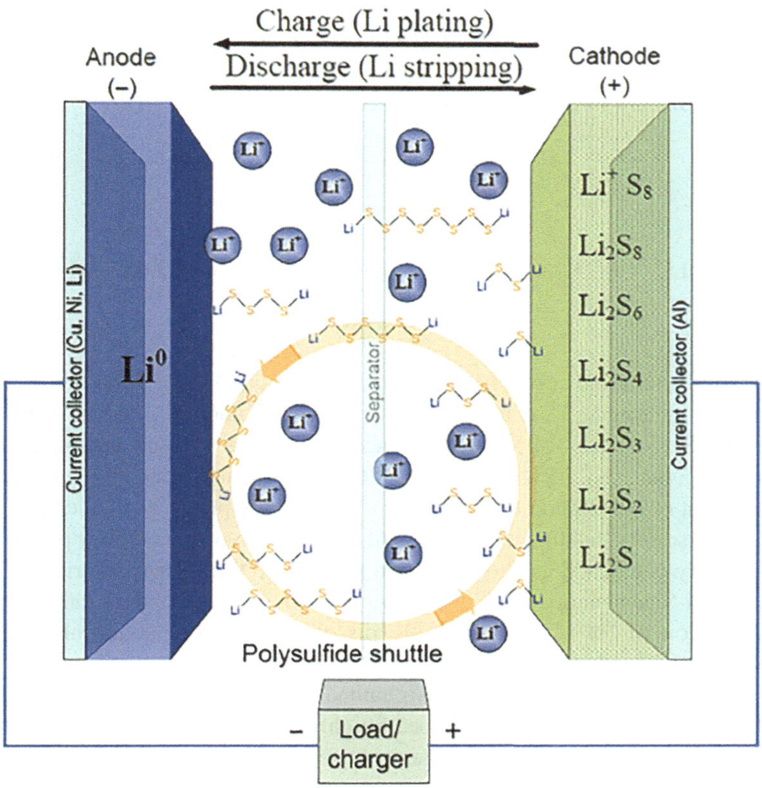

Fig. 2.3 Arrangement of lithium-sulfur battery with liquid electrolyte (Dunn et al. 2011)

greatly reduced by the same number of electrons emitted from electrochemical processes. Even though lithium-sulfur batteries have twice as much electromotive force (EMF) capacity as typical materials of cathode, sulfur can reach a maximum power density of 2500 $Whkg^{-1}$ (2800 WhL^{-1}), predicting total S reduction. Basic 2S (Mikhaylik and Akridge 2004; Yamin and Peled 1983; Rao 1968). typed Sulfur is also plentiful, inexpensive, and harmless in nature. In 1962, the lithium-sulfur battery was introduced. Much of the lithium-sulfur battery research took place in late 1900s. Throughout this time, all the knowledge of the chemistry of Li-S batteries was achieved. However, the lithium-sulfur was only occasionally explored for the storage of the energy due to its low cycle potential (Herbert and Ulam 1962; Shim et al. 2002; Marmorstein et al. 2000; Jeon et al. 2002; Rao and Shropshire 1981). In addition, the polysulfide shuttle effect is a well-documented phenomenon that occurs in Li-S cells when liquid electrolyte (Mikhaylik and Akridge 2003; Cheon et al. 2003; Yamaki et al. 1998) is used, resulting in a "chemical short circuit." Electrochemical cell can produce the polysulfide shuttle effect which reduces sulfur consumption and Coulombic efficiency, and induces solid precipitation at both the insoluble cathode and the anode (Barchasz et al. 2012) of Li_2S and/or Li_2S_2, resulting in two electrodes, formed electrochemically, resulting in unnoticed and loss of capacity. All of these issues have hampered the commercialization of Li-S batteries. Sulfur lawsuit may be summarized in four steps (2.2)–(2.5):

$$S_8 + 2e^- \rightarrow S_8^{2-} \tag{2.2}$$

$$S_8^{2-} + 2e^- \rightarrow 2S_4^{2-} \tag{2.3}$$

$$(2x + 2\gamma)Li^+ + 0.25\ (2x + \gamma)S_4^{2-} + 0.5(2x + 3\gamma)e^- \rightarrow xLi_2S_2 + \gamma Li_2S \tag{2.4}$$

$$2Li + Li_2S_2 + 2e \rightarrow 2Li_2S \tag{2.5}$$

Two solid and liquid components of the crown reaction like a solid ring lower the S8 to a more melting $S_8{}^{-2}$, with a 2.3 V plate. In the cathode area, the molten $S_8{}^{2-}$ is reduced to $S_4{}^{2-}$ in the following sequence, as well as various intermediates for instance $S_6{}^{2-}$, $S_3{}^{2-}$, $S_3{}^{-}$ (Zhang and Read 2012; Yamin and Peled 1983). Because of the increment of polysilicon (PS) ion, this methodology produces a quick rise in the electrolyte viscosity, leading to a significant decrease in voltage to a low peak when a high viscosity is reached as shown second place. The third stage contributes to the significant Li-S cell energy value, with a wavelength of 2.1 V, which is accompanied by a reduction of PS of low solubility dissolved in dipole solids almost Li_2S_2 dissolved or Li_2S, mentioned in Eq. (2.4). The following is a strong one-phase reaction to reduce Li_2S_2 to Li_2S (Barghamadi et al. 2013; Bresser et al. 2013). Despite the many benefits that have been highlighted, lithium-sulfur batteries continue to present challenges to their practical use:

(a) Sulfur is degraded by the properties of the electrical and ionic sulfur used as well as by its discharge products.
(b) The "shuttle effect" occurs when polysulfides – mediators of the cathodic reaction in a conventional liquid organic electrolyte – dissolve, resulting in a significant reduction of the active cathode material and lithium corrosion.
(c) The cathode structure becomes unstable when the cathode size changes from S to Li_2S by 76%.
(d) Due to the production and ignition of lithium dendrites in the liquid organic electrolyte, the use of a metallic lithium anode reduces the potential for safety.

Several attempts have been made to design sulfur cathodes to increase electronic conductivity and reduce dissociation (Bresser et al. 2013; Evers and Nazar 2013; Manthiram et al. 2013; Song et al. 2013; Chen and Shaw 2014; Xu et al. 2014). The activation of activated sulfur forms is typical of the conductive porous matrix design of nanostructured carbon and conductive polymers, as well as the dissolution of the physical or chemical barrier and the use of PS to suppress active material damage and shuttle effect circulation. The subject of these papers. In addition to the development of cathodes for Li-S batteries, Li-S electrolytes have been extensively studied due to their unique importance (Parfitt 2012).

Cathode Materials

Presently the most ordinarily taken on methodology in planning sulfur cathodes which will be momentarily presented beneath.

Sulfur-Carbon Composites

The low insulation property of the sulfur, a wide range of sulfur-carbon composites are used due to the high conductivity of carbon. A large amount of sulfur can be stacked on the surface pores of the carbon materials. The construction of pores on the carbon might give adequate space for the charge and release process of the sulfur, and relieve the disintegration of polysulfides; thus, the usage of sulfur and cycle execution can be moved along. Simultaneously, the permeable design could improve the response energy by giving the electrolyte particle transport channels.

Nazar gathering has created CMK-3/S compound that contains sulfur by a soften dissemination system as depicted in Fig. 2.4a (Liang et al. 2009). The nano-sized complex release limit is 1005 mAh g^{-1} at 168 mA g^{-1}. After 20 cycles, the limit actually kept up with at around 800 mAh g^{-1}. Liang and so forth detailed a sort of bimodal permeable carbon with mesopores of size 7.3 nm and under 2 nm mesopores through a delicate layout amalgamation technique joining with the potassium hydroxide actuation, as displayed in Fig. 2.4b (Xin et al. 2012; Zhang et al. 2010).

Gao proposed a carbon-sulfur composite hollow structures by incorporating sulfur into the carbon hollow circles. Firstly, sulfur was sublimed and mixed with the carbon structures under the thermal treatment at 155 °C. Small diameter of complex (0.7 nm) allows the penetration of the sulfur into the hollow circles of carbon, which limits the polysulfide dissemination and offers the great cathode conductivity (Fig. 2.4c). This sulfur-based cathode offers long cyclic life of up to 500 cycles at a maintenance rate of 85% (Li et al. 2014).

As of late, Huang group additionally examined the electrochemical component of the $S_{2–4}$ cathode confined in a microporous carbon lattice in different electrolytes through consolidated test results and hypothetical computation (Fig. 2.4d) (Wang et al. 2011b).

Dai's group has presented a PEG-wrapped material on the graphene/sulfur material, as displayed in Fig. 2.5. Sheets of graphene and dark carbon enhance the conductivity capability of sulfur, whereas PEG increases the volume of sulfur during the charge and release process. The PEG and graphene successfully dissolve the polysulfide, thus offering higher conductivity, and make the composite stable even after 100 cycles with a release limit of 750 mAh g^{-1} at 0.2 C (1 C = 1672 mA g^{-1}) (Cao et al. 2011; Zhang et al. 2012; Huang et al. 2013; Lv et al. 2014; Zhao et al. 2012).

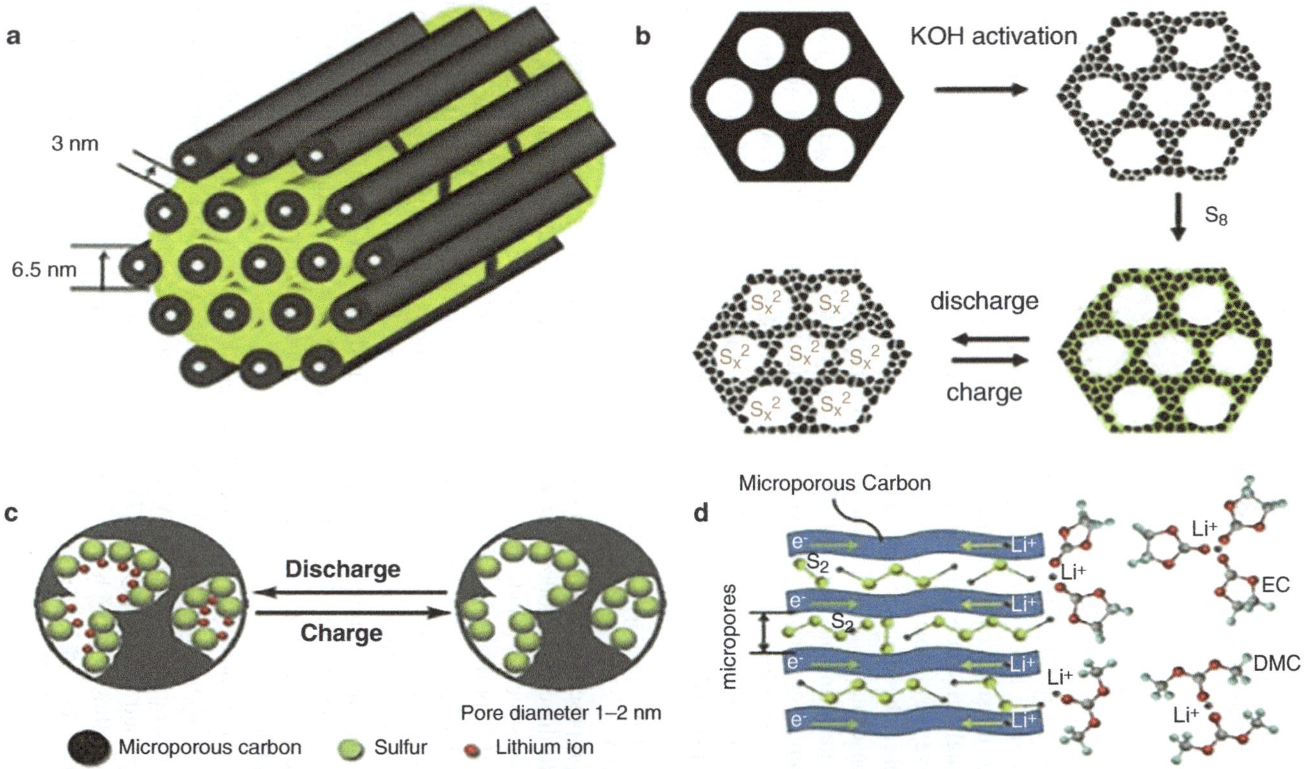

Fig. 2.4 (**a**) Sulfur interconnected with carbon in a porous structure (Liang et al. 2009). (**b**) Sulfur-carbon porous composite (Xin et al. 2012). (**c**) Electrochemical reaction beneath the S/C cathode composite (Li et al. 2014). (**d**) Process of lithiation in a carbon cathode (Wang et al. 2011b)

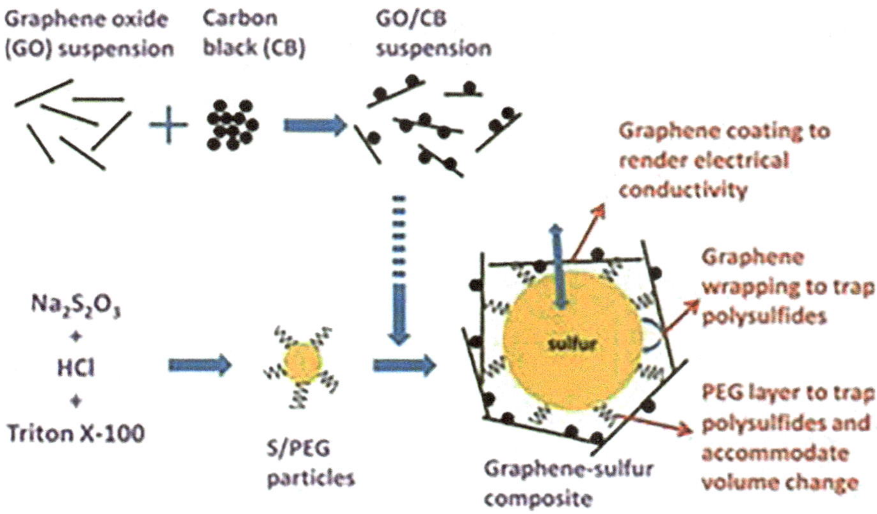

Fig. 2.5 Synthesis methodology of PEG-graphene-sulfur material (Cao et al. 2011)

Conducting Sulfur-Based Polymer Composites

The previously mentioned carbon covering strategies, for example, CVD covering of carbon (Zheng et al. 2011; Guo et al. 2011), ordinarily require treatment at a raised temperature to build the electrical conductivity of the covering layer (Yin et al. 2013). Nonetheless, with low temperature of sublimation of sulfur, the immediate covering should be completed at lower temperature that definitely restricts the quantity of materials (Wu et al. 2011, 2016; Wang et al. 2002; Fu and Manthiram 2012; Xiao et al. 2012).

Liu presented the utilization of the polyaniline (PANI) nanotubes as a S epitome grid (Xiao et al. 2012). The nanotubes were thermally heated with S at a temperature of about 275–280 °C, which forms a 3D shape, cross-connected, primarily stable S/PANI composite (Xiao et al. 2012), as displayed in Fig. 2.6a. The as-prepared cathode shows a long cyclic length of 500 cycles with a release limit of about 568 mAh g^{-1} after 100 cycles at 1 C rate (Fig. 2.6b).

Chen shows a film using the precipitation synthesis method to combine S nanoparticles (10–20 nm) and covered the films with the coating of 3,4-(ethylenedioxy) thiophene (PEDOT) to form S-PEDOT nanostructures (Fig. 2.6c) (Chen et al. 2013b). The nanocomposite shows usage of 50 cycles with a release limit of 1117 mAh g^{-1} and a steady limit of 930 mAh g^{-1}.

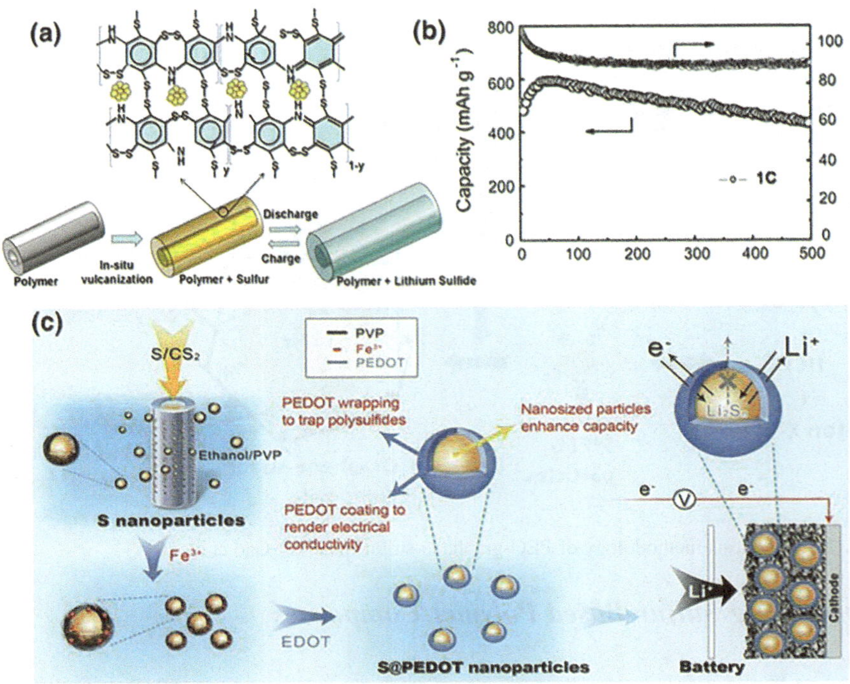

Fig. 2.6 (**a**) Synthesis and charge transfer process of the S/PANI nanotube complex. (**b**) Coulombic and cycling performance activity of the S/PANI composite (Xiao et al. 2012). (**c**) Synthesis of the S/PEDOT nanostructures for lithium-sulfur battery (Chen et al. 2013b)

Metal Oxide-Sulfur Complexes

Because of the large surficial area and solid polysulfide adsorption ability, metal oxides are utilized as practical sponges or covering layers to stifle polysulfide carrying in lithium-sulfur batteries. A few metal oxides are, for example, Al_2O_3 (Choi et al. 2007), TiO_2 (Evers et al. 2012), SiO_2 (Ji et al. 2011b), La_2O_3 (Zheng et al. 2006; Sun et al. 2013), and $Mg_{0.6}Ni_{0.4}O$ (Pastorino and Gamba 2000; Walus et al. 2013; Kulisch et al. 2014; Villevieille and Nova'k 2013; Moon et al. 2013; Liu et al. 2011a; Ji et al. 2009b, 2011b; Mikhaylik and Akridge 2003, 2004; Yamin and Peled 1983; Rao 1968; Herbert and Ulam 1962; Shim et al. 2002; Marmorstein et al. 2000; Jeon et al. 2002; Rao and Shropshire 1981; Cheon et al. 2003; Yamaki et al. 1998; Barchasz et al. 2012; Zhang and Read 2012; Barghamadi et al. 2013; Bresser et al. 2013; Evers and Nazar 2013; Manthiram et al. 2013; Song et al. 2013; Chen and Shaw 2014; Xu et al. 2014; Parfitt 2012; Liang et al. 2009; Xin et al. 2012; Zhang et al. 2010, 2012; Li et al. 2014; Wang et al. 2002, 2011b; Cao et al. 2011; Huang et al. 2013; Lv et al. 2014; Zhao et al. 2012; Zheng et al. 2006, 2011; Guo et al. 2011; Yin et al. 2013; Wu et al. 2011, 2016; Fu and Manthiram 2012; Xiao et al. 2012; Chen et al. 2013b; Choi et al. 2007; Evers et al. 2012; Sun et al. 2013;

Seh et al. 2013) as sponges have shown to be profoundly successful adsorbent of lithium polysulfides.

For example, Song incorporated nanostructure powder of $Mg_{0.6}Ni_{0.4}O$ by the sol-gel strategy and incorporated it with the sulfur cathode (Seh et al. 2013). The powder of $Mg_{0.6}Ni_{0.4}O$ shows to be an excellent adsorbent for the polysulfides, thus providing higher porosity of the cathode, and eventually enhances the capacity of the cathode. Thus, the underlying charge-release limit expanded from 741 (without added substances) to 1185 mAh g^{-1} (with added substances), and the cyclic strength of the cell improved from 76% of the underlying release ability to 85% at the 50th cycle after the option of nano-estimated $Mg_{0.6}Ni_{0.4}O$ particles (Seh et al. 2013). Ji proposed an idea to relieve the above issues; the methodology utilizes a permeable silica inserted inside the C-S composite that adsorbed the polysulfide intermediates, resulting of about 95% Coulombic efficiency (Fig. 2.7a) (Zheng et al. 2006).

Yolk-shell nanostructure based of sulfur-TiO_2 was proposed by Cui group. The yolk-shell structure was synthesized by the hydrolysis of S with TiO_2 (Fig. 2.7b) (Seh et al. 2013). The battery with the yolk-shell configuration exhibits an underlying release limit of 1030 mAh g^{-1}, and the limit maintenance was viewed as 67% after 1000 cycles (Fig. 2.7c).

Lithium Sulfide Materials

In lithium-sulfur batteries, sulfur ought to be matched with a lithium anode. Supplanting sulfur with lithium sulfide (Li_2S) is a promising technique. The hypothetical limit of Li_2S (1166 mAh g^{-1}) is a lot higher than that of currently used cathode materials, and the completely lithiated state (i.e., Li_2S state) evades the issue of volume development during charging (Yang et al. 2012; Zu et al. 2014; Son et al. 2015). Other than metallic lithium, it tends to be coordinated with other no lithium-containing anode materials like silicon, germanium, tin, and carbonaceous materials with more dependable well-being properties (Yang et al. 2010; Hassoun and Scrosati 2010).

Numerous researchers have made research to combine several conductive materials with lithium sulfide to enhance the electrical conductivity. Low-energy ball-processing technique was applied by Scrosati for blending lithium sulfide powder with carbon to act as a cathode (Li_2S/C), and tin-carbon composite as anode in the presence of the gel electrolyte to form $Sn/C//Li_2S/C$ (Fig. 2.8a) (Hassoun and Scrosati 2010). Lithium sulfide and carbon composite as cathode and a silicon nanowire structure as anode were presented by Cui for lithium battery without any metal (Fig. 2.8b) (Yang et al. 2010). The nanostructured plan of the two anodes helps with defeating the problems related with utilizing S mixtures and silicon in lithium-ion battery.

The above explores on little estimated Li_2S as dynamic materials; nonetheless, mass Li_2S particles actually experience the ill effects of low electrochemical action and a huge starting actuation energy hindrance (enormous overpotential) (Meini

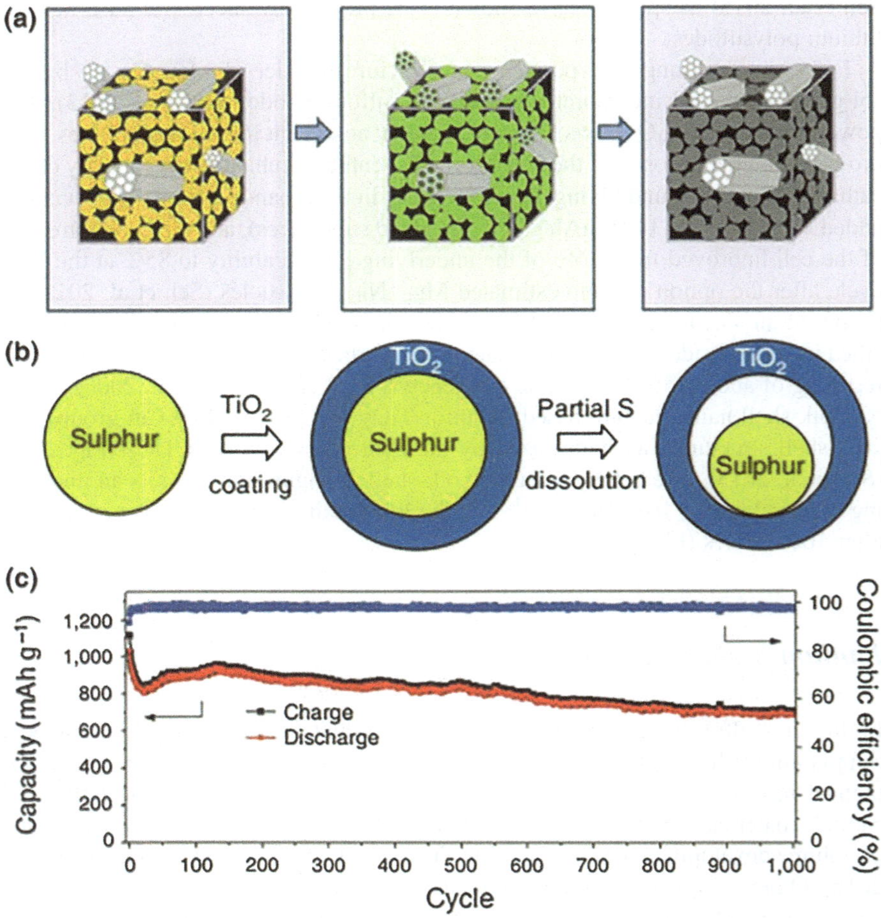

Fig. 2.7 (**a**) SBA-15-supported S electrodes (Zheng et al. 2006). (**b**) Manufacturing scheme for the yolk-shell sulfur-titanium dioxide nanostructure. (**c**) Coulombic efficiency and charging/discharging capability of the nanostructure (Seh et al. 2013; Tao et al. 2014)

et al. 2014; Wang et al. 2015). Numerous systems were projected to overcome the aforementioned problems, for example, utilizing high-energy ball-processed or beat laser affidavit strategies to manufacture Li_2S-metal composites (Li_2S-Fe (Obrovac and Dahn 2002), Li_2S-Cu (Hayashi et al. 2008), and Li_2S-Co (Zhou et al. 2007) composites) to work on the electrical conductivity of Li_2S. Li_2S-carbon composites are one more generally taken on technique used to defeat the current difficulties with utilizing Li_2S as a cathode. For instance, carbon-covered Li_2S center shell circles of shifting sizes have been combined through a two-venture CVD strategy (Nan et al. 2014) and had the option to exhibit a high introductory release limit and stable cycling execution north of 400 cycles at 0.5 C rate without extra carbon-added substances. Li_2S-C composite with lithium-nitrile connection has been mentioned by

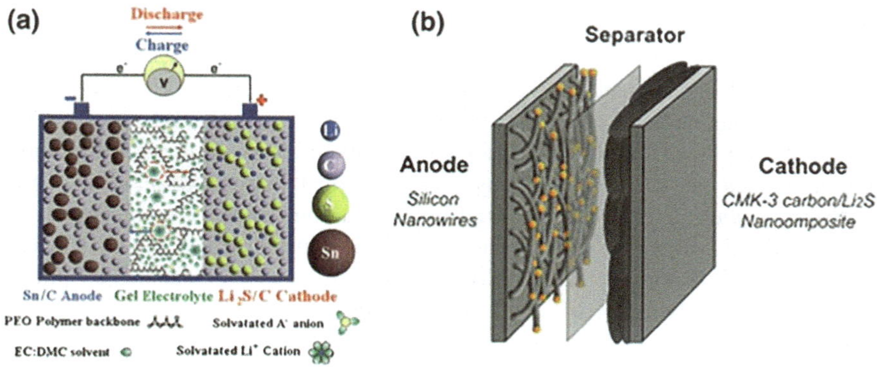

Fig. 2.8 (**a**) Sn/C/CGPE/Li_2S/C battery schematics (Hassoun and Scrosati 2010). (**b**) Li_2S/silicon battery (Yang et al. 2010)

Table 2.2 Various sulfur cathode material's advantages and challenges

Cathodes	Advantages	Challenges
Sulfur-carbon composites	Materials that contain carbon offer high surface area. Porous structure and excellent electric conductance and plentiful useful gatherings, which assist with working on the electrical conductivity and trap polysulfides	Streamline size of pore size, further develop the sulfur stacking
Sulfur-conducting polymer composites	Low-temperature combination, surface useful gatherings limit and trap polysulfides	Work on electrical conductivity and the sulfur stacking
Sulfur/metal oxide composite	Huge surface region adsorbs polysulfides and limits the van impact, catalyzes the sulfur decrease and Li_2S oxidation	Metal oxides offer less electrical conductivity, weighty mass of oxides diminishing the energy thickness of Li-S batteries
Lithium sulfide-based cathodes	Keep away from lithium dendrite issues and enormous measure of electrolyte utilization	Li_2S easily reacts with H_2O releasing hazardous H_2S

consistently blending lithium sulfide with PAN to shape a lithium sulfide-PAN cross-connected lattice (Guo et al. 2013), empowering great cycling dependability and high Coulombic efficiency. Besides carbon covering, the exemplification of Li_2S particles utilizing two-layered materials with high conductivity and solid restricting proclivity to Li_2S/Li_2S x species has as of late been understood and very much clarified by hypothetical computations (Seh et al. 2014). These strategies exhibited the practicality of utilizing Li_2S cathodes. In any case, the simple response of Li_2S with H_2O delivering unsafe H_2S actually restricts the activity of Li_2S in the dry climate or glove box. Advantages and challenges that are encountered by using various S-based cathodes are given in Table 2.2.

Anode Materials

Lithium-based anode materials show certain problems with the electrolytes and polysulfide intermediates, such as weathering of the Li metal, high usage of the electrolyte, and low Coulombic efficacy. Surface modification of the nanostructured lithium anodes can be used to increase the usage of lithium-based anodes such as synthesis of the interconnected layered hollow carbon nanospheres (Cheon et al. 2003) or ultrathin BN/graphene (Zheng et al. 2014; Yan et al. 2014; Lin et al. 2016). One more methodology detailed by Liu's gathering is that they planned a lithium-sulfur battery involving electrically associated graphite and lithium metal as a half-breed anode to control bothersome surface responses on lithium side (Liang et al. 2016) as mentioned in Fig. 2.9a.

Separators and Interlayers

Much work had been performed to highlight the impact of the cathode and anode materials. However, the separator has gotten less consideration in the lithium-sulfur battery. Separator is a significant part in lithium-sulfur battery with the capacity to isolate cathode and anode, and forestall inside hamper keeping up with the dispersion pathway of particles. Separator must have high wettability and should be thermally stable (Brückner et al. 2014). Zhang and Huang et al. detailed an efficient polysulfide particle safeguard by the consolidation of slender Nafion layer onto ordinary polypropylene (PP) separator (Fig. 2.10a) (Huang et al. 2015).

Unique in relation to the modification of the business separators, Manthiram's gathering proposed to embed an interlayer among cathodes and separators in a Li-S

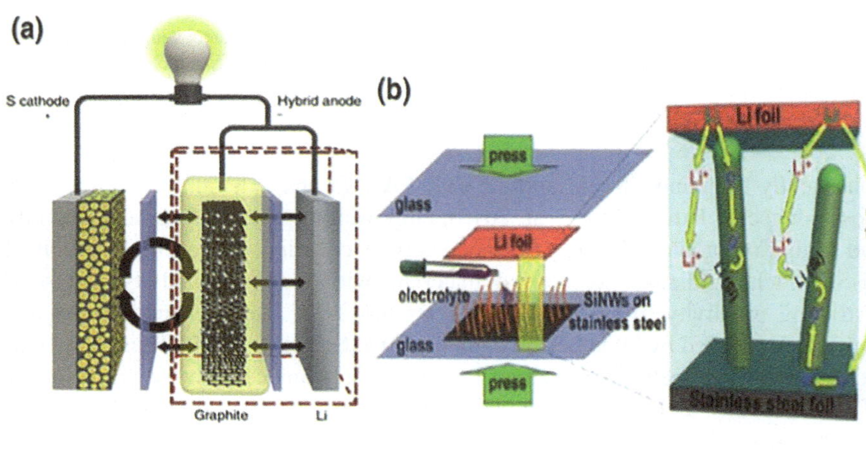

Fig. 2.9 (**a**) Hybrid anode on lithium-sulfur battery (Liang et al. 2016). (**b**) Pathways for the electron and Li ions during the prelithiation process (Huang et al. 2014a; Liu et al. 2011b)

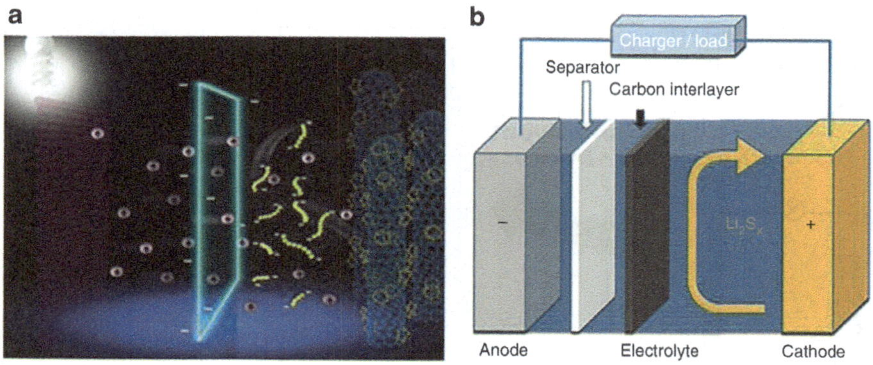

Fig. 2.10 (**a**) Lithium-sulfur batteries with membranes allowing selective ion transfer (Huang et al. 2015). (**b**) Lithium-sulfur battery with a carbon layer between the separator and cathode (Huang et al. 2014b)

cell to fill in as an upper current authority, which was compelling in bringing down the charge move opposition of the sulfur cathode and catching of polysulfides in the cathode (Fig. 2.10b) (Huang et al. 2014b). These highlights manage the cost of a superior electrochemical exhibition of the batteries in both high specific limit and brilliant soundness. A high beginning release limit of 1367 mAh g^{-1} was conveyed, and a high-rate limit of 846 mAh g^{-1} after 150 cycles at 2 C was accomplished, like multi-walled CNTs (MWCNT) interlayer (Su and Manthiram 2012a, b; Zu et al. 2013).

Electrolyte

Electrolyte is the passage for the exchange of lithium+ among anodes and cathodes, and the electrochemical practices can be significantly disparate in different electrolytes for Li-S batteries, because of the different dissolvability and compound cooperation of polysulfides with different electrolyte solvents. Subsequently, determination of reasonable electrolytes is basically significant for Li-S battery frameworks. Right now, fluid natural electrolytes are the principal electrolytes applied in Li-S batteries. $LiClO_4$, $LiPF_6$, Li $(CF_3SO_2)_2N$, and LiTFSI are ordinarily utilized Li salts in lithium-sulfur frameworks. Contrasted and the normally utilized LiPF6 salt in LIBs, LiTFSI shows higher warm and hydrolytic security in Li-S batteries. The principal solvents used in Li-S batteries incorporate ether and carbonate. Ether dissolvable shows great affinity with carbon and sulfur (Wang et al. 2013b; Shim et al. 2002) (Fig. 2.11).

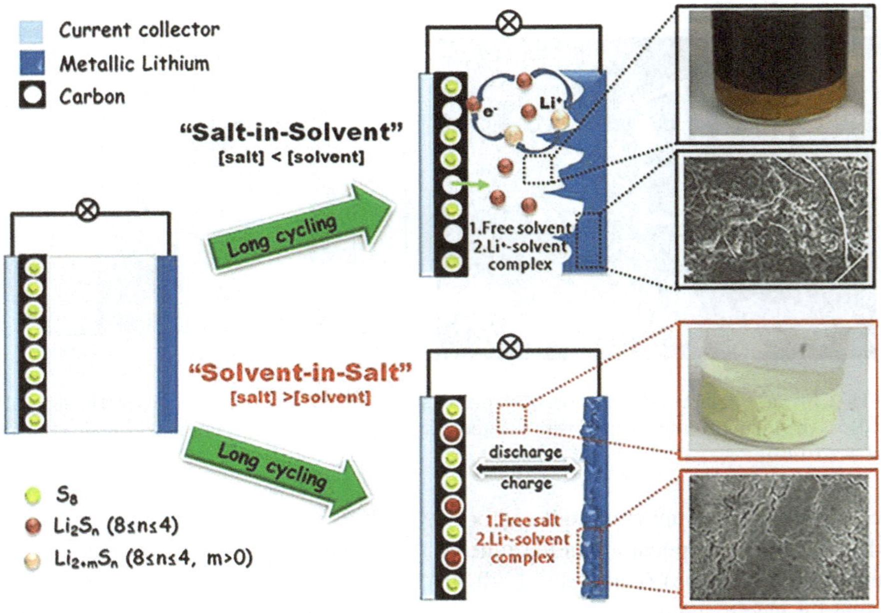

Fig. 2.11 Mechanism of the different electrolytes in the lithium-sulfur batteries (Gao et al. 2011)

References

Barchasz C, Molton F, Duboc C, Lepretre JC, Patoux S, Alloin F (2012) Lithium/sulfur cell discharge mechanism: an original approach for intermediate species identification. Anal Chem 84:3973–3980. https://doi.org/10.1021/ac2032244

Barghamadi M, Kapoor A, Wen C (2013) A review on Li-S batteries as a high efficiency rechargeable lithium battery. J Electrochem Soc 160:A1256–A1263. https://doi.org/10.1149/2.096308jes

Bresser D, Passerini S, Scrosati B (2013) Recent progress and remaining challenges in sulfur based lithium secondary batteries—a review. Chem Commun (Camb) 49:10545–10562. https://doi.org/10.1039/c3cc46131a

Bruce PG, Freunberger SA, Hardwick LJ, Tarascon JM (2012) Li-O2 and Li-S batteries with high energy storage. Nat Mater 11:19–29

Brückner J et al (2014) Carbon-based anodes for lithium sulfur full cells with high cycle stability. Adv Funct Mater 24(9):1284–1289

Cao YL et al (2011) Sandwich-type functionalized graphene sheet-sulfur nanocomposite for rechargeable lithium batteries. Phys Chem Chem Phys 13(17):7660–7665

Chang DR, Lee SH, Kim SW, Kim HT (2002) Binary electrolyte based on tetra(ethylene glycol) dimethyl ether and 1,3-dioxolane for lithium-sulfur battery. J Power Sources 112(2):452–460

Chen L, Shaw LL (2014) Recent advances in lithium-sulfur batteries. J Power Sources 267:770–783. https://doi.org/10.1016/j.jpowsour.2014.05.111

Chen Y, Li X, Park K, Song J, Hong J, Zhou L, Mai Y-W, Huang H, Goodenough JB (2013a) Hollow carbon-nanotube/carbon-nanofiber hybrid anodes for Li ion batteries. J Am Chem Soc 135(44):16280–16283

Chen HW et al (2013b) Ultrafine sulfur nanoparticles in conducting polymer shell as cathode materials for high performance lithium/sulfur batteries. Sci Rep 3:1910

Cheon SE, Ko KS, Cho JH, Kim SW, Chin EY, Kim HT (2003) Rechargeable lithium sulfur battery I. Structural change of sulfur cathode during discharge and charge. J Electrochem Soc 150: A796–A799

Choi YJ et al (2007) Electro chemical properties of sulfur electrode containing nano Al_2O_3 for lithium/sulfur cell. Phys Scr T129:62–65

Choi N-S, Chen Z, Freunberger SA, Ji X, Sun Y-K, Amine K, Yushin G, Nazar LF, Cho J, Bruce PG (2012) Challenges facing lithium batteries and electrical double-layer capacitors. Angew Chem Int Ed 51(40):9994–10024

Dixitar A (2020) Cathode materials for lithium ion batteries (LIBs): a review on materials related aspects towards high energy density LIBs. Xiv preprint arXiv:2008.10896. arxiv.org

Dunn B, Kamath H, Tarascon J-M (2011) Electrical energy storage for the grid: a battery of choices. Science 334(6058):928–935

Ellis BL, Lee KT, Nazar LF (2010) Positive electrode materials for Li-ion and Li-batteries. Chem Mater 22(3):691–714. https://doi.org/10.1021/cm902696j

Etacheri V, Marom R, Elazari R, Salitra G, Aurbach D (2011) Challenges in the development of advanced Li-ion batteries: a review. Energy Environ Sci 4:3243. https://doi.org/10.1039/c1ee01598b

Evers S, Nazar LF (2013) New approaches for high energy density lithium-sulfur battery cathodes. Acc Chem Res 46:1135–1143. https://doi.org/10.1021/ar3001348

Evers S, Yim T, Nazar LF (2012) Understanding the nature of absorption/adsorption in nanoporous polysulfide sorbents for the Li–S battery. J Phys Chem C 116(37):19653–19658

Fu Y, Manthiram A (2012) Ortho rhombic bi pyramidal sulfur coated with polypyrrole nano layers as a cathode material for lithium–sulfur batteries. J Phys Chem C 116(16):8910–8915

Gao J, Lowe MA, Kiya Y, Abruna HD (2011) Effects of liquid electrolytes on the charge-discharge performance of rechargeable lithium/sulfur batteries: electrochemical and in-situ X-ray absorption spectroscopic studies. J Phys Chem C 115(50):25132–25137

Ge Y, Jiang H, Fu K, Zhang C, Zhu J, Chen C, Lu Y, Qiu Y, Zhang X (2014) Copper doped Li4Ti5O12/carbon nanofiber composites as anode for high-performance sodium ion batteries. J Power Sources 272:860–865

Goodenough JB, Kim Y (2010) Challenges for rechargeable Li batteries. Chem Mater 22:587–603. https://doi.org/10.1021/cm901452z

Greenwood NN, Earnshaw A (1997) Chemistry of the elements, 2nd edn. Butterworth-Heinemann, Oxford

Gu M, Li Y, Li X, Hu S, Zhang X, Xu W, Thevuthasan S, Baer DR, Zhang J-G, Liu J, Wang C (2012) In situ TEM study of lithiation behavior of silicon nanoparticles attached to and embedded in a carbon matrix. ACS Nano 6(9):8439–8447

Guo JC, Xu YH, Wang CS (2011) Sulfur-impregnated disordered carbon nanotubes cathode for lithium–sulfur batteries. Nano Lett 11(10):4288–4294

Guo J, Yang Z, Yu Y, Abruña HD, Archer LA (2013) Lithium-sulfur battery cathode enabled by lithium-nitrile interaction. J Am Chem Soc 135(2):763–767

Hassoun J, Scrosati B (2010) A high-performance polymer tin sulfur lithium ion battery. Angew Chem Int Ed 49(13):2371–2374

Hayashi A, Ohtsubo R, Ohtomo T, Mizuno F, Tatsumisago M (2008) All-solid-state rechargeable lithium batteries with Li_2S as a positive electrode material. J Power Sources 183(1):422–426

Herbert D, Ulam, J (1962). Electric dry cells and storage batteries. US Patent 3,043, 896

Huang J-Q et al (2013) Entrapment of sulfur in hierarchical porous graphene for lithium– sulfur batteries with high rate performance from −40 to 60 °C. Nano Energy 2(2):314–321

Huang C et al (2014a) Manipulating surface reactions in lithium-sulphur batteries using hybrid anode structures. Nat Commun 5:3015

Huang JQ et al (2014b) Ionic shield for polysulfides towards highly-stable lithium-sulfur batteries. Energy Environ Sci 7(1):347–353

Huang J-Q, Zhang Q, Wei F (2015) Multi-functional separator/interlayer system for high-stable lithium-sulfur batteries: progress and prospects. Energy Storage Mater 1:127–145

Jeon BH, Yeon JH, Kim KM, Chung IJ (2002) Preparation and electrochemical properties of lithium–sulfur polymer batteries. J Power Sources 109:89–97
Ji X, Nazar LF (2010) Advances in Li-S batteries. J Mater Chem 20:9821. https://doi.org/10.1039/b925751a
Ji L, Zhang X (2009a) Fabrication of porous carbon nanofibers and their application as anode materials for rechargeable lithium-ion batteries. Nanotechnology 20(15):155705
Ji L, Zhang X (2009b) Electrospun carbon nanofibers containing silicon particles as an energy storage medium. Carbon 47(14):3219–3226
Ji L, Zhang X (2009c) Manganese oxide nanoparticle-loaded porous carbon nanofibers as anode materials for high-performance lithium-ion batteries. Electrochem Commun 11(4):795–798
Ji L, Zhang X (2010) Evaluation of Si/carbon composite nanofiber-based insertion anodes for new-generation rechargeable lithium-ion batteries. Energy Environ Sci 3(1):124–129
Ji L, Lin Z, Medford AJ, Zhang X (2009a) Porous carbon nanofibers from electro spun poly acrylonitrile/SiO2 composites as an energy storage material. Carbon 47(14):3346–3354
Ji XL, Lee KT, Nazar LF (2009b) A highly ordered nanostructured carbon-sulphur cathode for lithium-sulphur batteries. Nat Mater 8(6):500–506
Ji L, Yao Y, Toprakci O, Lin Z, Liang Y, Shi Q, Medford AJ, Millns CR, Zhang X (2010) Fabrication of carbon nanofiber-driven electrodes from electro spun poly acrylonitrile/poly pyrrole components for high-performance rechargeable lithium-ion batteries. J Power Sources 195(7):2050–2056
Ji L, Lin Z, Alcoutlabi M, Zhang X (2011a) Recent developments in nanostructured anode materials for rechargeable lithium-ion batteries. Energy Environ Sci 4(8):2682–2699
Ji X, Evers S, Black R, Nazar LF (2011b) Stabilizing lithium–sulphur cathodes using polysulphide reservoirs. Nat Commun 2:325–331
Kolosnitsyn VS, Karaseva EV (2008) Lithium-sulfur batteries: problems and solutions. Russ J Electrochem 44(5):506–509
Kulisch J, Sommer H, Brezesinski T, Janek J (2014) Simple cathode design for Li-S batteries: cell performance and mechanistic insights by in operando X-ray diffraction. Phys Chem Chem Phys 16:18765–18771
Li H, Wang Z, Chen L, Huang X (2009) Research on advanced materials for Li-ion batteries. Adv Mater 21:4593–4607. https://doi.org/10.1002/adma.200901710
Li Z et al (2014) Insight into the electrode mechanism in lithium-sulfur batteries with ordered microporous carbon confined sulfur as the cathode. Adv Energy Mater 4(7):1301473
Liang CD, Dudney NJ, Howe JY (2009) Hierarchically structured sulfur/carbon nanocomposite material for high-energy lithium battery. Chem Mater 21(19):4724–4730
Liang Z et al (2014) Sulfur cathodes with hydrogen reduced titanium dioxide inverse opal structure. ACS Nano 8(5):5249–5256
Liang Z et al (2016) Composite lithium metal anode by melt infusion of lithium into a 3D conducting scaffold with lithiophilic coating. Proc Natl Acad Sci U S A 113(11):2862–2867
Lin D et al (2016) Layered reduced graphene oxide with nanoscale interlayer gaps as a stable host for lithium metal anodes. Nat Nano 11(7):626–632
Liu Z, Fu W, Liang C (2011a) Lithium–sulfur batteries. In: Daniel C, Besenhard JO (eds) Handbook of battery materials, 2nd edn. Wiley-VCH Verlag GmbH & Co. KGaA. Chapter 24
Liu NA, Hu LB, McDowell MT, Jackson A, Cui Y (2011b) Prelithiated silicon nanowires as an anode for lithium ion batteries. ACS Nano 5(8):6487–6493
Lv W et al (2014) Tailoring microstructure of graphene-based membrane by controlled removal of trapped water inspired by the phase diagram. Adv Funct Mater 24(22):3456–3463
Manthiram A, Fu Y, Su Y-S (2012) Challenges and prospects of lithium-sulfur batteries. Acc Chem Res 46(5):1125–1134
Manthiram A, Fu Y, Su YS (2013) Challenges and prospects of lithium-sulfur batteries. Acc Chem Res 46:1125–1134. https://doi.org/10.1021/ar300179v
Manthiram A, Fu YZ, Chung SH, Zu CX, Su YS (2014) Rechargeable lithium-sulfur batteries. Chem Rev 114(23):11751–11787

Marmorstein D, Yu TH, Striebel KA, McLarnon FR, Hou J, Cairns EJ (2000) Electrochemical performance of lithium/sulfur cells with three different polymer electrolytes. J Power Sources 89:219–226

Meini S, Elazari R, Rosenman A, Garsuch A, Aurbach D (2014) The use of redox mediators for enhancing utilization of Li_2S cathodes for advanced Li–S battery systems. J Phys Chem Lett 5(5):915–918

Mikhaylik YV, Akridge JR (2003) Low temperature performance of Li/S batteries. J Electrochem Soc 150:A306–A311

Mikhaylik YV, Akridge JR (2004) Polysulfide shuttle study in the Li/S battery system. J Electrochem Soc 151:A1969–A1976

Mikhaylik Y, Kovalev I, Schock R, Kumaresan K, Xu J, Affinito J (2010) High energy rechargeable Li-S cells for EV application. Status, remaining problems and solutions. ECS Trans 25:23–34. https://doi.org/10.1149/1.3414001

Moon S, Jung YH, Jung WK, Jung DS, Choi JW, Kim do K (2013) Encapsulated monoclinic sulfur for stable cycling of Li-S rechargeable batteries. Adv Mater 25:6547–6553

Nan C et al (2014) Durable carbon-coated Li_2S core-shell spheres for high performance lithium/ sulfur cells. J Am Chem Soc 136(12):4659–4663

Obrovac MN, Dahn JR (2002) Electrochemically active lithia/metal and lithium sulfide/metal composites. Electrochem Solid State Lett 5(4):A70–A73

Pang Q, Liang X, Kwok CY, Nazar LF (2016) Advances in lithium-sulfur batteries based on multifunctional cathodes and electrolytes. Nat Energy 1:16132

Parfitt CE (2012) Characterization, modelling and management of lithium-sulphur batteries for spacecraft applications. Ph.D. thesis, University of Warwick

Pastorino C, Gamba Z (2000) Test of a simple and flexible molecule model for α-, β-, and γ-S8 crystals. J Chem Phys 112:282–286

Pastorino C, Gamba Z (2001) Toward an anisotropic atom–atom model for the crystalline phases of the molecular S8 compound. J Chem Phys 115:9421–9426

Qie L, Chen W-M, Wang Z-H, Shao Q-G, Li X, Yuan L-X, Hu X-L, Zhang W-X, Huang Y-H (2012) Nitrogen-doped porous carbon nanofiber webs as a nodes for lithium ion batteries with a super high capacity and rate capability. Adv Mater 24(15):2047–2050

Rao MLB (1968) Organic electrolyte cells. US Patent 3,413,154,

Rao BML, Shropshire JA (1981) Effect of sulfur impurities on Li/TiS2 cells. J Electrochem Soc 128:942–945

Seh Z et al (2013) Sulphur–TiO_2 yolk–shell nano architecture with internal void space for long-cycle lithium–sulphur batteries. Nat Commun 4:1331

Seh ZW et al (2014) Two-dimensional layered transition metal disulphides for effective encapsulation of high-capacity lithium sulphide cathodes. Nat Commun 5:5017

Shim J, Striebel KA, Cairns EJ (2002) The lithium/sulfur rechargeable cell. Effects of electrode composition and solvent on cell performance. J Electrochem Soc 149(10):A1321–A1325

Son Y, Lee JS, Son Y, Jang JH, Cho J (2015) Recent advances in lithium sulfide cathode materials and their use in lithium sulfur batteries. Adv Energy Mater 5(16):1500110

Song MK, Cairns EJ, Zhang Y (2013) Lithium/sulfur batteries with high specific energy: old challenges and new opportunities. Nanoscale 5:2186–2204. https://doi.org/10.1039/c2nr33044j

Su Y-S, Manthiram A (2012a) Lithium–sulphur batteries with a microporous carbon paper as a bifunctional interlayer. Nat Commun 3:1166

Su YS, Manthiram A (2012b) A new approach to improve cycle performance of rechargeable lithium–sulfur batteries by inserting a free-standing MWCNT interlayer. Chem Commun 48(70):8817–8819

Sun F et al (2013). A high-rate lithium-sulfur battery assisted by nitrogen-enriched mesoporous carbons decorated with ultrafine La_2O_3 nanoparticles. J Mater Chem A 1 (42):13283–13289. Song MS et al (2004) Effects of nano sized adsorbing material on electrochemical properties of sulfur cathodes for Li/S secondary batteries. J Electrochem Soc 151(6):A791–A795

Tao X et al (2014) Strong sulfur binding with conducting Magnéli-phase Ti_nO_{2n-1} nanomaterials for improving lithium-sulfur batteries. Nano Lett 14(9):5288–5294

Tarascon JM, Armand M (2001) Issues and challenges facing rechargeable lithium batteries. Nature 414(6861):359–367. https://doi.org/10.1038/35104644

Templeton LK, Templeton DH, Zalkin A (1976) Crystal structure of monoclinic sulfur. Inorg Chem 15:1999–2001

Villevieille C, Nova'k P (2013) A metastable β-sulfur phase stabilized at room temperature during cycling of high efficiency carbon fiber–sulfur composites for Li–S batteries. J Mater Chem A 1: 13089

Walus S, Barchasz C, Colin JF, Martin JF, Elkaim E, Lepretre JC, Alloin F (2013) New insight into the working mechanism of lithium-sulfur batteries: in situ and operando X-ray diffraction characterization. Chem Commun 49:7899–7901

Wang JL, Yang J, Xie JY, Xu NX (2002) A novel conductive polymer–sulfur composite cathode material for rechargeable lithium batteries. Adv Mater 14(13–14):963–965

Wang H, Cui L-F, Yang Y, Sanchez Casalongue H, Robinson JT, Liang Y, Cui Y, Dai H (2010) Mn3O4–graphene hybrid as a high-capacity anode material for lithium ion batteries. J Am Chem Soc 132(40):13978–13980

Wang H, Yang Y, Liang Y, Cui L-F, Sanchez Casalongue H, Li Y, Hong G, Cui Y, Dai H (2011a) LiMn1–xFexPO4 nanorods grown on graphene sheets for ultrahigh-rate-performance lithium ion batteries. Angew Chem 123(32):7502–7506

Wang HL et al (2011b) Graphene-wrapped sulfur particles as a rechargeable lithium-sulfur battery cathode material with high capacity and cycling stability. Nano Lett 11(7):2644–2647

Wang D-W et al (2012) A microporous-mesoporous carbon with graphitic structure for a high-rate stable sulfur cathode in carbonate solvent-based Li–S batteries. Phys Chem Chem Phys 14(24): 8703–8710

Wang D-W et al (2013a) Carbon-sulfur composites for Li–S batteries: status and prospects. J Mater Chem A 1(33):9382–9394

Wang X, Wang Z, Chen L (2013b) Reduced graphene oxide film as a shuttle-inhibiting interlayer in a lithium–sulfur battery. J Power Sources 242:65–69

Wang C et al (2015) Slurryless Li_2S/reduced graphene oxide cathode paper for high-performance lithium sulfur battery. Nano Lett 15(3):1796–1802

Wu F et al (2011) Sulfur/polythiophene with a core/shell structure: synthesis and electrochemical properties of the cathode for rechargeable lithium batteries. J Phys Chem C 115(13):6057–6063

Wu F, Lee JT, Zhao E, Zhang B, Yushin G (2016) Graphene–Li_2S–Carbon nanocomposite for lithium-sulfur batteries. ACS Nano 10(1):1333–1340

Xiao LF et al (2012) A soft approach to encapsulate sulfur: polyaniline nanotubes for lithium–sulfur batteries with long cycle life. Adv Mater 24(9):1176–1181

Xin S et al (2012) Smaller sulfur molecules promise better lithium-sulfur batteries. J Am Chem Soc 134(45):18510–18513

Xiong L, Xu Y, Tao T, Song J, Goodenough JB (2012) Excellent stability of spinel LiMn2O4-based composites for lithium ion batteries. J Mater Chem 22(47):24563–24568

Xu GY, Ding B, Pan J, Nie P, Shen LF, Zhang XG (2014) High performance lithium sulfur batteries: advances and challenges. J Mater Chem A 2:12662–12676. https://doi.org/10.1039/C4ta02097a

Yamaki JI, Tobishima SI, Sakurai Y, Saito KI, Hayashi K (1998) Safety evaluation of rechargeable cells with lithium metal anodes and amorphous V2O5 cathodes. J Appl Electrochem 28:135–140

Yamin H, Peled E (1983) Electrochemistry of a non-aqueous lithium/sulfur cell. J Power Sources 9: 281–287. https://doi.org/10.1016/0378-7753(83)87029-3

Yan K et al (2014) Ultrathin two-dimensional atomic crystals as stable interfacial layer for improvement of lithium metal anode. Nano Lett 14(10):6016–6022

Yang Y et al (2010) New nanostructured Li_2S/silicon rechargeable battery with high specific energy. Nano Lett 10(4):1486–1491

Yang Y et al (2012) High-capacity micrometer-sized Li_2S particles as cathode materials for advanced rechargeable lithium-ion batteries. J Am Chem Soc 134(37):15387–15394
Yang Y, Zheng GY, Cui Y (2013a) Nanostructured sulfur cathodes. Chem Soc Rev 42(7): 3018–3032
Yang Y, Zheng G, Cui Y (2013b) A membrane-free lithium/polysulfide semi-liquid battery for large-scale energy storage. Energy Environ Sci 6:1552. https://doi.org/10.1039/c3ee00072a
Yin S-C, Grondey H, Strobel P, Huang H, Nazar LF (2003) Charge ordering in lithium vanadium phosphates: electrode materials for lithium-ion batteries. J Am Chem Soc 125(2):326–327
Yin YX, Xin S, Guo YG, Wan LJ (2013) Lithium-sulfur batteries: electrochemistry, materials, and prospects. Angew Chem Int Ed 52(50):13186–13200
Zhang SS, Read JA (2012) A new direction for the performance improvement of rechargeable lithium/sulfur batteries. J Power Sources 200:77–82. https://doi.org/10.1016/j.jpowsour.2011.10.076
Zhang X-W, Patil PK, Wang C, Appleby AJ, Little FE, Cocke DL (2004) Electrochemical performance of lithium ion battery, nano-silicon-based, disordered carbon composite anodes with different microstructures. J Power Sources 125(2):206–213
Zhang B, Qin X, Li GR, Gao XP (2010) Enhancement of long stability of sulfur cathode by encapsulating sulfur into micropores of carbon spheres. Energy Environ Sci 3(10):1531–1537
Zhang FF, Zhang XB, Dong YH, Wang LM (2012) Facile and effective synthesis of reduced graphene oxide encapsulated sulfur via oil/water system for high performance lithium sulfur cells. J Mater Chem 22(23):11452–11454
Zhang Q, Cheng XB, Huang JQ, Peng HJ, Wei F (2014) Review of carbon materials for advanced lithium-sulfur batteries. New Carbon Mater 29(4):241–264
Zhao M-Q et al (2012) Graphene/single-walled carbon nanotube hybrids: one-step catalytic growth and applications for high-rate Li–S batteries. ACS Nano 6(12):10759–10769
Zheng W, Hu XG, Zhang CF (2006) Electro chemical properties of rechargeable lithium batteries with sulfur-containing composite cathode materials. Electrochem Solid State Lett 9(7):A364–A367
Zheng GY, Yang Y, Cha JJ, Hong SS, Cui Y (2011) Hollow carbon nanofiber-encapsulated sulfur cathodes for high specific capacity rechargeable lithium batteries. Nano Lett 11(10):4462–4467
Zheng G et al (2014) Interconnected hollow carbon nanospheres for stable lithium metal anodes. Nat Nanotechnol 9(8):618–623
Zhou Y, Wu C, Zhang H, Wu X, Fu Z (2007) Electrochemical reactivity of Co–Li_2S nanocomposite for lithium-ion batteries. Electrochim Acta 52(9):3130–3136
Zu CX, Su YS, Fu YZ, Manthiram A (2013) Improved lithium-sulfur cells with a treated carbon paper interlayer. Phys Chem Chem Phys 15(7):2291–2297
Zu C, Klein M, Manthiram A (2014) Activated Li_2S as a high-performance cathode for rechargeable lithium-sulfur batteries. J Phys Chem Lett 5(22):3986–3991

Chapter 3
Chemistry and Operation of Li-S Batteries

Nimra Jabeen, Aneeqa Husnain, Umair Azhar, Muhammad Arif, Muhammad Sagir, Muhammad Tanseer Hussain, and Muhammad Bilal Tahir

Abstract Lithium-sulfur (Li-S) batteries are promising high-energy-density energy storage systems. It is generally agreed that shuttle of the polysulfides in a functional battery is slowed by intense anchoring of the intermediates. However, there is still a lack of knowledge regarding the chemistry involved. Here, using advanced quantum chemical computations, we examine the Li bond chemistry in Li-S batteries with concept of hydrogen bond. The Li bond, a strong dipole-dipole connection between Li-S cathode materials and Li polysulfides, is facilitated by electron-rich donors like pyridinic nitrogen (pN), and is further strengthened by the conjugative and inductive effect of scaffold materials having π-electrons like graphene. This research elucidates the importance of Li bond chemistry in Li-S cells and provides a comprehensive knowledge, which is helpful for rational choice of cathode materials and implementation of Li-S batteries in the practical applications.

Because of sulfur's natural insulation, substantial volume expansion, shuttling of the soluble polysulfides, and, most critically, slow conversion of polysulfide intermediates, Li-S battery performance is still far from theoretical prediction. The electrochemical performance of Li-S batteries (e.g., lifespan, rate capability,

N. Jabeen · A. Husnain
Institute of Chemistry, Khwaja Fareed University of Engineering and Information Technology, Rahim Yar Khan, Punjab, Pakistan

U. Azhar · M. Arif · M. Sagir
Institute of Chemical and Environmental Engineering, Khwaja Fareed University of Engineering and Information Technology, Rahim Yar Khan, Punjab, Pakistan
e-mail: dr.umairazhar@kfueit.edu.pk; muhammad.arif@kfueit.edu.pk; dr.msagir@kfueit.edu.pk

M. T. Hussain
Air University School of Management (AUSOM), Air University Islamabad, Islamabad, Punjab, Pakistan

M. B. Tahir (✉)
Institute of Physics, Khwaja Fareed University of Engineering and Information Technology, Rahim Yar Khan, Punjab, Pakistan
e-mail: m.bilaltahir@kfueit.edu.pk

M. S. Tahir et al. (eds.), *Lithium-Sulfur Batteries: Key Parameters, Recent Advances, Challenges and Applications*, Springer Tracts in Electrical and Electronics Engineering, https://doi.org/10.1007/978-981-99-2796-8_3

cyclability, etc.) can be influenced by manipulating the architectures and functions of polymer materials; hence, they play an essential role in solving these problems. In this chapter a brief introduction of Li-S battery followed by its fundamental electrochemistry and challenges has been given. The mechanisms of operation of the Li-S batteries are then considered, with special emphasis on the uses of several polymers in all components.

Keywords Lithium-sulfur batteries · Li bond chemistry · Dipole-dipole interaction · Polymers · Electrochemical performance · Operations of Li-S battery

Introduction

With Li-ion batteries along their energy density approaching the extent, recently researchers have shifted their interest toward systems away from Li-ion. Due to low cost and abundance of the elemental sulfur, the lithium-sulfur (Li-S) batteries have been considered an advantageous choice (Zhang 2013). Conventionally the Li-S batteries are comprised of a cathode and an anode made of sulfur and lithium metals, respectively, and an electrolyte based on indispensable ether (Demir-Cakan 2017). Among numerous forms of sulfur, the ring structural octasulfur S_8 remains their stability at ambient temperature. The general electrochemical cell reaction based on Li-S redox reaction is expressed as

$$16\ \mathrm{Li} + \mathrm{S8} \leftrightarrow 8\ \mathrm{Li2S} \tag{3.1}$$

In result of complete reduction from the elemental sulfur to lithium sulfide (Li_2S), sulfur is anticipated to deliver an energy density about 2600 Wh Kg^{-1} and a specific capacity of 1675 Ah Kg^{-1}, which are 3–5 times higher than those of aspects of Li-ion batteries (Zhang 2013). Li-S battery (LSB) configuration working at room temperature acts for a beneficial option as they can provide low cost (about $150 per ton), high capacity (1672 mAh/g), low equivalent weight, and the environmentally favorable factors (Demir-Cakan 2017). To meet the required areal capacity and energy density of L-S batteries, an immense sulfur storing is required. In view to this, certain insightful reviews were published studying LSB anodes, cathodes, electrolytes, and multisystem designs for future advancement of LSBs to attain high sulfur storing (Rana et al. 2019).

Though lithium-sulfur batteries have been launched into the market in just a few single cases until now, it depicts that barriers need to be defeated and major challenges have to be resolved for this technology. An abundance of outstanding review articles prevail, summarizing the current state of research in fields of anode and cathode materials, cell design, processing conditions, the separator design, electrolyte systems, and many others (Borchardt et al. 2016). In spite of these precautions, a few studies have found that polysulfides are favorable for the cell life. Self-healing of the LSBs can be aided by establishing an efficient equilibrium between the precipitation and dissolution of Li polysulfides at the electrode interfaces, as revealed by Xu et al. Therefore, studies in the field of Li-S batteries have gradually shifted from focusing on these sulfur internments to utilizing the chemically generated dissolved polysulfides, either in a redox flow composition or a static condition. However, these polysulfides are used as enhanced cycling performance enhancers in electrolytes (Demir-Cakan 2017).

During this stage an important evolution was description of the electrolyte formulation applicable for the Li-S batteries. Frequent salt/solvent combinations had been proposed consisting of sulfone based, dimethyl sulfoxide, dimethylformamide, ionic liquid, tetraethylene glycol dimethoxyethane, dimethoxyethane, mixture of these specified solvents at various ratios, dioxolane or polyethylene oxide (PEO)-based polymer electrolytes, that is now safety measures of the Li-S batteries along an effective elimination of Li dendrite production (Demir-Cakan 2017).

The Li-S class of battery techniques these days already attains 160–350 Wh kg^{-1} in manufacturing cells, but it is currently restricted to niche small power applications demanding moderate cycle life. Solutions are needed to mark the unique characteristics of lithium-sulfur chemistry for broader consumption, which smother potential:

(i) Li-S cells have lithium-metal anodes, organic electrolytes, and active intermediates, all of which are sensitive to breakdown in the extreme conditions of a conventional battery.
(ii) The shuttle phenomenon at which point long-chain polysulfides disperse to anode and then react to design shorter polysulfides which in return diffuse toward cathode. That polysulfide shuttle wears down the lithium anode and devotes to self-discharge and the small Coulombic efficiency.
(iii) Solubility of mobile species directing to variable inner resistance and the increased exploitative mass over cell fabrication from binders, additives, and conductive carbons required to enhance cathode conductivity. Sulfur and its reductive product, Li-S, both are insulating and insoluble, resulting in dissolution chemical precipitation toward end of both sulfur (charge) and lithium sulfide (discharge) depending on their voltage. During cycling it results over active material waste (both the reversible and irreversible); furthermore, cathode and the separator decline, especially degradation of porous structure.
(iv) Charge/discharge characteristics like the number of the plateaus revealed in typical discharge curve, on contrary to the typical Li-ion curve. In consequence, applications engineers need new tools for understanding and to work with LSBs in applications.

Traditional research has directed on materials for addressing the challenges underlined in i–iii, and the numbers of latest publications are accessible summarizing loads of that struggle (Wild et al. 2015).

The Li-S system has a complex chemistry that includes many constituents and species within cell; thereby, its design necessarily requires a comprehensive approach that employs various characterization tools to understand why changes from one of the components affect its electrochemical behavior (Demir-Cakan 2017).

Chemistry of Li-S Batteries

Basic Property of Sulfur

There are more than 30 different solid allotropes of sulfur, making it the 17th most prevalent element in Earth's crust. In nature, cyclic octasulfur (S8) predominates, followed by cyclic tetrasulfur (S12). There are three different types of octasulfur, each with its own properties and uses. Thermodynamic stability means that b- and g-sulfur will always convert to a-sulfur when kept at normal temperature. A-sulfur begins to melt at 115.21 °C, boils at 444.6 °C, and quickly sublimes. When molten, sulfur exhibits a unique viscosity-temperature behavior. At temperatures above 190 ° C, sulfur begins to depolymerize and the viscosity drops, after having increased noticeably at roughly 160 °C as a result of polymerization and opening of S8 rings. This results in a minimum viscosity value of around 160 °C. This quality has been widely exploited for the creation of sulfur composites by penetrating sulfur into porous materials.

Above 160 °C, allotropes with 9–5, 18, and 20 sulfur atom rings emerge, and the quenching of all this melt produces amorphous sulfur, which is flexible and metastable. This amorphous sulfur returns to a crystalline allotrope when stored. Only vapor above 720 °C or somewhat lower temperatures at lower pressure contains allotropes with 2–5 sulfur atoms. When the temperature drops, these low atomic sulfur allotropes return to a solid allotrope (S_8).

In several polar electrolytic solvents, sulfur is only mildly soluble. Smaller polysulfide anions (Sn^{-2} with $n = 2$–5) are easily and consistently produced in solution, unlike elemental sulfur. Polysulfide solution (Li_2Sn with $n > 2$), for example, can be easily made by reacting elemental sulfur with lithium sulfide (Li_2S). Polysulfide anions can coexist in solution through a series of chemical equilibriums since their Gibbs free energies are so close to one another (Zhang 2013).

Li Bond Theory

Li bond theory was developed previously to explain the strong interaction between the lithium polysulfides and hosts consisting of the heteroatoms like nitrogen (N) and oxygen. Li bond between the Li and pyridinic N (pN) is consequently viewed as a dipole-dipole interaction, especially for pyridinic N (pN) in Fig. 3.1. However, a better understanding and use of this Li bond theory is critical for reasonable experimentation and material development.

To investigate the characteristics of Li bond and its significance in the Li-S chemistry, an intrinsic bond orbital analysis was carried out. To begin, the ionic LiN connection of the Li_3N crystal stands in stark contrast to the negligible negative electron transfer of 0.07 e- from PD to Li_2S_8. The transfer orientations of ionic bond

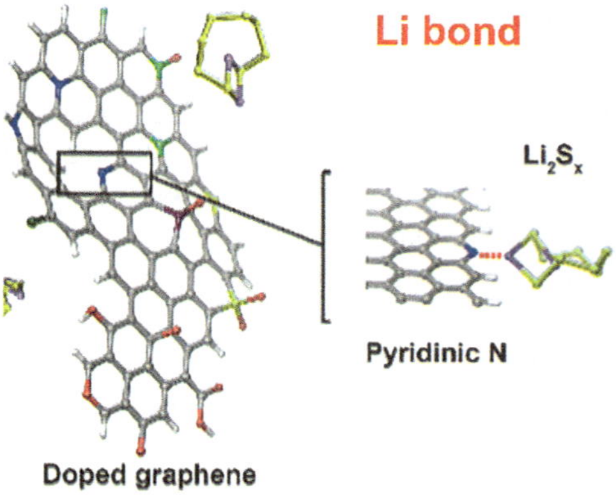

Fig. 3.1 Schematic diagram of Li bond in Li-S batteries (Zhang 2013; https://dornsife.usc.edu/labs/narayan/lithium-sulfur-batteries)

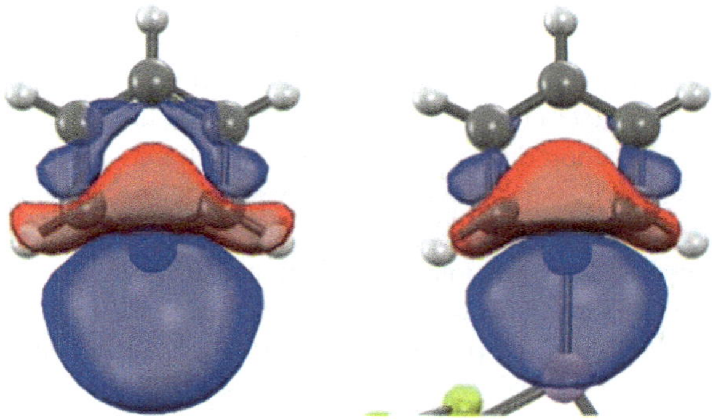

Fig. 3.2 Li Natural bond study of the Li_2S_8 before and after combining with PD, Li bond analysis in Li-S system (positive and the negative phases of orbital are shown by blue and red, correspondingly. C, N, H, Li, and S atoms are represented by gray, blue, white, purple, and yellow spheres, respectively) (Liu et al. 2019)

(LiN, Li to N) and the Li bond (Li…pN) are polar opposites, despite the large magnitude of the charge transfer. Li_2S_8's terminal Li acts as the acceptor for an electron donated by pN, which forms the Li bond. Therefore, the interaction between Li bonds is distinct from ionic bonds. Besides the dipole-dipole electrostatic interaction, the Li bond also possesses inductive and conjugative effects, as shown by the natural bond analysis (Fig. 3.2). The π-backdonation of metal-ligand complexes is remarkably reflected in the Li…pN bond's attraction to the π- electrons in the conjugated system of PD. Not only does the Li bond change the aromatic ring's electron dispersion, but it also changes overall bonding type of N. In particular, after PD binding with Li_2S_8, Wiberg bond ordering of the NC bond shifts between the ranges of 1.423 and 1.385. Li bond (Li…pN) has a bond order of 0.108. Finally, nature of the Li bond has been discovered to be a dipole-dipole interaction, with the dipole providing the majority of the bond strength, as well as additional inductive and conjugative effects (Hou et al. 2017).

Sulfur Cathode

Elemental sulfur was initially studied as a cathode substance for lighter metal (Li, Na)-dependent electrochemical cells in the 1960s, but it has lately piqued attention. Because of its benefits as a cathode substance for Li chargeable batteries, such as a great conceptual specific capacity, relatively inexpensive, an enormous amount, and little atmospheric effect, it has been extensively explored. The Li-S cell has piqued the curiosity of academics, as evidenced by the numbers of scientific publications and studies issued since 2010. The electrochemical process outlined below allows sulfur cathode in Li-S cell scheme to satisfy two Li ions per S atom:

$$S_8 + 16Li^+ + 16e^- \leftrightarrow 8Li_2S \left(E^0 = 2.15\ V\ vs.Li/Li^+\right) \quad (3.2)$$

The sulfur cathode's huge theoretical specific capacity (1675 mAh/gS) compensates for its reduced redox potential when contrasted to transition metal oxide cathodes, allowing Li-S cell to have theoretical specific energy of 2600 Wh/kg and a theoretical power density of 2800 Wh/L under the supposition of comprehensive Li_2S creation during the emission procedure, which is much higher than the Li-ion cell ($LiMO_2$/graphite system). Despite the encouraging benefits, numerous research and technological obstacles for the sulfur cathode must be resolved in terms of developing a functional Li-S cell; the establishment and degradation of Li_2S during the discharge and charge procedures result in a significant volumetric transformation of about 80%, respectively, leading in the sulfur cathode's structural breakdown.

The electrically shielding properties of S and Li_2S prevent total sulfur to Li_2S conversion. Essential elements for the development of the sulfur cathode may be determined by knowing the reactivity and breakdown processes of the sulfur cathode:

(i) Mechanical strength of the sulfur cathode throughout cycling
(ii) Advancement of the sulfur cathode's ionic and electronic conductivities
(iii) Safety to prevent Li polysulfides from dissolving into the electrolyte solution
(iv) Physical or chemical trapping of Li polysulfides to prevent the shuttle effect

These issues must be resolved to rationally design a sulfur cathode for a Li-S cell with large specific energy, and the cathode's design should not be limited to producing sulfur-based compounds. However, additional electrode elements like the binder, additives, and current collector should be taken into account.

When nanoengineered sulfur develops distinctive composite materials with complementary substances such as other mediums, the electrochemical efficiency of the sulfur cathode may be increased even further. Sulfur/porous substance nanocomposites are one of the best common patterns because nanopores not only give room for sulfur to expand during lithiation but also physically trap polysulfides in their space, a notion known as the "polysulfide reservoir."

With this approach, the cell may have a long cycle life while maintaining high Coulombic efficiency. Porous surfaces are often made of porous carbons or metal oxides. Because of the structural advantages of nanocomposites, most sulfur cathodes that have demonstrated encouraging cell performance use sulfur-based nanocomposites as the active ingredient rather than pure sulfur powder (Cairns and Hwa 2017).

Binder

All the sulfur species have been absorbed into the liquid electrolyte after discharge Region I, leaving many gaps in the cathode, whereas soluble PS deposited back onto cathode in both kinds of Li_2S and Li_2S_2 in discharge Region III. This necessitates a binder competent for keeping the cathode's extremely porous structure during the Li/S cell's cycling process. Because of gelling or expanding caused by electrolyte solvents, traditional binders like poly(ethylene oxide) (PEO) and poly(vinylidene fluoride) (PVDF) are unable to satisfy this need. Moreover, the intermediate of sulfur reduction is anionic polysulfide radical, which interacts through a wide range of organic polymers. A sign of an interaction between elemental sulfur and PVDF-HFP molecules, a potent thiol odor is produced when poly(vinylidene fluoride-cohexafluoropropene) (PVDF-HFP) is ball-milled with sulfur in acetonitrile solvent. Polymers containing unsaturated >C] C< bonds, like natural and synthetic rubbers, undergo vulcanization when PS and sulfur react with them. Therefore, an ideal cathode binder for a Li/S cell has to be chemically stable toward all the sulfur species and expand in the presence of liquid electrolyte (Zhang 2013).

Lithium Anode

Since the reduction potential of most commonly used anode (negative electrode) materials is near to that of reference (Li^+/Li) electrode potential, the positive electrode (cathode) determines the majority of a lithium battery's operating potential. Thus, positive electrode materials are considered to be the vital component of a rechargeable cell, with the majority of current applied and basic scientific research devoted to them. Extensive efforts are made to look for components that have high volumetric and gravimetric energy densities at a low price (Piątek et al. 2021).

Because of its low electrochemical potential (3.040 V versus a normal hydrogen electrode) and light weight (molecular weight 6.94 g mol^1, density 0.534 g cm^3) compared to other metals, Li metal is selected as an anode in the Li-S batteries (Kim et al. 2013). Dendritic and mossy Li-metal deposits occur over continuous charge/discharge cycles, causing low CE, limited cycle life, and security risks when using lithium metal as an anode in the nonaqueous electrolyte cells. A passivation layer rapidly forms on metallic Li anode surface due to the presence of polysulfides along with electrolyte additives within Li-S batteries. This coating will cause the Li-S battery to collapse due to higher cell resistivity (Cao et al. 2015).

Several solutions have been devised to limit Li anode corrosion and lessen their rise in impedance under working circumstances, involving usage of the electrolyte additives, cosolvents, interlayers, and polymer electrolytes (Su and Manthiram 2012). Concentrated electrolytes have recently been shown to have decreased polysulfide solubility, which might contribute to longer cycle stability in the Li-S batteries. Stable cycling and high CE for Li-S batteries were claimed to be provided by a new "solvent-in-salt" electrolyte family consisting of lithium bis (trifluoromethanesulfonyl)imide [LiTFSI or $(LiN(SO_2CF_3)_2]$ in 1,2-dimethoxyethane (DME) and 1,3-dioxalane (DOL) at high salt concentrations (up to 7 M). By reducing the number of unwanted side reactions that occur between Li-metal and polysulfide species, electrolytes with a greater salt concentration were able to reduce corrosion to the Li anode and prevent the growth of Li dendrites. Lithium bis(fluorosulfonyl)imide [$(LiN(SO_2F)_2$ or LiFSI)], a concentrated electrolyte, has been described as being used in Li-S devices to create an insulating barrier on both the anode and cathode sides (Kim et al. 2015). Furthermore, lithium salt anions have been discovered to play critical roles in forming solid electrolyte interphase layer and, to a degree, regulation of the electrochemical efficiency (Cao et al. 2016).

Electrolyte

Beyond guaranteeing strong ionic conductivity, the electrolyte serves a key function in Li-S battery. As already stated, the type of discharge in the Li-S batteries is dependent on solubility of the redox-active LiPS, which mediates charge transfer reaction kinetics. When it comes to designing and improving Li-S batteries' electrolytes, this poses a distinct set of problems and possibilities. Li-S electrolyte should be able to solvate and tolerate a wide range of lithium polysulfide species due to solution-mediated reaction processes. While the breakdown of lithium polysulfides is necessary, it also has the side effect of a parasitic shuttling of the active material to anode. During riding, that results in high active material loss. Another factor to consider is the reactive lithium-metal anode's long-term stability in the availability of selected electrolytes. As a result, the electrolyte must provide adequate ionic conductance while also ensuring good LiPS solubility, minimizing LiPS shuttle effect, and the long-term stability with Li-metal anode. Most often utilized ether-based electrolyte is presently a finest equilibrium among these several factors.

Ionic conductivity and LiPS solubility are both quite good in this electrolyte. When it comes to polar aprotic solvents, ethers are among the most stable over Li metal. The endeavor to extend Li-metal's long-term stability and reduce LiPS shuttling gives substantial prospects for liquid electrolytes, but it will need continuing research into novel electrolyte solvents, salts, and additives.

Investigation of novel electrolyte structures, such as solid-state electrolytes and sparsely solvating electrolytes, is also possible. To inhibit LiPS dissolution in sparsely solvating electrolytes, the salt coordinates the solvent completely. In all methods, battery must rely upon solid-state reduction from the elemental sulfur to Li_2S to avoid LiPS dissolution. This might result in considerable savings in terms of LiPS shuttling and the long-term Li-metal stabilization. The low carbon percentage required for practical use, along with the high sulfur loadings, presents substantial kinetic problems. Despite the abundance of options, it is essential to keep in mind that every new electrolyte framework should strive to fulfill the parameters discussed in this article. Ongoing investigation into fully solvating, partially solvating and the solid-state electrolytes may eventually help to meet the stringent design requirements of the Li-S battery (Bhargav et al. 2020).

Solid electrolytes shown in lithium-ion, sodium-ion, and notably Li-S batteries are regarded as fundamental benefit for realistic battery solutions toward attaining high safety, high power density, and high energy density in all solid-state lithium batteries. Lithium-ion mobility in the solid electrolyte has influenced the electrochemical efficiency of solid-state batteries. With the advancement of solid electrolyte technology, solid electrolytes with high lithium-ion conductivity in the range of 104 S cm^1 at room temperatures, such as oxide-type and sulfide-type electrolytes, have shown tremendous promise in all solid-state lithium batteries (Yuan et al. 2019).

To counterbalance low working voltage (2.1 V vs. Li/Li^+) and equal high capacity of the sulfur, high-energy-density Li-S cells require employment of the Li-metal-based anode.

Therefore, one of most difficult difficulties and major possibilities that require development in the Li-S batteries has permitting steady cycling of Li-metal anode. Abundant Li, anode design, and electrolyte deterioration are all critical factors in unlocking Li-S performance in this area (Bhargav et al. 2020). In comparison with LIB technology, the particular Li-S battery requirements on the electrolyte are higher. Li-metal anode's stability vs. polysulfides (PSs) and elemental sulfur—both radicals and anions—and known regulated solubility of the PSs are the three main criteria.

Each of these requirements will be discussed in detail below since they are the primary reasons for our current emphasis on creating innovative Li-S battery electrolytes. The PS redox shuttle mechanism is the most frequently stated aspect of the Li-S batteries, and this is mostly connected to the last criteria listed above. In summary, PSs generated at the sulfur-containing cathode spread across the electrolyte and are converted to intractable Li-PS species, which coat the Li-metal anode (Johansson et al. 2017).

Lithium Polysulfide Shuttle

In a basic Li-S system, soluble LiPS moving back and forth between electrodes is referred to as the "LiPS shuttle," which causes irreversible sulfur loss, continual anode corrosion, and amplified electrolyte breakdown, resulting in low discharge capacity and battery cycle stability (Yang et al. 2020). The entire influence of both electric field and concentration gradients of LiPSs across electrodes causes such shuttle behavior in LiPSs, posing a severe challenge in the areas of thermodynamics and kinetics. As a result, enhanced LiPS shuttle mitigation is required to achieve a good battery life (Yu et al. 2021).

Carbonaceous substances, including porous carbon, carbon nanotubes (CNTs), and graphene (oxide) are intensively investigated as a second media for sulfur cathode due to their reasonable conductivity, Li permeability, and other advantages based on their physical and chemical composition. Polymers that are elastic or conducting are also being studied as a secondary medium to increase the sulfur cathode's electrochemical efficiency. Inorganic compounds such as oxides and sulfides, as well as metal-organic frameworks (MOFs), have been researched in addition to carbonaceous substances and polymers. Chemical surface functionalization that can chemically trap polysulfide species to reduce the polysulfide shuttle effect might improve these physical combination ideas even further (Cairns and Hwa 2017).

Even though shuttle effect in the Li-S batteries could be reduced by covering and making composites with the carbon-based materials, these methods would not address inherent material deterioration that occurs in these batteries. During repeated redox reaction, the substantial expansion of volume generated by conversion of the S into Li_2S causes the coating components to decompose (Li et al. 2019). Because of its general specific capacity of 1166 $mAhg^1$, Li_2S, the last discharge product of the sulfur, has piqued attention as pure S replacement in recent times (Son et al. 2015). When Li_2S is used as cathode for the Li-S batteries, it is expected that substantial volume variations may be avoided. During the charging period, Li_2S particles shrink, leaving many blank spaces and gaps for sulfur to fill (Qiu et al. 2015).

Operation of Li-S Battery

Lithium-sulfur batteries are the most favorable energy storage systems for future generations. Li-S batteries have acquired vast attention due to their great theoretical energy density. The battery is an appliance that enables its retained chemical energy in its active substance to be changed directly into electric-powered energy through an electrochemical oxidation and reduction reaction. In contrast with Li-ion technologies, lithium-sulfur batteries supply an increase of threefold in energy density. The lithium-sulfur (Li-S) batteries' unique specific energy density theoretical value of 2600 W h kg^{-1}has displayed it as superior future battery technology (Lopez et al. 2020).

Primary Characteristics

A few terms like functional capacity, power, and energy are characteristic of battery it can provide, and these above features are defined by the chemical effects of cell per component. The electrical energy of the battery is measured by either per unit of volume (W hL^{-1}) or per unit of weight (W h kg^{-1}); battery internal resistance is an operation that regulates Coulombic efficiency (%), capacity (Ah kg^{-1}), and cell potential (V). The gravimetric energy density is expressed by per unit of weight (W h kg^{-1}) that resembles the quantity of per unit mass energy stored by the battery. An essential parameter volumetric energy density widely is expressed in W hL^{-1}. The energy density depends on the accumulator voltage discharge and the mass storage capability of the electrodes. The power density of a battery is the ratio of the capacity gained to the volume in W L^{-1}, while the power-to-weight ratio is the ratio of the amount of power generated by a cell or battery to its weight in W kg^{-1} (Yermukhambetova 2017).

Performance Measures

When the lithium-sulfur battery cell begins to operate, its active substance converts its solid form into a soluble phase. This process means that the material does not bound inside the structure of the electrode. After dissolving in the electrolyte, it can also impact its properties, e.g., conductivity, viscosity, and resistance, and the active substance returns to its solid form Li_2S at the discharge end. These characteristics exhibit the contrast between the conventional lithium-ion batteries, where Li^+ ions are de-intercalated between the host electrodes (cathode and anode), while active substance remains in electrode system and the electrolyte structure is not involved (Walus 2015).

Lithium-sulfur battery is distinguishable from the standard Li-ion because Li-S displays an active mechanism inside the cell that is essential and of great significance to have its detailed understanding. Lithium-sulfur cells' organic components that enter the cell are an additional, significant benefit beyond their energy storage capabilities. In comparison with Li-ion cathodes based on heavy metals, sulfur is a safer and more cost-effective option for the environment. Lithium-sulfur could be generated at a smaller scale (around $100 per kWh), making it difficult to assess the cost deduction compared with Li-ion at this time. In usage, the lithium-sulfur battery state of charge can be discharged at 0% and does not demand recharging when in storage (Fotouhi et al. 2017).

The following measures must be satisfied previous successful commercialization of the battery:

(i) **Specific energy**

The total quantity of energy that is stored per unit volume or mass.

(ii) **High power density**

The discharge of the highest energy rate per unit volume or mass. High current density at the electrode requires optimal ionic and electronic conductivity.

(iii) **Protection**
Specific battery cell parts at high temperatures disintegrate and go through an exothermic reaction that directs reduction for safe function.
(iv) **Cycle strength**
In order to meet the durability standards of a wide range of applications, specific energy and power density must be stable under constant cycling.
(v) **Expense**
Must contend with different existing energy storage devices.
(vi) **Chemical and mechanical stability**
The chemical structure and mechanical properties, its composition, temperature, and surface over period in the functional conditions are maintained by the electrodes. The chemical and mechanical stability constraints come from the phase change, electrolyte reaction, low conductivity, and corrosion of materials in their discharged condition.
(vii) **Degree of operation**
(viii) For automotive and service use, a battery can withstand temperatures between −50 and 80 °C. The temperature range for typical use is between 0 and 40 °C. The ability of batteries to function in harsh environments depends on the interfaces between their electrolytes.
(ix) **Shape of discharge curve**
For the function of an electronic appliance, a low-voltage plateau is advantageous. A diagonal curve is favored for applications when specifying the state of charge, and the curve figure shows sloping and two-phase reactions (plateau) (Yermukhambetova 2017).

Mechanism of Li-S Batteries

A standard lithium-sulfur cell is made up of a metallic lithium negative electrode and the sulfur-based positive electrode, split within a porous polymer separator embedded in an organic electrolyte. The binding agent used in lithium-sulfur cells is composed of synthetic polymers that are electronically and ionically conducting materials.

The purpose of the binding agent is to actively bind the conductive agent and an active positive material to current collector and improve its mechanical probity of electrode. The main conditions to be fulfilled by a binding agent include simple wet-ability of the electrolyte and vast solubility in the solvent. Sulfur at the cathode initiates a conductive system between the conductive agent and active material in polysulfides which retains the porous form. For cathode preparation, the organic solvent is based on the high binding agent and low sulfur solubility to prevent the settling of dense sulfur particles that might assemble an insulating coating on the current collector (Walus 2015).

A Li-S battery system is established on electrochemical reaction of lithium with sulfur to produce the final reaction product, lithium sulfide (Li_2S), which is the final reaction product. Sulfur is favorable as a cathode element because of its ability to intercalate two electrons at once, moreover to its lower expense and instinctive abundance. The optimization of sulfur cathodes has two major classifications, changing the structure form and integrating strategies of conductive chemical addition, and cathodes initiate from its polysulfides. The ring structure of eight sulfur atoms formed an S_8 molecule having strong symmetry at standard pressure and temperature.

The sulfur cathode develops an insulating character that demands a supportive electron-conductive additive like carbon black. To prevent Li^+ ions backup in the pathway and for the advantage of fair electrical contact electron transport is reduce, cathode stable structural dependability in the advancement of a cell becomes essential aspect mainly to conduct highly homogeneous and highly permeable diffusion of sulfur (Baloch 2016).

Lithium-sulfur cell experiences an electrochemical redox reaction and begins with a release which happens because of sulfur charged state. The cell function starts from discharge when sulfur is in its charged state, while oxidation reaction happens at metallic lithium at the negative electrode and the Li^+ ion formed with loss of an electron. The diffused lithium-ions move toward the positive electrode (cathode) via the electrolyte, and electrons travel through an external source. The cell components in a typical Li-S and its charge and discharge operation happen in conducting polar electrolyte with metallic lithium at the anode and produce lithium-ion.

At the negative electrode, the oxidation reaction occurs as given below:

$$16\,Li^0 \rightarrow 16\,Li^+ + 16\,e^- \tag{3.3}$$

The lithium-sulfur battery has large practical and theoretical specific energy which is the main benefit for its application. Specific energy can be obtained by integrating a sulfur cathode with a lithium-metal anode. The theoretical value of a lithium cell at the cathode is 2500 $Wh.kg^{-1}$ in Watt-hour and has an energy capacity of 1675 $mAh.g^{-1}$ by the reduction of 16 electrons from the S_8 molecule. The reduction reaction happens at the positive electrode; the sulfur atom will be reduced by accepting an electron from lithium metal (Fotouhi et al. 2017).

According to the following the equation

$$S_8 + 16\,Li \rightarrow 8\,Li_2S \tag{3.4}$$

Inside the lithium-sulfur battery, the overall reaction occurs at the voltage of 2.15 V which is shown in Eq. (3.5):

$$S_8 + 16Li + 16e \rightleftharpoons 8Li_2S \tag{3.5}$$

Li_2S is insoluble in the electrolyte and electrically insulating material, and Li_2S is formed near the end while discharging. The few sulfur species that remain during reduction have high solubility in aprotic solvents.

A principal challenge to the lithium-sulfur battery is due to multiple reactions that happen from S_8 to Li_2S. In contrast with the typical process of intercalation in Li-ion batteries, sulfur goes through a series of electrochemical oxidation-reduction reactions. The following series of redox reactions are given below:

$$2Li^+ + S_8 + 2e^- \leftrightarrow 2Li_2S_8 \quad (a)$$

$$2Li^+ + Li_2S_8 + 2e- \leftrightarrow 2Li_2S_6 \quad (b)$$

$$2Li^+ + Li_2S_6 + 2e- \leftrightarrow 2Li_2S_4 \quad (c)$$

$$2Li^+ + Li_2S_4 + 2e- \leftrightarrow 2Li_2S_2 \quad (d)$$

$$2Li^+ + Li_2S_2 + 2e- \leftrightarrow 2Li_2S \quad (e)$$

The operation (charge and discharge) of lithium-sulfur batteries relies on the successful transition of each reaction. Similarly, the physical characteristics of these species diverge remarkably, e.g., Li_2S and S_8 are solid and not soluble in standard electrolytes. Correspondingly, the intermediate reduction species of polysulfide has a high solubility in organic solvents and also in ionic polymers and liquids. Upon discharge, the molecules of sulfur atom (S_8) gain electrons and show a reduction reaction which conducts high-order lithium polysulfides Li_2Sn ($6 < n \leq 8$) at the upper plateau having voltage around 2.3–2.4Li2S V (Baloch 2016) (Fig. 3.3).

As the reduction persists, further discharge of polysulfides is carried out which steps down voltage at 2.1 V, and formation of lower-order lithium polysulfides exists having a chain reaction of Li_2Sx ($2 < \times \leq 6$). In ether-based liquid electrolytes, the plateau of 2.3 V–2.1 V occurs due to discharge and represents transformations of a sulfur molecule (S_8) to Li_2S_4 and Li_2S_4 to Li_2S, respectively (Demir-Cakan 2017).

The opposing reaction charging and oxidation occur when enforced with an exterior electric field, which has a definite potential difference that leads to decay of the Li_2S to Li and S. In this technique, the electrochemical potential at cathode slowly expands. The molecular weight of sulfur is moderately light and has an advantage over positive electrode transition metal oxide material because of its electrons gaining per one atom of sulfur.

A specificity of this technique mainly lies in its complex functional mechanism. The general reaction between lithium and element sulfur formed lithium sulfide product, and it does not show intercalation. It is not a direct reaction as the solid sulfur reduction is followed by the arrangement of a species family, called lithium polysulfides (Li_2S_n, $3 < n \leq 8$); they are readily dissolved in organic electrolyte. The distinct polysulfides are deeply correlated with the chain length and features based upon their color and solubility. We can determine polysulfide groups on the basis of their chain length:

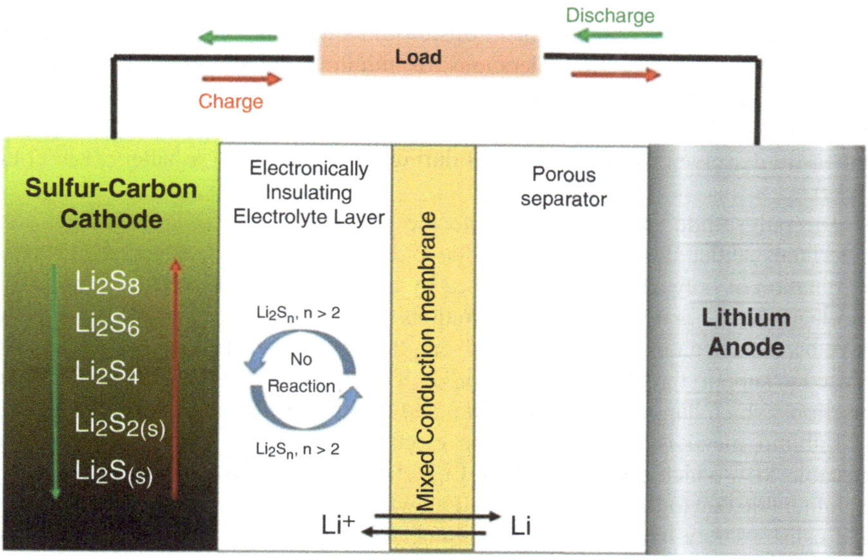

Fig. 3.3 Polysulfides species in Li-sulfur battery (https://dornsife.usc.edu/labs/narayan/lithium-sulfur-batteries)

(i) "Long-chain" and "high-order" polysulfides: the species which have the long chain length, for example, Li_2S_8 and Li_2S_6.
(ii) "Intermediate order" is particularly Li_2S_4, fairly and easily solubilized.
(iii) "Short-chain" and "low-order" polysulfides are the following species: Li_2S_3, Li_2S_2, and Li_2S.

The high-order polysulfides (Li_2S_n, n = 4–8) are dissolved into electrolyte and low-order polysulfides, for example, Li_2S_2 and Li_2S adsorbed as solid species onto the electrode.

The medium lithium polysulfides can cause the degeneration of lithium at the anode, and it is the primary hindrance for efficient application of the lithium-sulfur batteries. The stable generation of solid electrolyte interphase layer stops production of the lithium dendrite and lithium precipitation. This formation of the SEI layer directs the liquid phase into a solid phase as the cycle proceeds in Li-S batteries. The following techniques are specified to prevent hindrance, and different characterization procedures have been practiced to increase the intermediate lithium polysulfides' performance by the SEI layer formation method.

Limitations

The method limitations appear from various cell parts: an electrolyte, negative, and positive electrode. Lithium-sulfur batteries have many challenges that slow cell operation. The sulfur molecule (S_8) and lithium sulfide (Li_2S) are electrical insulators, having 5×10^{-30} S cm^{-1} of intrinsic conductivity at 25 °C. This feature has many significances for the operations of lithium sulfide cells. Lithium-sulfur rechargeable batteries upon reduction of sulfur show a formation of Li polysulfides (Li_2S_n, $n \geq 2$) which get dissolved in the aprotic solvents. These polysulfide species are easily dissolved in the electrolytes during the operation of a battery (Liu et al. 2019).

The polysulfide species migrate into the separator from the cathode electrode. Later, they settled down at the anode due to the loss of active substances. The toxic elimination of polysulfides is established by the following techniques to protect the Li electrode, e.g., solid-state configurations of polymer membranes and ceramic have been carried out. In an aprotic solvent, the electrolyte additive (lithium nitrate) is used to support the SEI layer at the anode. The electrolyte additive avoids parasitic reactions which happen between lithium and polysulfide (Demir-Cakan 2017).

Lithium metal is reactive, and the electrolyte reacts at the anode to initiate unstable SEI, which directs to utilize Li metal and electrolyte. Similarly, Li metal is vulnerable due to the dendrite formation, which potentially conducts short circuits and causes safety hazards. The development of dendrites also guides to constant reforming and breaking of the SEI layer, which also destroys electrolyte and Li metal. The broker dendrites will be detached from the anodes and isolated Li will reduce the Coulombic efficiency (Fig. 3.4).

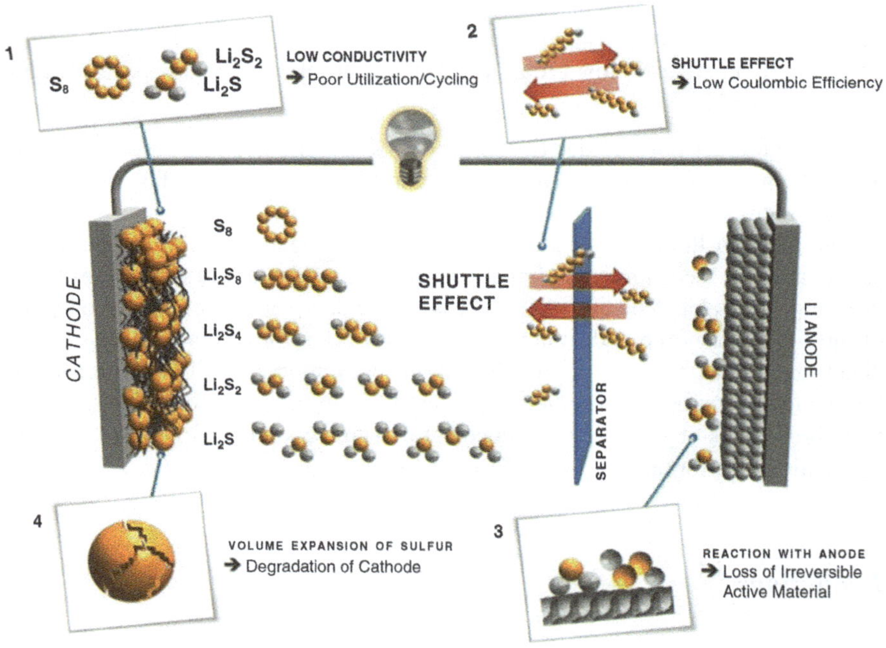

Fig. 3.4 Limitations of Li-sulfur battery (Yang 2020)

The shuttle effect is another biggest issue encountered by the Li-S battery because of soluble species formation. It happens due to the high solubility of long-chain lithium polysulfide species Li_2S_n ($4 < n \leq 8$). During charging, short-chain species experience a sequence of oxidation-reduction reactions and increase their solubility in solution to long-chain species. At cathode, the concentration of the long-order species is vast, so they disperse through the concentration gradient at the anode. The long-chain polysulfides reach the anode and go through parasitic reaction Li metal to break down the chain into a lower order. Within the growth of short-chain Li species, the concentration increases and one cycle of shuttle mechanism is completed when species dispersed back to the sulfur electrode. The self-discharge in lithium-sulfur batteries caused by shuttle effect and parasitic reaction creates insoluble lithium sulfide, which deposits at the anode. The deposition of Li_2S on the electrode surface results in permanent loss of capacity and active material (McCreary et al. 2021).

The gels and aqueous-based systems minimize the high-order polysulfide formation and increase solubility capacity. The lithium electrode modifies with surface chemistry to preserve it from the polysulfide species shuttle effect and sustain its electrolytes interface. The current compositions of high sulfur-content materials and supported cathode structures are chemically tuned, and enough mechanical power has displaced sulfur elements (Lopez et al. 2020).

References

Baloch M (2016) Strategies towards performance enhancement in lithium-sulphur batteries. Doctoral dissertation, Universidad del País Vasco-Euskal Herriko Unibertsitatea

Bhargav A, He J, Gupta A, Manthiram A (2020) Lithium-sulfur batteries: attaining the critical metrics. Joule 4(2):285–291

Borchardt L, Oschatz M, Kaskel S (2016) Carbon materials for lithium sulfur batteries—ten critical questions. Chemistry Eur J 22(22):7324–7351

Cairns EJ, Hwa Y (2017) Sulfur cathode. In: Li-S batteries: the challenges, chemistry, materials, and future perspectives. World Scientific, London, p 31

Cao R, Xu W, Lv D, Xiao J, Zhang JG (2015) Anodes for rechargeable lithium-sulfur batteries. Adv Energy Mater 5(16):1402273

Cao R, Chen J, Han KS, Xu W, Mei D, Bhattacharya P, . . . Zhang JG (2016) Effect of the anion activity on the stability of Li metal anodes in lithium-sulfur batteries. Adv Funct Mater 26(18): 3059–3066

Demir-Cakan R (2017) Introduction to rechargeable lithium–sulfur batteries. Li-s batteries: the challenges, chemistry, materials, and future perspectives, 1

Fotouhi A, Auger DJ, O'Neill L, Cleaver T, Walus S (2017) Lithium-sulfur battery technology readiness and applications—a review. Energies 10(12):1937

Hou TZ, Xu WT, Chen X, Peng HJ, Huang JQ, Zhang Q (2017) Lithium bond chemistry in lithium–sulfur batteries. Angew Chem 129(28):8290–8294

https://dornsife.usc.edu/labs/narayan/lithium-sulfur-batteries

Johansson P, Demir-Cakan R, Hayashi A, Tatsumisago M (2017) Lithium–sulfur battery electrolytes. In: Li-S batteries: the challenges, chemistry, materials, and future perspectives. World Scientific, London, p 149

Kim H, Jeong G, Kim YU, Kim JH, Park CM, Sohn HJ (2013) Metallic anodes for next generation secondary batteries. Chem Soc Rev 42(23):9011–9034

Kim H, Wu F, Lee JT, Nitta N, Lin HT, Oschatz M, ... Yushin G (2015) In situ formation of protective coatings on sulfur cathodes in lithium batteries with LiFSI-based organic electrolytes. Adv Energy Mater 5(6):1401792
Li T, Bai X, Gulzar U, Bai YJ, Capiglia C, Deng W, ... Proietti Zaccaria R (2019). A comprehensive understanding of lithium–sulfur battery technology. Adv Funct Mater 29(32):1901730
Liu Y, Peng B, Wang X, Xie Y, Schaefer HF (2019) The nature of lithium bonding in C2H2Li2, C6Li6, and lithium halide dimers. Organometallics 38:4708–4716
Lopez CV, Maladeniya CP, Smith RC (2020) Lithium-sulfur batteries: advances and trends. Electrochem 1(3):226–259
McCreary C, An Y, Kim SU, Hwa Y (2021) A perspective on Li/S battery design: modeling and development approaches. Batteries 7(4):82
Piątek J, Afyon S, Budnyak TM, Budnyk S, Sipponen MH, Slabon A (2021) Sustainable Li-ion batteries: chemistry and recycling. Adv Energy Mater 11(43):2003456
a) Qiu Y, Rong G, Yang J, Li G, Ma S, Wang X, ... Zhang Y (2015) Highly nitridated graphene–Li2S cathodes with stable modulated cycles. Adv Energy Mater 5(23):1501369. b) Sun D, Hwa Y, Shen Y, Huang Y, Cairns EJ (2016) Li2S nano spheres anchored to single-layered graphene as a high-performance cathode material for lithium/sulfur cells. Nano Energy 26:524–532
Rana M, Ahad SA, Li M, Luo B, Wang L, Gentle I, Knibbe R (2019) Review on areal capacities and long-term cycling performances of lithium sulfur battery at high sulfur loading. Energy Storage Mater 18:289–310
Son Y, Lee JS, Son Y, Jang JH, Cho J (2015) Recent advances in lithium sulfide cathode materials and their use in lithium sulfur batteries. Adv Energy Mater 5(16):1500110
Su YS, Manthiram A (2012) Lithium–sulphur batteries with a microporous carbon paper as a bifunctional interlayer. Nat Commun 3(1):1–6
Walus S (2015) Lithium/sulfur batteries: development and understanding of the working mechanisms. Doctoral dissertation, Université Grenoble Alpes
Wild M, O'Neill L, Zhang T, Purkayastha R, Minton G, Marinescu M, Offer GJ (2015) Lithium sulfur batteries, a mechanistic review. Energy Environ Sci 8(12):3477–3494
Yang L·(2020) A review of cathode materials in lithium-sulfur batteries. Ionics. Springer. https://link.springer.com
Yang C, Li P, Yu J, Zhao LD, Kong L (2020) Approaching energy-dense and cost-effective lithium–sulfur batteries: from materials chemistry and price considerations. Energy 201:117718
Yermukhambetova A (2017) Development of lithium sulphur battery and insights into its failure mechanism. Doctoral dissertation, UCL (University College London)
Yu S, Cai W, Chen L, Song L, Song Y (2021) Recent advances of metal phosphides for Li–S chemistry. J Energy Chem 55:533–548
Yuan H, Peng HJ, Huang JQ, Zhang Q (2019) Sulfur redox reactions at working interfaces in lithium–sulfur batteries: a perspective. Adv Mater Interfaces 6(4):1802046
Zhang SS (2013) Liquid electrolyte lithium/sulfur battery: fundamental chemistry, problems, and solutions. J Power Sources 231:153–162

Chapter 4
Advanced Characterization Techniques for Li-S Batteries

Ayesha Tariq, Muhammad Sagir, Lubna Jahangeer, Shahzad Murtaza, and M. B. Tahir

Abstract Since the creation of their first prototype in the 1960s, lithium-sulfur (Li-S) batteries have been anticipated to be a viable high-energy-density secondary battery system. By addressing the issues at the lab-level model systems, significant progress has been made over the past 10 years in promoting the performance of Li-S batteries. Li-S battery applications are receiving more attention; however, the bottleneck is being caused by brand-new difficulties at realistic cell sizes. In this review, we first introduce the lithium-sulfur batteries and describe their principle. Then, we elaborate all the recent advances and finally concluded along with future research objectives.

Keywords Lithium-sulfur (Li-S) batteries · High-energy-density · Energy applications

Literature Review

Because of overabundance of natural resources which has seriously harmed ecosystem, it's crucial to create sustainable and innovative fuels alongside highly effective storage mechanisms. Discovering cutting-edge electrically powered lithium batteries having suitable thermal and electrical capacities over use in mobile gadgets, smart

A. Tariq (✉) · L. Jahangeer · M. B. Tahir
Institute of Physics, Khawaja Fareed University of Engineering and Information Technology, Rahim Yar Khan, Punjab, Pakistan
e-mail: m.bilaltahir@kfueit.edu.pk

M. Sagir
Institute of Chemical and Environmental Engineering, Khawaja Fareed University of Engineering and Information Technology, Rahim Yar Khan, Punjab, Pakistan
e-mail: dr.msagir@kfueit.edu.pk

S. Murtaza
Institute of Chemistry, Khawaja Fareed University of Engineering and Information Technology, Rahim Yar Khan, Punjab, Pakistan

M. S. Tahir et al. (eds.), *Lithium-Sulfur Batteries: Key Parameters, Recent Advances, Challenges and Applications*, Springer Tracts in Electrical and Electronics Engineering, https://doi.org/10.1007/978-981-99-2796-8_4

networks, as well as hybrid cars is of tremendous concern. High capacity along with inexpensive stuff for electrodes is crucial for continuing lithium-ion battery advancements in real life. The present article intends to provide a consideration of new developments in lithium-sulfur battery packs as well as related tactics to earn their operational sustainability, i.e., efficient means to improve capacity of lithium-ion batteries include raising end result execution potential along with stuff for electrode capability (Xu et al. 2023).

According to research by Ohta et al., a buffer film of $LiNbO_3$ with a nanoscale measurement has been embedded across $LiCoO_2$ as well as sulfur compound solid salt to been approach to lessen friction via electrode as well as improve features regarding tangible lithium batteries. This offered promising opportunities to ensure secure and viable use of lithium battery backups (Ohta et al. 2007). Being able to detect lithium mobility within small-scale batteries while power pack is operating is provided by neutron depth profiling (NDP), according to subsequent research by Oudenhoven et al. Incorporation with an overloaded $6LiCoO_2$ electrode provides additional information into ways in which transport occurs taking place beneath every layered power source which is working on a small scale, while it is shown that such NDP outcomes are generally compatible to outcomes of electrolytic experiments (Oudenhoven et al. 2011).

In years since 2013, we witnessed advances in our understanding of the physical as well as physiological development of components of electrodes due to in situ underprocess of transmission electron microscopy (TEM) investigations of battery packs containing lithium ion employing a martyr open-cell format. According to research by Gu et al., the salt solution within a typical open-cell arrangement comprises a type of ionized fluid or hard lithium oxide. To address such a requirement, they created an operando TEM electrolytic fluid cell, which provides an authentic battery arrangement as well as a pertinent fluid mineral, a point that is proven via the fact that only one Si nanowire lithiation or de-lithiation conduct whether they use a fluid cell or not matches up about the findings compared to open-cell examinations (Gu et al. 2013). In order to tackle vital problems, recent advancements within Li-S power sources, which involve S-negative electrode; Li-positive electrode, along with fluid; as well as novel forms of Li-S battery packs featuring a conductive Li-free electrode, are discussed by Y.-X. Yin et al. They further describe the chemical makeup about Li-S power sources and go over challenges leading of electrical activity, like electro-activity as well as polysulfide disintegration. A potential technique regarding manufacturing sector within decades to come is to build S compounds that are enclosed in conducting small pore containing carbon compounds to increase durability of lithium-ion battery cells (Yin et al. 2013). Sulfur has been found to be a desirable cathode substrate alongside a large specific power about 1673 $mAhg^{-1}$, as well as Wei Seh et al. have shown how to devise a sulfur-TiO_2 yolk-shell nanoscale engineering using inner gaps for sulfur's volumetric growth while maintaining an unaffected TiO_2 wrap to reduce polysulfide bankruptcy. Around 0.5 C, the starting particular power about 1030 $mAhg^{-1}$ is attained, along with a 1000 round Coulombic yield of 98.4%. Additionally, endurance of large round time battery packs with lithium-sulfur has improved, with their power degradation over 1000 rounds being as low as 0.033% each round (Wei Seh et al. 2013). Lithium-sulfur (Li-S) packs have a potential electrical power that is 3–5 times greater compared to lithium-ion ones, as reported by S. S. Zhang et al. Because of multiple problems, involving tiny lifespan, low assurance, small going productivity, as well as an excessive autonomy velocity that contributes with disintegration regarding lithium polysulfide (PS), an assortment of reduced sulfur precursors, within electrolytic solution as well as ensuing invading processes via lithium electrode alongside fluid elements, marketing of Li-S batteries has remained unsuccessful. However, it is additionally necessary to ensure Li-S cell's operation since a Li-S cell can't function gradually because natural sulfur as well as its reduced merchandise is insulator (Zhang 2013).

Yao et al. provide convincing graphic proof which shows upon electrocatalytic recharging or empty for aqueous polysulfide, tangible sulfur as well as lithium sulfide entities predominantly lodge upon Sn-doped indium oxide opposed to carbon. They additionally construct carbon nanofibers having Sn-doped indium oxide nanoscale species adorning their outer layer like blended 3D separators to enhance variety about installation points in order to weave that idea about controlled space as better power source separators. Blended electrode can be loaded to 4.3 mg cm^{-2} before reaching permitted amount of Li_2S_8 capability, which is 1470 $mAhg^{-1}$ (sulfur density), with an intake of C/5 (Yao et al. 2014). Q. Wang et al. use in situ electron paramagnetic resonance (EPR) technology to show that sulfur ions may be directly seen. Subsequently shown chemistry as well as electric reactions in Li-S cells stimulate themselves to continue via sulfur radicals, causing both entirely separate responses avenues while released as well as charged. This depends upon variations in amount regarding sulfur ions based on various powers along with electrolytic features within compartment. It offers fresh angles on study of relations among sulfur molecules along with fluid, igniting creative approaches to creation of Li-S cells (Wang et al. 2015).

In 2017, investigation into lithium (Li) metallic science has accelerated owing to an alluringly large hypothetical potential as well as its essential function as part of future-oriented Li-S as well as Li-air packs. Through conventional Freon gas R134a as product, Lin et al. create a ubiquitous LiF film on the external layer of Li. Using balanced cells, it was demonstrated that a LiF coating on 3D stacked Li-rGO conductors greatly decreased adverse effects but improved cycling reliability lacking excessive potential enhancement during more than 200 rounds. Additionally, Li-S batteries having LiF-shielded Li-rGO show greatly enhanced durability along with Coulombic yield whereas preserving rapid growth potential (800 $mAhg^{-1}$ at 2 C) (Lin et al. 2017a). New advances into large-sulfur filling lithium-ion (Li-S) cells are highlighted by H.-J. Peng et al. by using multi-scale organization approaches. Basic understanding of interface processes, tiny fabrication techniques, original topologies, as well as structural creativity within electrodes, along with isolator, has a high importance. It is suggested that multi-scale architecture order will direct subsequent research on large-sulfur filling Li-S cells (Peng et al. 2017). Usable electrical power obtained from a Li-S cell is governed via a variety of architectural criteria; many of those are frequently disregarded, resulting in the inability to satisfy business demands. In their discussion on how to develop dependable Li-S power sources, Fang et al. analyze how different factors affect useable energy content (Fang et al. 2017). Among increasingly popular options to future-oriented Li power sources, notably Li-S as well as Li-air infrastructure involves usage of metal-based Li. Metal-based Li is finally set on a comeback while spending many years in obscurity due to safety issues, owing to advancement of forensic techniques along with nanotechnology-based approaches. By researching existing knowledge of Li electrodes, latest significant developments in material engineering along with sophisticated methods of characterization, as well as prospects alongside possible next steps for Li electrodes in usage, are created by Lin et al. (2017b). Among the most important concern about modern life is how to improve electrolytic preservation of energy. There are difficulties in hunt for novel supplies for batteries as well as desire to increase efficiency while reducing expense of both new and old cells. Recognizing foundational techniques of substance along with relationships among each element concerned is unquestionably an initial step in achieving fulfillment in such fields. In their study about cell investigation, Pecher et al. emphasize use on solid form nuclear magnetic resonance (NMR) spectral analysis, a method that may be very effective in identifying regional patterns in materials used for batteries throughout excessively disorganized environments (Pecher et al. 2017). Due to its strong electrical properties and outstanding processing ability, argyrodite Li_6PS_5Cl was found by Auvergniot et al. to be a promising hard electrolyte over massive lithium-ion cells. Li_6PS_5Cl is oxygenated onto fundamental lithium core on connection alongside anode influential supplies, according to their investigation through connection durability of argyrodite regarding multiple anode supplies, including $LiCoO_2$, $LiNi_{1/3}Co_{1/3}Mn_{1/3}O_2$, and $LiMn_2O_4$ (Auvergniot et al. 2017). Fundamentally sulfides generate low overpotential because Li_2S breakdown energy barrier is connected to interaction between separated ions. Some of those compounds can be useful as activation catalysts to speed up conversion of Li_2S into sulfur, according to Zhou et al. (2017). Koerver et al. synthesized $LiNi_{0.8}Co_{0.1}Mn_{0.1}O_2$, NCM-811. Chemo-mechanical contraction of active material upon de-lithiation causes contact loss between solid electrolyte and active material particles, further increasing interfacial resistance and capacity loss where majority of interphase formation takes place (Koerver et al. 2017).

According to research work performed in 2018 by B. Li et al., solid electrolyte interphase is unstable and inhomogeneous. Additionally, from the viewpoint of anode, efficient solutions to these issues are presented, such as surface engineering, controlling electric field, and controlling lithium-ion flow (Li et al. 2018). In their discussion of how atomic force microscopy may be utilized technique to help researchers understand common electrode materials (Wang et al. 2018). Sulfide materials are viewed as enabling material for high-energy bulk-type all-solid-state batteries due to superionic conductivities (as high as 10^2 S cm^{-1}) and deformability of Li-ion batteries. In their analysis of recent developments in liquid-phase synthesis and solution of SEs, Park et al. emphasized potential for maintaining strong ionic connections and creating sheet-type electrodes (Park et al. 2018). By adjusting various concentrations of solid electrolytes and mass loading, Nam et al. thoroughly analyze electrochemical performance of traditional dry-mixed electrodes and wet-slurry produced electrodes for ASLBs. When cut using scissors and heated at 110 °C, pouch-type 80 60 mm^2 all-solid-state $LiNi_{0.6}Co_{0.2}Mn_{0.2}O_2$/graphite complete cells show high cell-based energy density (184 $Whkg^{-1}$ and 432 WhL^{-1}) (Nam et al. 2018).

While CNTs and graphene do not interact electrochemically with lithium ions and electrons, they do alter behavior of a particular electroactive material while storing lithium, expanding spectrum of uses for a lithium battery. Fang et al. came at this conclusion in 2019 (Fang et al. 2019). More and more scientists are concentrating on triggering anionic redox process to increase energy and power density of Na-ion batteries. A new viewpoint on Na-rich materials is provided by X. Zhang et al.'s discovery of Mn-based $Na_{1.2}Mn_{0.4}Ir_{0.4}O_2$ (NMI) (Zhang et al. 2019). Li/CuO@Ti-mesh anodes with a high Coulombic efficiency (CE) of 94.2% at 10 $mAcm^{-2}$ over 90 cycles are created by Huang et al. They felt that limital anode will be able to go on to beneficial applications due to ultrahigh rates and extraordinarily extended cycle life of LCTM anode (Huang et al. 2019). Stability of metallic Li via interphase design or microstructure development is examined by Guo et al. together with benefits and limitations of first 3D scaffolds. Given complexity of dynamic problem of high-rate cycling of metallic Li, a scaffold with high mixed ionic and electronic conductivity could be an effective remedy (Guo et al. 2019). Tan et al. noted reversible cycling and an electrolyte capacity of 965 $mAhg^{-1}$ in half-cells made of Li-In | Li_6PS_5Cl | Li_6PS_5Cl-C. Sulfur (S) and phosphorus pentasulfide (P_2S_5) are produced by Li_6PS_5Cl during charging; however, during discharge, these products are first reduced to a Li_3PS_4 intermediate and then produce lithium sulfide (Li_2S) and lithium phosphide (Li_3P) (Tan et al. 2019).

Coulombic efficiency (CE) values of at least 99.7%, and preferably >99.9%, are required for Li-metal cells to live up to claims made by Zheng et al. regarding highly reversible Li anodes, which raises prospects for significant advancement toward commercially relevant Li-metal batteries. A detailed examination of literature reveals that realistic CE values have plateaued at roughly 99.5%, which is troubling (Zheng et al. 2020). Densely linked interfaces created TiO_2 on Ni_3S_2 surface, according to R. Wang et al., which are essential for bidirectional catalysis. TiO_2 gathers polysulfides for reduction, while Ni_3S_2 catalyzes conversion. Both TiO_2 and Ni_3S_2 exhibit catalytic activity when Li_2S is dissolved during oxidation, refilling catalyst surface (Wang et al. 2020b). A reduced graphene oxide "host" framework for Li-metal anodes was created by H. Wang et al., and it was improved by adding silicon (Si) nanoparticles in between graphene layers. In the absence of pre-stored Li, they act as Li nucleation seeds to encourage Li deposition inside framework. Li_xSi alloy particles act as "pillars" or supports between graphene layers after metallic Li has been fully removed, preventing thickness shrinkage. 99.4% Coulombic efficiency over 600 cycles was achieved when paired with a Li-compatible electrolyte, and constant cycling of a Li||NMC532 complete cell over 380 cycles with minimal capacity loss was realized (Wang et al. 2020a). Due to its advantageous characteristics for Li-metal storage, Hyeon et al. suggest using porous carbon materials produced from zeolitic imidazolate frameworks (ZIFs) as host materials. This approach, which is quite practical, focuses on exploiting host materials to store Li-metal among other ways (Hyeon et al. 2020). Progress of useful Li-metal batteries is currently acknowledged as requiring creation of a SEI on Li anodes by Zhai et al. Recent advances in analytical techniques have increased our understanding of SEI's composition and fine structure, which has sped up creation of stable Li-metal anode interface engineering (Zhai et al. 2020). Development of ASSBs has significantly advanced identification highly conductive SSEs. Tan et al. explored a range of different SSEs utilizing conductivity and chemical stability as assessment criteria, assessed SSEs beyond traditional criteria, and provided some insight into a number of bulk, interfacial, and nanoscale phenomena (Tan et al. 2020). According to Krauskopf et al., solid electrolyte separators can serve as a mechanical barrier to prevent lithium from dispersing dendritically throughout cell, which holds particular potential for development of high-performing lithium metal anodes. However, contact loss and uneven electrodeposition make employing a lithium metal anode in solid-state batteries generally difficult (Krauskopf et al. 2020). Contrary to what is often believed, W. Yin et al. disclose a throughout crystallites as opposed to simply at surface, extending cycle $Li_{1.2}Ni_{0.13}Mn_{0.54}Co_{0.13}O_2$ (Yin et al. 2020). Near-atomic structural examinations of Li dendrites and solid electrolyte interphases are made possible by high spatial resolution offered by scanning probe imaging and electron microscopy. Analysis of Li microstructure formation in huge cells is made possible by high beam penetration depth offered by X-ray-based technologies. Neutron imaging based on 6Li is more sensitive than X-ray imaging and can show Li-ion concentration in relation to sample depth (Foroozan et al. 2020). Y.-G. Lee et al. demonstrate how thin Ag-C layer may effectively control deposition of Li, resulting in a battery with a really long electrochemical cycle life (Lee et al. 2020).

Fundamental problems preventing Li-S batteries from working properly are high migration and leakage of soluble lithium polysulfide intermediates out of cathodes during cycling. Methods for anchoring polysulfides are categorically divided into physical confinement and chemical bonding based on electrochemistry of Li-S batteries (Zhou et al. 2021). Horstmann et al. address probable pathways for reversible cycling of two-dimensional lithium metal. They also talk about how our knowledge of lithium metal nucleation, deposition, and stripping at nanoscale has improved. At electrode level, morphological instabilities can be reduced, and significant shape change can be prevented by using artificial interphases and three-dimensional host frameworks (Horstmann et al. 2021). Theoretically, smooth Li are present in polymorphous Li deposits on Li-metal anodes. These Li polymorphs mostly form in reaction to various physical fields, including temperature field, etc. (Zou et al. 2021). Understanding dynamic development of electrodes, solid electrolytes, and their interfaces and interphases during cycling of ASSLBs is essential for improving performance and attaining practical applications of ASSLBs. According to Cheng et al., in situ TEM is frequently helpful for revealing underlying processes and offering design guidance for creation of high-performance ASSLBs (Cheng et al. 2021). A development strategy for an effective oxide and sulfide-based ASSLB is presented, with an emphasis on interfacial challenges and consideration of five factors. This is accomplished by considering the historical evolution of technology

(Kim et al. 2021). Recently, interest in rechargeable metal (Li, Na, Mg, Al)-sulfur batteries for electrical energy storage has increased due to low cost of cathode and ease of obtaining elemental sulfur. Development of metal-sulfur batteries is hampered by a number of visible barriers. Primary problems with Li-S and Na-S batteries are cathode conductivity problems, polysulfides shuttle, and anode dendritic problems. There are insufficient accessible appropriate electrolytes for Mg-S and Al-S batteries (Liu et al. 2021). A 1-A-h level pouch cell can run for 80 cycles with >95% Coulombic efficiency and produce >300 Whkg^{-1} with just a 100% excess of lithium (Zhao et al. 2021a). In order to study oxygen redox reactions in cathode materials from atomistic level to microstructure regime, G.-H. Lee et al. use a multi-scale methodology (Lee et al. 2021). N. Li et al. create a novel idea for layered-rocksalt intergrown structures by integrating benefits of each phase. This idea led to development of lithium nickel ruthenium oxide, which has a high capacity and an intergrown rocksalt primary layered structure (Li et al. 2021). Due to our growing reliance on high-energy-density lithium-ion batteries, interest in layered lithium-rich cathode materials is predicted to revive after a lengthy period of slumber (Zhao et al. 2021b).

$LiNi_{0.8}Mn_{0.1}Co_{0.1}O_2$ and Li-metal anode are examined by W. Li et al. in the present year. Due to weak oxidation stability of ether-based electrolyte, transition metal ions dissolve in substantial amounts, which causes dendrite growth on anode and quick cell death. An ester-based electrolyte, on the other hand, demonstrates robust electrochemical performance at 4.5 V with little crossover impact (Li et al. 2023). In this chapter, we briefly review all work done for enhancement of Li-S batteries, i.e., advanced characterization regarding lithium-sulfur batteries.

References

Auvergniot J, Cassel A, Ledeuil J-B, Viallet V, Seznec V, Dedryvère R (2017) Interface stability of argyrodite Li6PS5Cl toward LiCoO2, LiNi1/3Co1/3Mn1/3O2, and LiMn2O4 in bulk all-solid-state batteries. Chem Mater 29(9):3883–3890. https://doi.org/10.1021/acs.chemmater.6b04990

Cheng Y, Zhang L, Zhang Q, Li J, Tang Y, Delmas C, . . . Huang J (2021) Understanding all solid-state lithium batteries through in situ transmission electron microscopy. Mater Today 42:137–161. https://doi.org/10.1016/j.mattod.2020.09.003

Fang R, Zhao S, Sun Z, Wang D-W, Cheng H-M, Li F (2017) More reliable lithium-sulfur batteries: status, solutions and prospects. Adv Mater 29(48):1606823. https://doi.org/10.1002/adma.201606823

Fang R, Chen K, Yin L, Sun Z, Li F, Cheng H-M (2019) The regulating role of carbon nanotubes and graphene in lithium-ion and lithium–sulfur batteries. Adv Mater 31(9):1800863. https://doi.org/10.1002/adma.201800863

Foroozan T, Sharifi-Asl S, Shahbazian-Yassar R (2020) Mechanistic understanding of Li dendrites growth by in-situ/operando imaging techniques. J Power Sources 461:228135. https://doi.org/10.1016/j.jpowsour.2020.228135

Gu M, Parent LR, Mehdi BL, Unocic RR, McDowell MT, Sacci RL, . . . Wang C-M (2013) Demonstration of an electrochemical liquid cell for operando transmission electron microscopy observation of the lithiation/delithiation behavior of Si nanowire battery anodes. Nano Lett 13(12):6106–6112. https://doi.org/10.1021/nl403402q

Guo W, Liu S, Guan X, Zhang X, Liu X, Luo J (2019) Mixed ion and electron-conducting scaffolds for high-rate lithium metal anodes. Adv Energy Mater 9(20):1900193. https://doi.org/10.1002/aenm.201900193

Horstmann B, Shi J, Amine R, Werres M, He X, Jia H, . . . Latz A (2021) Strategies towards enabling lithium metal in batteries: interphases and electrodes. Energy Environ Sci 14(10): 5289–5314. https://doi.org/10.1039/D1EE00767J

Huang K, Li Z, Xu Q, Liu H, Li H, Wang Y (2019) Lithiophilic CuO nanoflowers on Ti-mesh inducing lithium lateral plating enabling stable lithium-metal anodes with ultrahigh rates and ultralong cycle life. Adv Energy Mater 9(29):1900853. https://doi.org/10.1002/aenm.201900853

Hyeon Y, Lee J, Qutaish H, Han SA, Choi SH, Moon SW, . . . Kim JH (2020) Lithium metal storage in zeolitic imidazolate framework derived nanoarchitectures. Energy Storage Mater 33:95–107. https://doi.org/10.1016/j.ensm.2020.07.015

Kim KJ, Balaish M, Wadaguchi M, Kong L, Rupp JLM (2021) Solid-state Li–metal batteries: challenges and horizons of oxide and sulfide solid electrolytes and their interfaces. Adv Energy Mater 11(1):2002689. https://doi.org/10.1002/aenm.202002689

Koerver R, Aygün I, Leichtweiß T, Dietrich C, Zhang W, Binder JO, . . . Janek J (2017) Capacity fade in solid-state batteries: interphase formation and chemomechanical processes in nickel-rich layered oxide cathodes and lithium thiophosphate solid electrolytes. Chem Mater 29(13): 5574–5582. https://doi.org/10.1021/acs.chemmater.7b00931

Krauskopf T, Richter FH, Zeier WG, Janek J (2020) Physicochemical concepts of the lithium metal anode in solid-state batteries. Chem Rev 120(15):7745–7794. https://doi.org/10.1021/acs.chemrev.0c00431

Lee Y-G, Fujiki S, Jung C, Suzuki N, Yashiro N, Omoda R, . . . Han IT (2020) High-energy long-cycling all-solid-state lithium metal batteries enabled by silver–carbon composite anodes. Nat Energy 5(4):299–308. https://doi.org/10.1038/s41560-020-0575-z

Lee G-H, Lau V W-h, Yang W, Kang Y-M (2021) Utilizing oxygen redox in layered cathode materials from multiscale perspective. Adv Energy Mater 11(27):2003227. https://doi.org/10.1002/aenm.202003227

Li B, Wang Y, Yang S (2018) A material perspective of rechargeable metallic lithium anodes. Adv Energy Mater 8(13):1702296. https://doi.org/10.1002/aenm.201702296

Li N, Sun M, Kan WH, Zhuo Z, Hwang S, Renfrew SE, . . . Tong W (2021) Layered-rocksalt intergrown cathode for high-capacity zero-strain battery operation. Nat Commun 12(1):2348. https://doi.org/10.1038/s41467-021-22527-z

Li W, Jie Y, Chen Y, Yang M, Chen Y, Li X, . . . Jiao S (2023) Crossover effects of transition metal ions in high-voltage lithium metal batteries. Nano Res 16(6), 8417–8424. https://doi.org/10.1007/s12274-022-5334-y

Lin D, Liu Y, Chen W, Zhou G, Liu K, Dunn B, Cui Y (2017a) Conformal lithium fluoride protection layer on three-dimensional lithium by nonhazardous gaseous reagent Freon. Nano Lett 17(6):3731–3737. https://doi.org/10.1021/acs.nanolett.7b01020

Lin D, Liu Y, Cui Y (2017b) Reviving the lithium metal anode for high-energy batteries. Nat Nanotechnol 12(3):194–206. https://doi.org/10.1038/nnano.2017.16

Liu X, Li Y, Xu X, Zhou L, Mai L (2021) Rechargeable metal (Li, Na, Mg, Al)-sulfur batteries: materials and advances. J Energy Chem 61:104–134. https://doi.org/10.1016/j.jechem.2021.02.028

Nam YJ, Oh DY, Jung SH, Jung YS (2018) Toward practical all-solid-state lithium-ion batteries with high energy density and safety: comparative study for electrodes fabricated by dry- and slurry-mixing processes. J Power Sources 375:93–101. https://doi.org/10.1016/j.jpowsour.2017.11.031

Ohta N, Takada K, Sakaguchi I, Zhang L, Ma R, Fukuda K, . . . Sasaki T (2007) LiNbO3-coated LiCoO2 as cathode material for all solid-state lithium secondary batteries. Electrochem Commun 9(7):1486–1490. https://doi.org/10.1016/j.elecom.2007.02.008

Oudenhoven JFM, Labohm F, Mulder M, Niessen RAH, Mulder FM, Notten PHL (2011) In situ neutron depth profiling: a powerful method to probe lithium transport in micro-batteries. Adv Mater 23(35):4103–4106. https://doi.org/10.1002/adma.201101819

Park KH, Bai Q, Kim DH, Oh DY, Zhu Y, Mo Y, Jung YS (2018) Design strategies, practical considerations, and new solution processes of sulfide solid electrolytes for all-solid-state batteries. Adv Energy Mater 8(18):1800035. https://doi.org/10.1002/aenm.201800035

Pecher O, Carretero-González J, Griffith KJ, Grey CP (2017) Materials' methods: NMR in battery research. Chem Mater 29(1):213–242. https://doi.org/10.1021/acs.chemmater.6b03183

Peng H-J, Huang J-Q, Cheng X-B, Zhang Q (2017) Review on high-loading and high-energy lithium–sulfur batteries. Adv Energy Mater 7(24):1700260. https://doi.org/10.1002/aenm.201700260

Tan DHS, Wu EA, Nguyen H, Chen Z, Marple MAT, Doux J-M, . . . Meng YS (2019) Elucidating reversible electrochemical redox of Li6PS5Cl solid electrolyte. ACS Energy Lett 4(10):2418–2427. https://doi.org/10.1021/acsenergylett.9b01693

Tan DHS, Banerjee A, Chen Z, Meng YS (2020) From nanoscale interface characterization to sustainable energy storage using all-solid-state batteries. Nat Nanotechnol 15(3):170–180. https://doi.org/10.1038/s41565-020-0657-x

Wang Q, Zheng J, Walter E, Pan H, Lv D, Zuo P, . . . Xiao J (2015) Direct observation of sulfur radicals as reaction media in lithium sulfur batteries. J Electrochem Soc 162(3):A474. https://doi.org/10.1149/2.0851503jes

Wang S, Liu Q, Zhao C, Lv F, Qin X, Du H, . . . Li B (2018) Advances in understanding materials for rechargeable lithium batteries by atomic force microscopy. Energy Environ Mater 1(1):28–40. https://doi.org/10.1002/eem2.12002

Wang H, Cao X, Gu H, Liu Y, Li Y, Zhang Z, . . . Cui Y (2020a) Improving lithium metal composite anodes with seeding and pillaring effects of silicon nanoparticles. ACS Nano 14(4):4601–4608. https://doi.org/10.1021/acsnano.0c00184

Wang R, Luo C, Wang T, Zhou G, Deng Y, He Y, . . . Yang Q-H (2020b) Bidirectional catalysts for liquid–solid redox conversion in lithium–sulfur batteries. Adv Mater 32(32):2000315. https://doi.org/10.1002/adma.202000315

Wei Seh Z, Li W, Cha JJ, Zheng G, Yang Y, McDowell MT, . . . Cui Y (2013) Sulphur–TiO2 yolk–shell nanoarchitecture with internal void space for long-cycle lithium–sulphur batteries. Nat Commun 4(1):1331. https://doi.org/10.1038/ncomms2327

Xu J, Cai X, Cai S, Shao Y, Hu C, Lu S, Ding S (2023) High-energy lithium-ion batteries: recent progress and a promising future in applications. Energy Environ Mater 6(5):e12450. https://doi.org/10.1002/eem2.12450

Yao H, Zheng G, Hsu P-C, Kong D, Cha JJ, Li W, . . . Cui Y (2014) Improving lithium–sulphur batteries through spatial control of sulphur species deposition on a hybrid electrode surface. Nat Commun 5(1):3943. https://doi.org/10.1038/ncomms4943

Yin Y-X, Xin S, Guo Y-G, Wan L-J (2013) Lithium–sulfur batteries: electrochemistry, materials, and prospects. Angew Chem Int Ed 52(50):13186–13200. https://doi.org/10.1002/anie.201304762

Yin W, Grimaud A, Rousse G, Abakumov AM, Senyshyn A, Zhang L, . . . Tarascon J-M (2020) Structural evolution at the oxidative and reductive limits in the first electrochemical cycle of Li1.2Ni0.13Mn0.54Co0.13O2. Nat Commun 11(1):1252. https://doi.org/10.1038/s41467-020-14927-4

Zhai P, Liu L, Gu X, Wang T, Gong Y (2020) Interface engineering for lithium metal anodes in liquid electrolyte. Adv Energy Mater 10(34):2001257. https://doi.org/10.1002/aenm.202001257

Zhang SS (2013) Liquid electrolyte lithium/sulfur battery: fundamental chemistry, problems, and solutions. J Power Sources 231:153–162. https://doi.org/10.1016/j.jpowsour.2012.12.102

Zhang X, Qiao Y, Guo S, Jiang K, Xu S, Xu H, . . . Zhou H (2019) Manganese-based Na-rich materials boost anionic redox in high-performance layered cathodes for sodium-ion batteries. Adv Mater 31(27):1807770. https://doi.org/10.1002/adma.201807770

Zhao C, Xu G-L, Yu Z, Zhang L, Hwang I, Mo Y-X, . . . Zhao T (2021a) A high-energy and long-cycling lithium–sulfur pouch cell via a macroporous catalytic cathode with double-end binding sites. Nat Nanotechnol 16(2):166–173. https://doi.org/10.1038/s41565-020-00797-w

Zhao S, Yan K, Zhang J, Sun B, Wang G (2021b) Reaction mechanisms of layered lithium-rich cathode materials for high-energy lithium-ion batteries. Angew Chem Int Ed 60(5):2208–2220. https://doi.org/10.1002/anie.202000262

Zheng J, Kim MS, Tu Z, Choudhury S, Tang T, Archer LA (2020) Regulating electrodeposition morphology of lithium: towards commercially relevant secondary Li metal batteries. Chem Soc Rev 49(9):2701–2750. https://doi.org/10.1039/C9CS00883G

Zhou G, Tian H, Jin Y, Tao X, Liu B, Zhang R, . . . Cui Y (2017) Catalytic oxidation of Li_2S on the surface of metal sulfides for Li–S batteries. Proc Natl Acad Sci 114(5):840–845. https://doi.org/10.1073/pnas.1615837114

Zhou L, Danilov DL, Eichel R-A, Notten PHL (2021) Host materials anchoring polysulfides in Li–S batteries reviewed. Adv Energy Mater 11(15):2001304. https://doi.org/10.1002/aenm.202001304

Zou P, Sui Y, Zhan H, Wang C, Xin HL, Cheng H-M, . . . Yang C (2021) Polymorph evolution mechanisms and regulation strategies of lithium metal anode under multiphysical fields. Chem Rev 121(10):5986–6056. https://doi.org/10.1021/acs.chemrev.0c01100

Chapter 5
Materials, Design Consideration, and Engineering in Lithium-Sulfur Batteries

Umair Azhar, Muhammad Arif, Muhammad Sagir, Nimra Jabeen, and Humayun Afzal

Abstract Groundbreaking researches in the field of electrochemistry have increased the longevity and electric power density of lithium-sulfur (Li-S) batteries in recent decades. Moreover, these batteries have been recognized as a viable alternative to the Li-ion battery and are expected to play a key part in high-energy-density electrochemical storage systems owing to its low cost, plentiful resources, environmental friendliness, and high energy density. In this chapter, particular focus is paid on research and developments pertaining to various issues in Li-S batteries such as sulfur cathode, separator, electrolyte, and Li anode. Finally, a future perspective on the development of high-performance Li-S batteries is discussed.

Keywords Lithium-sulfur (Li-S) batteries · Electrochemical storage · Electrolyte

---

U. Azhar (✉) · M. Sagir · H. Afzal
Institute of Chemical and Environmental Engineering, Khwaja Fareed University of Engineering and Information Technology, Rahim Yar Khan, Punjab, Pakistan
e-mail: dr.umairazhar@kfueit.edu.pk; dr.msagir@kfueit.edu.pk

M. Arif
Institute of Chemical and Environmental Engineering, Khwaja Fareed University of Engineering and Information Technology, Rahim Yar Khan, Punjab, Pakistan

Centre for Thermal and Renewable Energy Research, Khwaja Fareed University of Engineering and Information Technology, Rahim Yar Khan, Punjab, Pakistan
e-mail: muhammad.arif@kfueit.edu.pk

N. Jabeen
Institute of Chemistry, Khwaja Fareed University of Engineering and Information Technology, Rahim Yar Khan, Punjab, Pakistan

M. S. Tahir et al. (eds.), *Lithium-Sulfur Batteries: Key Parameters, Recent Advances, Challenges and Applications*, Springer Tracts in Electrical and Electronics Engineering, https://doi.org/10.1007/978-981-99-2796-8_5

Historical Background

Lithium-ion rechargeable battery technology has progressed significantly since the commercial manufacturing of the first Li-ion-based rechargeable battery in which cathode is made of lithium cobalt oxide and anode made of graphite. Nonetheless, recent developments have revealed that the batteries based on Li^+ have reached point of congestion, i.e., the total energy density (300–350 Wh/kg) is approaching the limit (Fotouhi et al. 2016; Marom et al. 2011; Turcheniuk et al. 2018). The electrode capacity and cell voltage are commonly used parameters to establish the energy density limit. The efficiency of current methods that use some insertion-type oxide for the construction of cathodes has reached its full potential value of ~250 mAh/g in current Li-ion battery technology, and the energy storage tendency of those anodes based on graphite has been likewise restricted at 372 mAh/g. The gap between the voltages of presently available organic electrolytes of liquid phase, on contrary, can scarcely work reliably above 4.3 V. As a result, new battery material design and combinations that may offer either larger storage tendency, better stability, or both are badly a need of the hour in order to fulfill ever increasing energy demands.

Li-S batteries, which work with the combination of readily available and negligibly harmful sulfur cathodes with anodes made of very light lithium element (0.534 g/cm^3), are being considered revolutionary in the field of energy storage. High energy density and their low cost are mainly two reasons for them being successful and advantageous (Nazar et al. 2014; Yang et al. 2018a; Manthiram et al. 2014; Fang et al. 2017; Yin et al. 2013; Bresser et al. 2013; Ma et al. 2015; Fotouhi et al. 2017). Thanks to the stripping/plating mechanism, that has made metallic lithium capable of giving the lowest potential value (3.04 V) for its oxidation-reduction reactions and also the maximum known value of specific capacity (3860 mAh/g) on anodic side. Theoretically speaking, the cost-effective active sulfur cathode may achieve a specific capacity value of somewhat around 1675 mAh/g, thanks to the multi-electron conversion process. Consequently, Li-S batteries have a potential gravimetric energy density of 2510 Wh/kg (depending entirely on a common release potential difference of 2.15 V), which is significantly beating the typical Li-ion-based batteries.

The invention of rechargeable batteries based on Li-S system can be traced back to 1962. The main cell models were initially proposed by Herbert and colleagues (Danuta et al. 1962), who used Li and some of its alloys as anode along with sulfur, selenium, and other halogens which were used as cathodic materials. The electrolyte is described in the invention as perchlorates of basic nature, halides, sulfocyanides, or chlorates which could dissolve in a saturated nonaromatic amine. Bhaskara developed high-energy-density metal-sulfur batteries in 1968, where light basic elements were used which include Li, Ca, Be, Mg, and Al with an organic electrolyte containing a blend of nonreactive conductors and sulfur. Some organic solvents to be suggested in the electrolytes were the likes of propylene carbonate, and these energy storage systems could produce cell potential difference of 1.16–2.52 V. Moreover, the suggested metal-sulfur batteries' theoretic gravimetric and volumetric energy densities were discussed in detail. Primary Li-S batteries constructed via several other solvents which are organic in nature, namely, tetrahydrofuran and dioxolane (DOL), were continuously established during the next 20 years (Peled et al. 1989a; Rauh et al. 1979; Martin et al. 1973). Another system of Li-S-based battery with a design of Li foil/0.1 M $Li2S_8$ in a combination of ethers/Teflon-bonded carbon materials was created by Peled (Peled et al. 1989b) in 1989, demonstrating the viability of building a rechargeable system of batteries based on Li-S in DOL-rich electrolyte. At discharge rate of 6 mA, the charge tendency is typically better as compared to the discharge capacity, with 50 numbers of maximum rounds to letdown. Even though the suggested Li-S cell model's reversibility is insufficient, it has spurred the researchers to devote significant time and effort to the study of rechargeable energy storage system based on Li-S in the future. A research in the 1990s on solid-phase electrolytes helped advances in effective employment electrolytes that are based on polymers and ceramics (Hayashi et al. 2003). In 1998, there was a patent awarded according to which solid-phase energy storage system of Li-S based had polyethylene electrolyte. Under 90 °C, this suggested system demonstrated to be giving an intense capacity value of somewhere around 1500 mAh/g.

Two major research articles published by Wang (Wang et al. 2002a, b) in 2002 first revealed that sulfur/activated carbon and sulfur/polyaniline nanocomposites are two extremely effective cathode materials to increase specific capacity and cycle stability. Based on the composite (sulfur: active carbon, 3:7), elemental sulfur integrated and trapped in the pores of high-specific-area active carbon has a reversible capacity of 440 mAh/g and a sulfur utilization of 90%. The molecular level mixed sulfur/polyaniline composite is made up of a highly electrically conductive polymer matrix in which active sulfur is homogeneously integrated. The prevention of aggregation of sulfur and suppression of polysulfide dissolution can be achieved by this structure. Consequently, the sulfur/polyaniline composite electrode can withstand 50 cycles of operation while maintaining a capacity of over 600 mAh/g. Nazar and colleagues made yet another big breakthrough in the prevention of polysulfide dissolution (Ji et al. 2009). They created the ordered mesoporous carbon (CMK-3)-sulfur composite with nanostructured polymer modifications and exhibited efficient confinement effects at the nanoscale. The mesopores in the CMK-3 carbon operate as multiple reaction microchambers, allowing active sulfur to be used more efficiently. Furthermore, the polymer coating aids in the retardation of polysulfide diffusion out of the composite, reducing active mass loss and boosting cycle stability (Sun et al. 2014; Lyu et al. 2015; Fang et al. 2016; Zhang et al. 2012; Chen et al. 2015). The "physical confinement" method advocated in the abovementioned studies has been widely adopted. Since 2000, several basic in-depth investigations on Li-S battery chemistry have been conducted, focusing on self-discharge characteristics, chemical structure development of sulfur species, and electrolyte system optimization, which includes organic solvent, additives, and Li salts (Jin et al. 2003; Hassoun and Scrosati 2010; Lee et al. 2003; Ji and Nazar 2010; Aurbach et al. 2009; et al. 2004; Mikhaylik and Akridge 2004). These foundational studies lay the groundwork for a comprehensive knowledge of rechargeable Li-S battery electrochemical processes.

Following 2010, additional resources were devoted to the study of rechargeable Li-S batteries, as seen by the rapid increase in the number of research papers. In the interim, several notable improvements have been made in the development of Li-S batteries with high specific capacity and extended cycle life (Li et al. 2015a; Huang et al. 2014a; Suo et al. 2013). In particular, some proposed strategies, such as the "small sulfur molecule," interlayer construction, and chemical adsorption enabled by oxides and heteroatom-doped carbons, were shown to be highly effective for improving the electrochemical performance of Li-S batteries, and shed light on future research directions on sulfur cathode performance enhancement (Evers et al. 2012; Xin et al. 2012; Su and Manthiram 2012; Sun et al. 2012). Arava and coworkers eventually reported another major technique of electrocatalytic conversion in Li-S batteries in 2015, motivated by the enhanced kinetics of aqueous polysulfides in photoelectrochemical solar cells and redox flow cells utilizing electrocatalytic electrodes (Al Salem et al. 2015). Polysulfide electrocatalytic conversion processes have been discovered to occur on surfaces including Pt, Au, and Ni, which can efficiently decrease polarization and boost specific capacity (Babu et al. 2015). In the meanwhile, significant progress has been achieved in the design and manufacture of functional separators, binders, and conductive additives to suppress the shuttle effect and enhance active sulfur utilization over the last 10 years (Tao et al. 2014; Xiao et al. 2019; Chen et al. 2018; Xu et al. 2017; Ghazi et al. 2017; Bai et al. 2016).

It's worth noting that, in addition to sulfur dissolving and sluggish conversion kinetics at the cathode, Li dendrite development and low Coulombic efficiency at the anode might potentially be major factors in Li-S battery performance degradation, although they've gotten little attention over the years. In 2014, Liu and colleagues developed a synthetic self-regulated solid-state electrolyte interface (SEI) layer of lithiated graphite on the surface of Li metal based on previous research on Li anode chemistry, which not only supplies Li-ion on demand but also minimizes direct contact between the Li anode and polysulfides (Huang et al. 2014b). The suggested surface engineering technique for employing Li metal anodes successfully inhibited the negative side reactions of Li anodes, resulting in significant performance improvements. The discovery has sparked a lot of interest in the protection of the Li anode, which is encouraging the development of Li-S battery systems (Cheng et al. 2017; Cao et al. 2015). Even though tremendous progress has been achieved in tackling the underlying scientific difficulties surrounding Li-S battery chemistry, researchers have discovered that there is still a huge gap between academic research and industrial manufacturing (Dörfler et al. 2020). To make Li-S batteries commercially viable, more effort must be put into material design and structural optimization under realistic operating circumstances. Luckily, current research on the production of high-mass-loading sulfur cathodes and solutions for cathode passivation in a lean electrolyte has shown promising results, boosting researchers' confidence in the commercialization of Li-S batteries (Fang et al. 2016; Zhao et al. 2020; Xue et al. 2019; Pan et al. 2018).

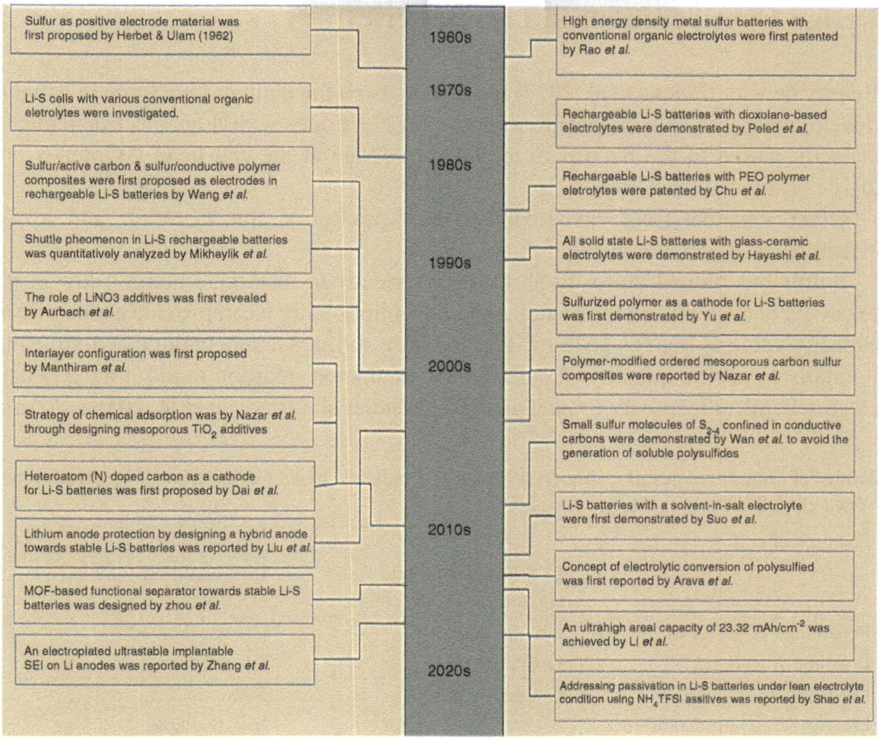

Fig. 5.1 A short historical record of the evolution of Li-S batteries, from the beginning to the present (Li and Guo 2021)

Taking a glance at the six-decade evolution in this field (Fig. 5.1), it is clear to assume that considerable progress has been made and that practical applications are becoming more common. The vital reaction processes and methodical obstacles in the creation of Li-S batteries are initially introduced in this article. We exhibit exemplary methodologies for material design and structure optimization based on a thorough grasp of Li-S battery chemistry to counter and tackle prevailing technical difficulties in Li-S battery systems. The main challenges with Li-S battery commercialization are then explored. To end with, we review the current state of Li-S batteries and offer some predictions for the progress of even better and more competent and application-specific Li-S batteries with high energy density and long-lasting permanency.

Design Considerations in Li-S Batteries

Energy density (Placke et al. 2017) has always been the main motive force in battery development, which translates to a search for electrodes with huge electrochemical capacity paired to the maximum attainable voltage at the cell level (Winter et al. 2018) (Table 5.1).

Other considerations are toxicity, the scarcity of the material, and cost-effectiveness as shown in Fig. 5.2.

Sulfur has piqued curiosity as a pristine electrode material since it's low-cost, excessively available on the planet, and, most importantly it can provide a theoretical capacity value as high as 1675 mAh/g. Because sulfur is a nonconductive nonmetal, it entails a specific quantity of carbon as a conductor, addition to aid it have better electrochemical reactivity (Yin et al. 2013; Wild et al. 2015; Yu et al. 2016a, 2018).

Table 5.1 Important design considerations for Li-S batteries based on liquid and solid electrolytes (Yang et al. 2018b)

Features of cell		
	Li-S based on liquid electrolyte	Li-S based on solid electrolyte
Percentage of sulfur in cathode	Greater than 50%	Greater than 30%
Loading of cathode	Greater than 3	Greater than 2
Li excess	Greater than 50%	50%
Electrolyte to sulfur ratio in ml/g	<4 for 300+ Wh/kg <3 for 400+ Wh/kg	Not available
Thickness of solid electrolyte	Not available	Less than 50
Specific capacity	Less than 1100	Less 1300
Utilization of sulfur	Greater than 66%	Greater than 78%

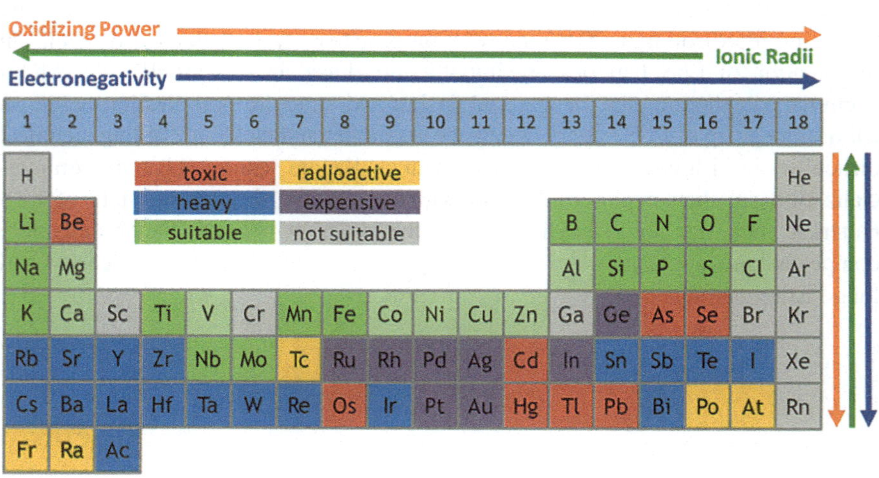

Fig. 5.2 Periodic table pallet for the design of electrode material (Palacin 2021)

During the early stages of research, the majority of the focus was on growing conductive sulfur using carbon with micropores for the reason that it could provide enhanced surface area as well as pore volume (Park et al. 2016; Jeong et al. 2016; Zhao et al. 2015; Pu et al. 2014; Li et al. 2014; Chen et al. 2014; Zhang et al. 2013; Wang et al. 2013). In general, a sulfur infiltration approach into carbon pores the usage of warmness treatment has been applied, the major bodily lure of the sulfur with inside the pores to spare you the dispersal of solvable lithium polysulfides (LiPSs) (Ji et al. 2009; Kim et al. 2020). At some point during the mid- and long-term cycling, the dissolution of polysulfides happens; however, this technique becomes less effective, and useless sulfur is easily formed from sulfur trapped in an ideal pore. Breakage of strong sulfur and decomposition of LiPSs are two processes that occur in lithium-sulfur batteries. The ability of the matrix/additive to show conductive behavior, as well as the number of C/S energetic sites, supports the breakdown of sulfur (S8) and the transition of strong sulfur into LiPSs (Park et al. 2018). On the contrary, because long-chain LiPSs are soluble in electrolytes rather than strong sulfur, the transition of long-chain LiPSs to short-chain LiPSs is greatly hampered by how well the LiPSs touch the conductive matrix. As a result, in recent years, a heteroatom-doped carbon, metallic oxide, or sulfide-like adsorbent containing LiPSs has been employed to prevent the LiPSs from diffusing away from the electrode (Hwang et al. 2016; Park et al. 2017; Tian et al. 2019; Li et al. 2015b; Yu et al. 2016b). Though this strategy can help to maintain the dissolved LiPSs close to the electrode, it still has a problem in that it's difficult to return the LiPSs to the electrode's floor after they've been pushed away (Kim et al. 2019). It is possible to deduce, in this regard, that the ease with which LiPSs go from the electrode's majority to the electrode's floor (or conversely) is critical to performance.

The variety of the preliminary substance and core constructions for accomplishing steady cyclic capability and rate performance of the lithium-sulfur battery should be well-thought-out grounded on the basic conformation with preliminary conditions of sulfur, and it is believed that the realistically planned structure initiates lithium-sulfur battery into large-scale manufacturing as the post-lithium-ion battery for overpowering the existing energy restraint of lithium-ion battery (Palacin 2021).

Design of the Materials in Li-S Batteries

As it is evident from the above illustration, namely, sulfur cathode, separator, electrolyte, and Li anode are four main components to be designed (Fig. 5.3, 5.4). A typical Li-S battery structure has been shown in the below diagram showing all the major components in working. Moreover, a reaction kinetic profile has also been shown alongside in a plot between potential *V* and electron transfer number per sulfur atom (Fig. 5.5).

	Concept Validation	Research	Applied Research	Development	Advanced Development
	An idea in a creative mind	Scale-up experiments	Lap/Prototype cells	Confirm research results	Design initial cell product
	Limited exploratory laboratory experiments	Characterize fundamental properties of the concept, chem, composition structure, etc.	Initial map of performance, rate, cycling, temperature, etc.	Establish initial product format	Design and construct unit operations
	Establish repeatability of performance	Evaluate the size of commercial opportunity	Scale-up of material preparation	Develop unit assembly operations	Scale-up prototype cell fabrication
	Is there a market?		Preliminary market scope	Make, test, and characterize 5 to 10 cell lots of 100 cells each	Run 3 to 5 sizable pilot line factory trials
Timing	1 to 3 years	1 to 3 years	3 to 4 years	3 to 5 years	3 to 4 years
Staffing	One	Two to four	Four to ten	Eight to sixteen	Twelve to thirsty
Materials Batch	Grams	10 to 50g	100g to 1kg	1kg to 10kg	10kg to 100kg

Fig. 5.3 General overview of the overall battery R&D process from conception to production (Palacin 2021)

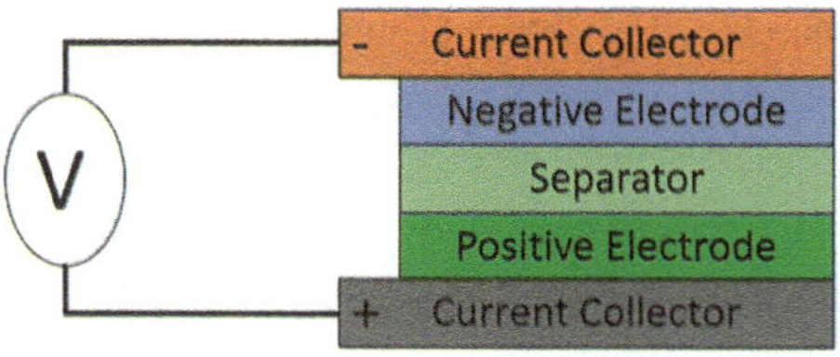

Fig. 5.4 Scheme of a battery cell (Palacin 2021)

Sulfur Cathode Design and Structuring

One of the initial ideas to get enhanced performance from sulfur cathodes was to encapsulate active sulfur in porous carbon hosts. A carbon host having micropores could prove an enhancement in the electrically conductive behavior of sulfur electrodes, accept large capacity variations in some classes of sulfur, and halt solvable polysulfides. As of now, carbons of microporous nature of different

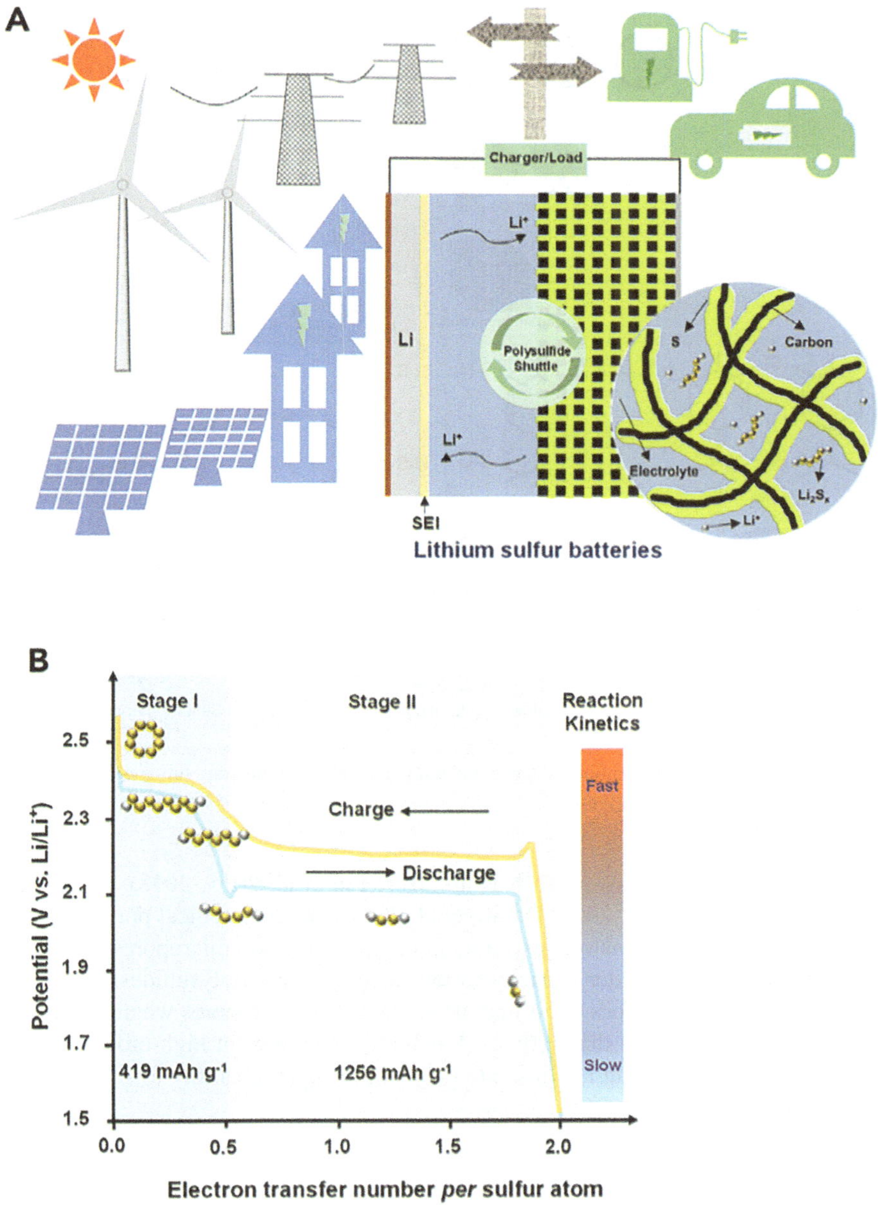

Fig. 5.5 (**a**) A diagram depicting the Li-S battery's construction. (**b**) General reaction kinetics profiles of Li-S battery (Li and Guo 2021)

dimensions such as three-dimensional (3D) carbon frameworks, two-dimensional (2D) grapheme and carbon sheets, one-dimensional (1D) tubes and carbon fibers, zero-dimensional (0D) hollow carbon spheres (Zhang et al. 2013, 2015; Yang et al.

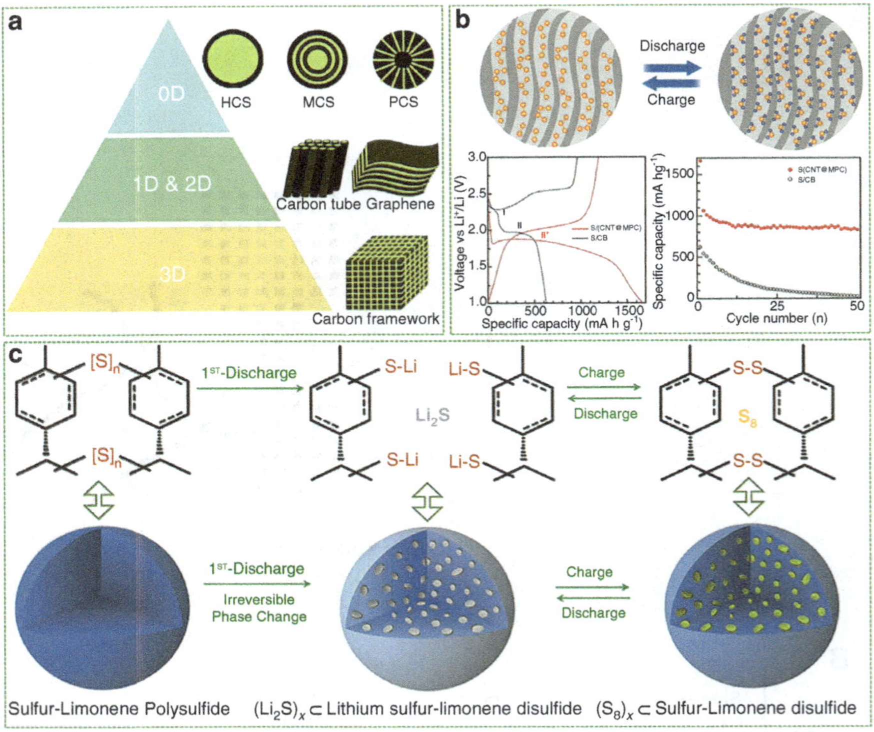

Fig. 5.6 Advanced sulfur cathode strategies including physical confinement, tiny molecular sulfur, covalent sulfur, and chemisorption

2019; Qu et al. 2013; Jayaprakash et al. 2011; Liu et al. 2015, 2018), multi-shell carbon spheres (Fu et al. 2019; Dhawa et al. 2019), and hierarchical porous carbon spheres have now been majorly explored in Fig. 5.6a. Carbon micropores have been discovered to have a better capacity to physically contain polysulfides. Only tiny metastable sulfur allotropes S2–4 may be contained in micropores when the size of the holes is lowered less than 0.6 nm. Micropores can prevent high-order solvable polysulfides from forming and insulate polysulfide approachability to solvent molecules (0.6–0.8 nm) Fig. 5.6b.

The disruption of long chains of those S-S happens to generate Li_2S nanoparticles within the limonene disulfide matrix during the first discharge, as illustrated in Fig. 5.6c. The embedded structures serve as ion sieves, allowing Li-ion transit while blocking solvent penetration. When the Li_2S nanoparticles are charged, the R-S-Li groups are delighted to form the R-S-S-R chains, and the Li_2S nanoparticles are reversibly converted back to S8. S8 and organosulfide molecules with R-S-S-R chains go through oxidation and reduction reactions separately in the following cycles. Resultantly, electrodes of such materials possess a high specific capacity of 1000 mAh/g at 0.5 C and a very little diminishing rate of 0.008% each round.

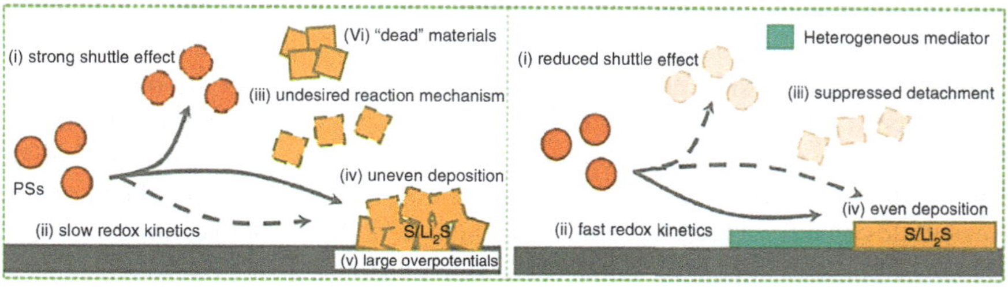

Fig. 5.7 The operating ideologies of Li-S batteries in the presence and absence of heterogeneous mediators

Even though the internment of sulfur in those carbon hosts which show no polarity might increase the specific capacity of Li-S batteries, not so strong connection between the polysulfides and their host materials makes long-term cycle stability problematic. Scientific scholars progressively came to realization that immobilizing polysulfides with durable polar bonds can excellently reduce the shuttle outcome and consequently momentously extend cycle life.

Two types of materials that have been explored for suitable host materials of polysulfides are:

- Metal oxides (e.g., oxide of titanium, aluminum, silicon, and cobalt)
- Carbon having heteroatomic doping

Most of the abovementioned oxides, unlike conductive carbon materials, have happened to be electronically very weak conductive, as well as substantial adsorption of polysulfide on the surface of insulated metal oxides can likely halt their dynamic sites and as a result impairing their electrochemical performance.

Nonetheless, several investigations have shown that the introduction of poor conductive metal oxides resulted in a significant improvement in electrochemical performance. Cui and colleagues were successful in developing the sense of competitiveness between chemical adsorption and dispersion of polysulfide adsorbates on these metal oxides to explain this behavior. On cerium oxide, aluminum oxide, lanthanum oxide, magnesium oxide, and calcium oxide, the geometrical optimization of the most stable Li_2S and Li_2 S8 is shown in Fig. 5.6d. Li_2S and Li_2S_8 have modest adsorption energies of 5.71 and 1.87 eV, respectively, on MgO (100). In their research, they looked at the dispersion of Li on the superficial of the prementioned five metal oxides using a density functional theory calculation. The Li on magnesium oxide sample has the bottommost distribution block of 0.45 eV among these samples.

Incorporating heterogeneous intermediaries into Li-S batteries is another effective technique for improving sulfur cathode performance. Heterogeneous facilitators, shown in Fig. 5.7, can attach to polysulfides and expedite polysulfide conversion by triggering various reaction pathways through their redox interactions with polysulfides.

In recent years, the notion of electrocatalytic polysulfide conversion has sparked a lot of attention in the ground of Li-S battery systems. So as by decreasing the energy barrier during conversion processes, polysulfide electrocatalysts are thought to be able to speed up the translation kinetics of polysulfides. Accelerating polysulfide conversion reduces polysulfide buildup and consequently suppresses polysulfide shuttle. The high-performance rate of Li-S batteries is further aided by the increased polysulfide conversion rate. Till now, a large range of electrocatalytic materials have been produced and used in Li-S batteries, which include:

- Carbon with heteroatomic doping
- Some nonreactive metals
- Oxides of metals and some nonmetals (N, C, S, Se, P)

The new idea of single-atom electrocatalysts has also been presented to increase sulfur consumption in Li-S batteries in order to fully use electrocatalysts. Unlike their names, these single-atom materials might act in a bifunctional manner for expedition of the formation and breakdown of Li_2S throughout the process of charging and discharging, according to experimental and theoretical data.

Separator Design and Structuring

Designing functionalized separators, in addition to enhanced sulfur cathodes, is another viable and successful technique for achieving new heights and improvements in the efficiency and performance of batteries. The capacity of conventional porous polyolefin separators to restrict the polysulfide shuttle and impede Li dendrite penetration is limited. Additional functional sheets, for added purpose, can be employed amid the Li anode/sulfur cathode and the separator to control polysulfide diffusion and Li stripping/plating conduct. In pursuance to improve the electrochemical performance of Li-S batteries, a self-supporting conductive multi-walled carbon nanotube (MWCNT) interlayer was inserted amid the sulfur cathode and the separator. This self-sufficient and self-supported interlayer could provide aid in improving active sulfur usage and reducing irreversible polysulfide diffusion. As a consequence, after 100 cycles, a high capacity of 804 mAh/g at a high rate of 1 C may be sustained.

Because the addition of an interlayer will unavoidably diminish the energy density of Li-S cells, greater emphasis is being paid to the creation of shallow and shrill covering sheets along with a unified interlayer and separator structure. Even though different interlayers for polysulfide immobilization have received a lot of attention, the interlayers between the anode and the separator for manipulating the stripping/plating behavior of Li anodes have received less attention. Advanced interlayers can be used to create a strong shield or to control the dispersal and concentration of near-surface Li-ions on the Li metal anode. Consequently, it plays a big part in improving the electrochemical performance of anodes.

In spite of the improved electrochemical performance of batteries, accumulation of interlayers to separators increases width and mass; this, to some extent, can cause intensifications of Li-ion diffusion resistance, causing electrode wetting capacity to deteriorate, and lowering energy density. Another simple and successful mode to increase the performance of Li-S batteries is to design new materials for separators with having no need of adding further interlayers to expand ionic discernment, thermal permanency, and mechanical qualities. The use of plentiful, ecologically benign, and biodegradable natural polymers for the production of separators has piqued researchers' attention.

Separators can also be made out of inorganic materials that have a good thermal permanency and a high Young's modulus. Glass fiber-based membranes, for example, have been efficaciously proven in Li-S batteries, with high permeability, improved thermal permanency, cheap price, and electrolyte's great wetting capability. These ideas might help us design new functional separators for high-performance Li-S batteries employing additional ordered mesoporous materials. Of late, there has been another development (Fig. 5.8a, b), in which cellulose-based porous materials, derivative from facial mask, have shown impressive performance. In another modern research, a fire-resistant and multipurpose separator organized via electro-spun nanofilaments of polyacrylonitrile combined with ammonium phosphate was documented for Li-S batteries for them being beneficial in terms of stability and safety (Fig. 5.8c). The linkage of these separators has been proved to be very strong, thanks to AAP for its high enrichment with amine groups (Fig. 5.8d). Moreover, the polyacrylonitrile separator is analyzed and found to be flameproof (Fig. 5.8e) due to combination with ammonium phosphate (APP) which is a fire-resistive material.

Electrolyte Design and Structuring

The stability and sustainability of the sulfur cathode and the Li metal anode equally in standard liquid organic electrolytes has been shown to be inadequate for long-lasting Li-S battery operation. Modern research has made significant progress in manipulating the constituents of electrolytes (solvents and Li salts) to optimize sulfur consumption and cyclic ability of Li anode. It is critical to utilize the right solvent to test the polysulfide solubility and stability of the Li metal anode. Many different solvents have been studied in Li-S batteries up to this point. Although carbonate-based electrolytes have been utilized successfully in Li-ion batteries, the vast majority of them are incompatible with Li metal anodes and S8 cathodes. Li metals/solvable polysulfides and carbonates have significant side reactions that stymie their development in Li-S batteries. Even though DOL-/DME-based electrolyte systems are extensively employed in Li-S batteries, their development is still hampered by excessive polysulfide dissolution and the instability of the Li anode, both of which have negative consequences for Li-S battery cycle stability. Using extremely concentrated electrolytes is one of the most successful component

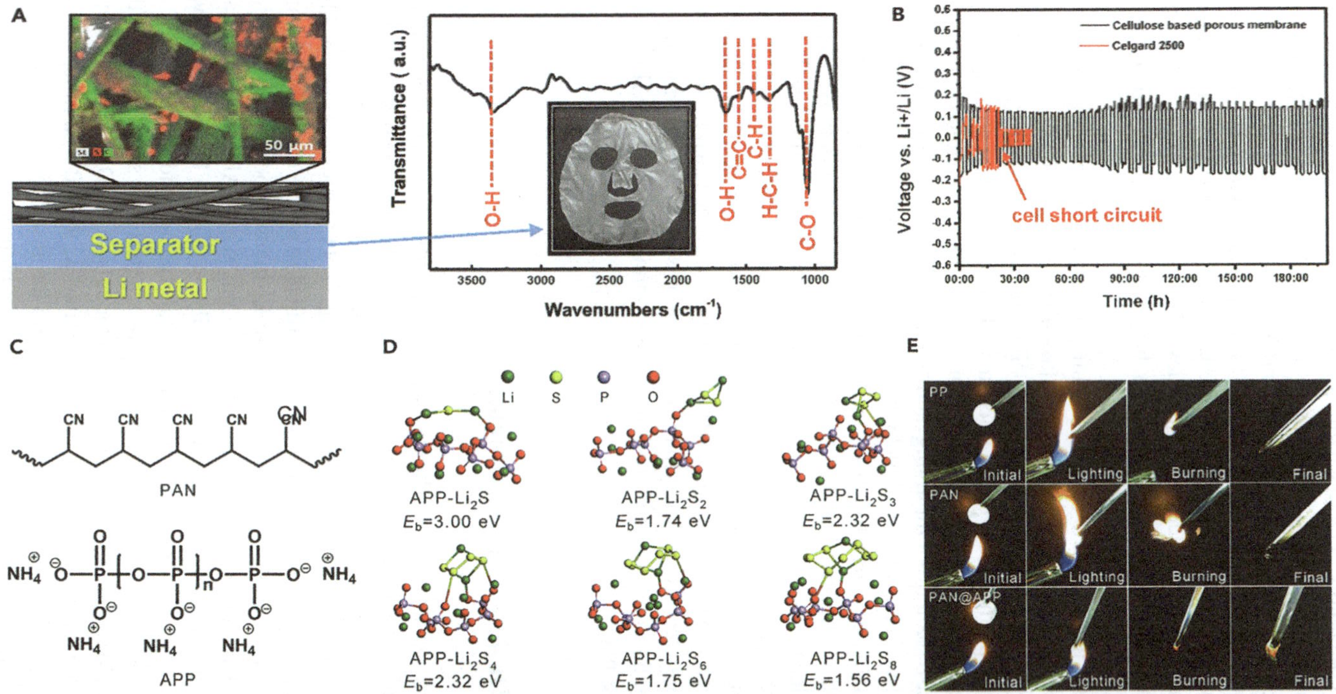

Fig. 5.8 Innovative separators under development: (**a**) A Li-S battery with a cellulose separator is shown schematically, along with the cellulose separator's Fourier transforms infrared spectrum. (**b**) Voltage profiles of Li/Li symmetric cells with a Celgard 2500 and a cellulose separator at a current density of 2.4 mA cm^2 (4.8 mAh cm^2), (**c**) PAN and APP chemical structures. (**d**) The energies of interaction between polysulfides and APP. (**e**) Images of PP, PAN, and PAN@APP separators demonstrating their fire-retardant qualities (Li and Guo 2021)

regulation techniques for improving the electrochemical performance of Li-S batteries (HCEs).

A unique and highly coordinated diethylene glycol DME system (G2:LiTFSI) has been designed to convert the sulfur dissolution-precipitation process to a quasi-solid-state conversion in order to decouple the electrolyte and sulfur usage along with restriction on the production of Li dendrites. Figure 5.9a, b, c shows the gradual change in G2:LiTFSI ratio and relates the ratio changes to the change in voltage plateau.

In addition to altering diluents and salts of lithium, adding additional materials into electrolytes has been suggested as a simple and active way to optimize sulfur cathode use and safeguard Li metal anodes. RMs have gotten a lot of interest as a sulfur-improvement addition because they can help with the electrochemical conversion of Li2S and so causing improvements in the tendency and rate of performance in Li-S battery systems (Fig. 5.9d).

Concerns regarding unstable and igneous liquid organic electrolytes, in particular, constitute significant hurdles for practical Li-S battery applications. The combination with fire-resistive chemicals to electrolytes has been considered a modest but active technique for addressing Li-S battery safety concerns. In order to improve compatibility and safety, few phosphorus-rich chemical additions have been tested and employed in Li-S systems in recent years.

In spite of significant development on batteries that are majorly based on liquid organic electrolytic material, the stability of electrode still needs to be enhanced. The replacement of the standard liquid-phase electrolytes with solid-phase electrolytes to gain better performance has been recognized as a potential technique. The three types of solid-phase electrolytes are gel polymer electrolytes, solid polymer electrolytes, and solid inorganic electrolytes.

At ambient temperature, electrolytes based on polymers have a high Young's modulus and low resistance at the interface, whereas inorganic ceramic-based electrolytes have shown a high Young's modulus and high ionic conductivity. The abovementioned forms of electrolytes, alongside anode and cathode made of Li and sulfur, respectively, produce various borderline characteristics. Constructing Li-S batteries that are full solid is expected to solve the problem of shuttling of polysulfides and development of lithium dendrites and lowers the flammability of the electrolyte. However, mutually building finely merged boundaries at sulfur cathodes and Li anodes remains a big difficulty. In order to establish a close contact point among the boundaries of sulfuric cathodes and SEIs, nano-engineering is considered the most successful and promising way. To create homogenous and dense electrode architectures, high-energy ball milling and high pressure have often been required. At the boundaries between SIE and Li anodes, an endwise connection between solid-state anode made of lithium and ceramic-based electrolytes (which are not flowable) results in a high interface resistance. Commonly, electrolytes that are based on gel polymerization have shown great ionic conductance and adaptability. As an excellent illustration, Fig. 5.10a, b shows the polymerization of very highly conductive electrolyte, namely, SHGP using a method called "ring-opening" polymerization.

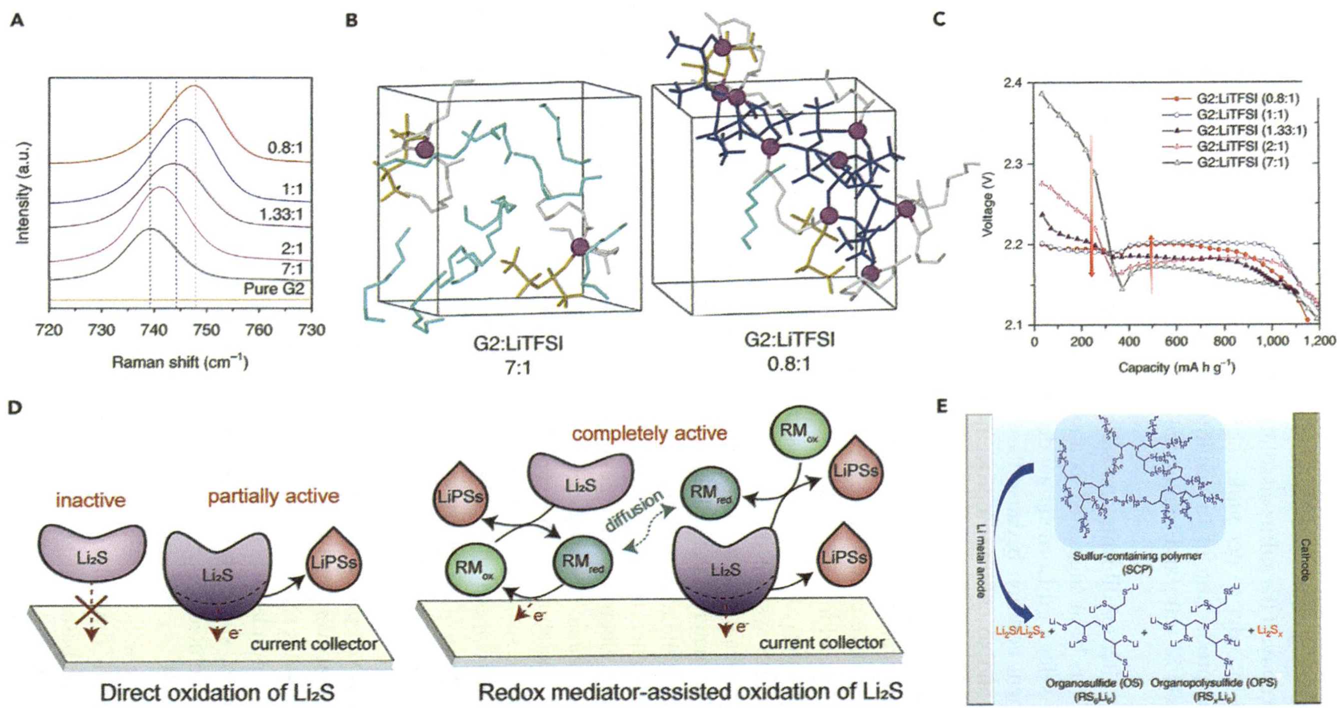

Fig. 5.9 Components and additions of liquid electrolytes and additives for Li-S batteries are regulated. (**a**) $TFSI^-$ band Raman spectra for pure G2 and various G2:LiTFSI electrolytes. (**b**) AIMD trajectories for the two end-member electrolytes (Li-ion, purple; free and coordinated G2 molecules, cyan and gray, respectively; contact-ion-pairs and aggregates of TFSI_ anions, gold and blue, respectively). (**c**) Distinct G2:LiTFSI electrolytes have different discharge patterns. (**d**) Direct Li2S oxidation and RM-assisted Li2S oxidation in Li-S batteries are depicted schematically. (**e**) The production of stable SEI layers including organic (organosulfide/organopolysulfide) and inorganic (Li2S/Li2S2) units is depicted schematically (Li and Guo 2021)

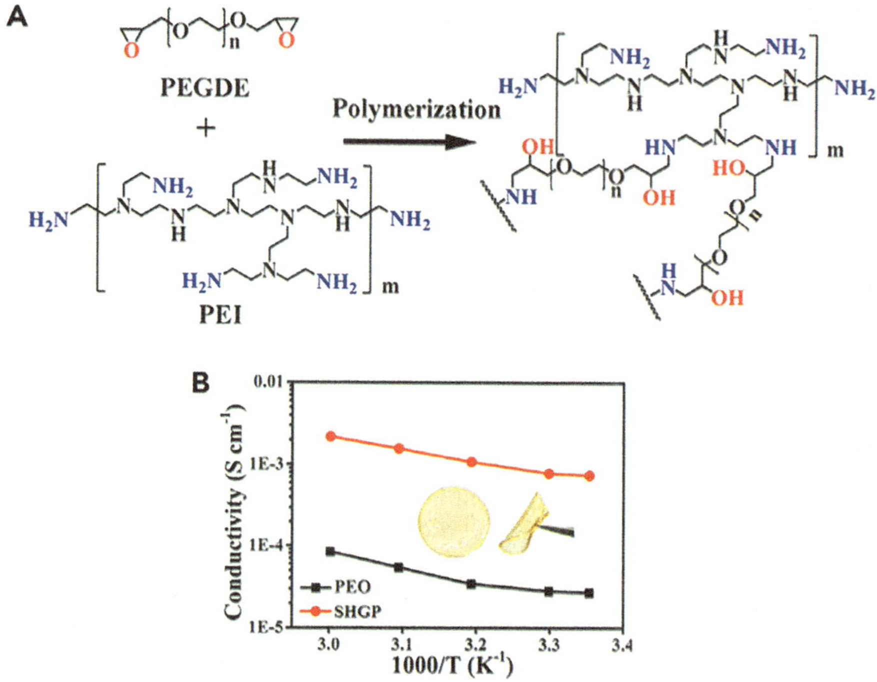

Fig. 5.10 *Advanced solid-state electrolytes for Li-S batteries are being developed.* (**a**) The polymerization of PEGDE and PEI is shown in this synthesis scheme. (**b**) Temperature-dependent ionic conductivities of the SHGP electrolyte and PEO gel polymer (Li and Guo 2021)

Even though battery systems that are wholly solid based upon SIEs could be a viable solution for addressing Li-S batteries' stability and safety issues, it is still challenging to fully exploit their capacity and rate capabilities. SIEs are frequently paired alongside liquid-phase electrolytes or GPEs (which are also termed as hybrid electrolytes) as a compromise method to provide adequate connection between the interfaces and wetting capability of electrode, particularly for those cathodes that are made of sulfur. Auspicious rate of performance and a high capacity under these conditions is a stern possibility. Still, it is very hard to overlook the fact that such type of electrolytes can cause difficulty in immobilization of the flowing electrolytes at battery topologies, increasing the structural design and preparation complexity. As a result, significant efforts should continue to be made to find better materials or novel manufacturing procedures that will enable Li-S batteries wholly based on solids to be realized and also not negotiating electrochemical performance.

Li Anode Design and Structuring

Metal-based lithium (an integral constituent of battery systems) has shown all the required necessities making it the "Holy Grail" of battery chemistry. Nevertheless, we need to consider various aspects which might affect battery performance before lithium can be considered a trustworthy option as an anodic material in batteries. Infinite form modification, characterization of substances, and structural refining for 3D hosts (Fig. 5.11), to accept the lithium metal and minimize thickness variation during cyclic roundup, have been considered potential methods to the real-world usage of lithium metal anodes, while extensive work is needed. Up till now, various 3D host materials which include carbons (graphene aerogels, carbon nanofibers, CNT sponges, carbon felts, carbon clothes, and carbonized wood), polymers (polyimide nanofibers, PAN nanofibers, and polyethyleneimine sponge), and conductive metals and metal compounds (Ni foams, Cu foams, and MXene aerogels) have been stated to normalize the stripping/plating behavior of lithium metal anodes.

Chemical alteration of the surface of hosts, in addition to regulating the microstructure of 3D hosts, is a key method used while developing Li-based anodes of very high stability. Making the surface functional with metal oxides, for example, has proved to be a very useful technique to increase the lithium attraction of carbon hosts which have Li stored beforehand by molten infusion before battery assembly that plays an important part in producing homogenous and unified fused Li-based anodes. Another good way to keep a check on the behavioral stripping and plating in the anode is to make alloys of Li metal. Surface coating using Li alloys, in addition to making bulk Li alloys, is seemingly a simple yet operative way to regulate the plating and stripping conduct of Li.

Artificially made SEI sheets, whether organic or inorganic, have their own benefits which are combined together to make fused SEI sheets, making them a more viable option for preserving Li metal anodes. As shown in Fig. 5.12, using a simple drop-casting method, a fused artificially made SEI sheet comprised of Cu_3N rooted in a styrene-butadiene rubber (SBR) medium has been recently produced.

The natural SEI-inspired multilayer design, in addition to the design of a composite with multiple components with artificially constructed SEI sheets, is another useful technique for constructing composite artificial SEI sheets which could serve

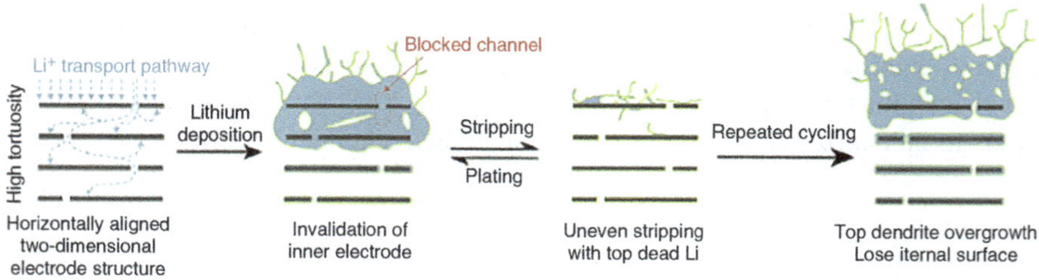

Fig. 5.11 Li's structure development in which a 3D host material is aligned in horizontal manner is seen schematically (Li and Guo 2021)

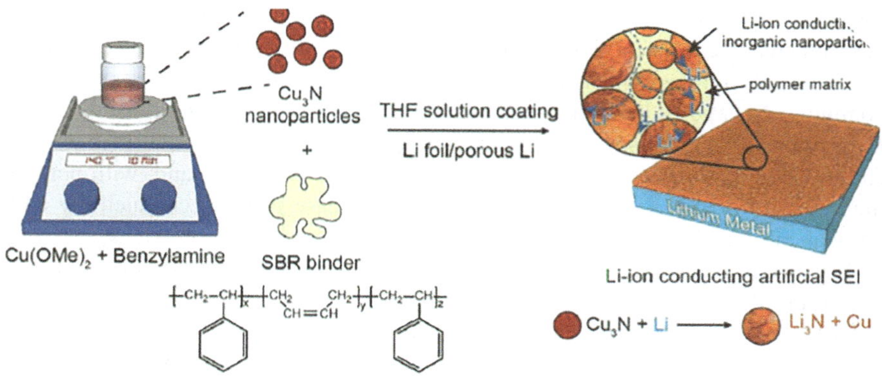

Fig. 5.12 The Cu_3N/SBR composite artificial SEI layers are created in a schematic illustration (Li and Guo 2021)

the purpose of very high performance. Off late, a merged artificially developed SEI sheet based on Cu_3N nanoparticles rooted in a medium of styrene butadiene rubber was established by employing a very ordinary drop molding method; Fig. 5.12. The findings further show that by inhibiting side reactions between polysulfides and Li anodes, the developed bilayered SEI sheets may considerably cause much improvements in the permanency of Li-based anodes and extend the lifespan of batteries.

Engineering Perspective

Because of its clearly higher power density than currently available Li-ion batteries, the rechargeable battery systems based on Li-S materials working on the complete conversion reactions are anticipated to be a great option in modern critical power garage applications, such as electric-driven motors and desk-bound power garages. A Li steel anode, a sulfur composite cathode, a natural electrolyte, and a porous polymer-primarily based completely separator are typical components of classic Li-S batteries. Unlike traditional Li-ion batteries, which rely vastly on the process of io-insertion, the reactions taking place in Li-S batteries are quite complex in nature. Even though the Li-S battery's working mechanism is complex and controversial, the commonplace location functioning principle may be stated. Because of the starting cloth on the cathode, the orthorhombic cyclic octasulfur also known as OCS or sulfur-8 (S8) has shown the highest strength at ambient temperature among the more over 30 strong sulfur allotropes. The very reason that Li-S based batteries start with a discharging process is that the electrode material of octasulfur is free of Li. The range of S8 molecules is reduced during the galvanostatic discharge process. Their very specific "crown shape" starts to open and unites with Li-ions. The S8 molecules are sequentially electrochemically lowered to construct excessive-order solvable lithium polysulfide intermediates (Li2Sx, $4 < x < 8$) at the early degree

(degree I), providing a larger voltage plateau at roughly 2.35 V than Li/Li + (0.5 electron switch consistent with a sulfur atom on the quilt of degree I). On contrary, solvable polysulfides which possess a very highly volatile pastime may need to continuously expose the response surfaces of the sulfur electrode, assisting in the acquisition of quick response kinetics and optimizing the utilization of energetic sulfur. The dissolution of high-order polysulfides, on the other hand, will increase the viscosity of the electrolyte, resulting in a decrease in ionic conductivity and, consequently, a shortage of energetic materials.

Because of the aggregate with additional Li-ions created during discharge, lower ranked unsolvable lithium polysulfides (Li_2Sx, $1 < x < 4$) are formed, resulting in a voltage plateau of roughly 2.15 V as contrasted to Li/Li + (1.5 electron switch consistent with a sulfur atom at some point of degree II). It's worth mentioning that at some time during the release process, an overpotential is often seen on the transition area between degrees I and II, which is attributable to the induction of diverse nucleation of unsolvable polysulfides on conductive substrates. Furthermore, solid-to-solid conversion reaction involving Li_2S_2 and Li_2S makes the kinetic response in phase II more complex and slower as compared to phase I. The Li_2S is successively thrilled to shape S8 via more than one polysulfide intermediate during the following fee method. Because polysulfide intermediates are frequently detected in ether-based electrolytes, detailed knowledge of the mechanism of complex sulfur chemistries may be obtained. Sulfur redox processes based on strong-liquid-strong conversion may be effectively regulated by fabric design and shape optimization. Geometry of charge and discharge curves can be variable in accordance with sulfuric cathodes and electrolytes. Molecular sulfur (of small molecular size) and polymers (based on sulfur), for example, have the most convenient single discharge plateau, suggesting solid-to-solid section conversion minus the creation of any kind of polysulfides that might be solvable.

Furthermore, when there are no naturally occurring solvents are available for polysulfides, all-country Li-S batteries predominantly often trail an opportune solid-to-solid conversion course amid sulfur and Li2S.

The following points should be the focus of future commercial Li-S battery research:

1. Metallic lithium needs to be modified with multifunctional and cost-effective current collectors or stabilizing coatings to buffer the infinite volume change of lithium and prevent it from corrosion. To avoid safety issues, lithium dendrites must be cleaned and eliminated as much as feasible.
2. Large-scale development of the equipment of cathode preparation based on conventional carbon (e.g., super P carbon, acetylene black carbon, and carbon nanotubes) and sulfur species are required. The cathode materials not only perform well electrochemically, but they are also inexpensive enough to compete with commercially available Li-S batteries.
3. Develop innovative separators to reduce lithium polysulfide transport, avoid the shuttle effect, and increase sulfur efficiency. Furthermore, the expense must be properly analyzed.

4. Choose the best electrolytes and chemicals for vesicle cells. Newly selected additives or electrolytes must be tested and validated under certain sulfur loading and excess limits.
5. Sulfur mass loading was optimized to increase vesicle cell capacity. It is impossible to aim for a sulfur concentration in vesicles more than 10 mg/cm2. To provide the competitive power density, endurance, and cheap cost of commercially available Li-S batteries, the sulfur mass charge must be properly tuned.
6. To determine the safety of Li-S batteries, reliable test procedures are required. More improved technologies in material handling, cell production, cell regulation, cell control, and management are required.

References

Al Salem H et al (2015) Electrocatalytic polysulfide traps for controlling redox shuttle process of Li–S batteries. J Am Chem Soc 137(36):11542–11545

Aurbach D et al (2009) On the surface chemical aspects of very high energy density, rechargeable Li–sulfur batteries. J Electrochem Soc 156(8):A694

Babu G et al (2015) Electrocatalysis of lithium polysulfides: current collectors as electrodes in Li/S battery configuration. Sci Rep 5(1):1–7

Bai S et al (2016) A long-life lithium–Sulphur battery by integrating zinc–organic framework based separator. J Mater Chem A 4(43):16812–16817

Bresser D, Passerini S, Scrosati B (2013) Recent progress and remaining challenges in sulfur-based lithium secondary batteries–a review. Chem Commun 49(90):10545–10562

Cao R et al (2015) Anodes for rechargeable lithium-sulfur batteries. Adv Energy Mater 5:1402273

Chen S et al (2015) Multi-chambered micro/mesoporous carbon nanocubes as new polysulfides reserviors for lithium–sulfur batteries with long cycle life. Nano Energy 16:268–280

Chen W et al (2018) A new hydrophilic binder enabling strongly anchoring polysulfides for high-performance sulfur electrodes in lithium-sulfur battery. Adv Energy Mater 8(12):1702889

Chen Y et al (2014) Sulfur encapsulated in porous hollow CNTs@ CNFs for high-performance lithium–sulfur batteries. J Mater Chem A 2(26):10126–10130

Cheng X-B et al (2017) Implantable solid electrolyte interphase in lithium-metal batteries. Chem 2(2):258–270

Danuta H et al (1962) Electric dry cells and storage batteries. US Pat, 3043896

Dhawa T et al (2019) Carbon@ carbon double hollow spheres as efficient cathode host for high rate LiS battery. Mater Chem Phys 225:309–315

Dörfler S et al (2020) Challenges and key parameters of lithium-sulfur batteries on pouch cell level. Joule 4(3):539–554

Evers S, Yim T, Nazar LF (2012) Understanding the nature of absorption/adsorption in nanoporous polysulfide sorbents for the Li–S battery. J Phys Chem C 116(37):19653–19658

Fang R et al (2016) 3D interconnected electrode materials with ultrahigh areal sulfur loading for Li–S batteries. Adv Mater 28(17):3374–3382

Fang R et al (2017) More reliable lithium-sulfur batteries: status, solutions and prospects. Adv Mater 29(48):1606823

Fotouhi A et al (2016) A review on electric vehicle battery modelling: From Lithium-ion toward Lithium–Sulphur. Renew Sust Energ Rev 56:1008–1021

Fotouhi A et al (2017) Lithium-sulfur battery technology readiness and applications—a review. Energies 10(12):1937

Fu A et al (2019) Recent advances in hollow porous carbon materials for lithium–sulfur batteries. Small 15(10):1804786

Ghazi ZA et al (2017) MoS2/celgard separator as efficient polysulfide barrier for long-life lithium–sulfur batteries. Adv Mater 29(21):1606817
Hassoun J, Scrosati B (2010) Moving to a solid-state configuration: a valid approach to making lithium-sulfur batteries viable for practical applications. Adv Mater 22(45):5198–5201
Hayashi A et al (2003) All-solid-state Li/S batteries with highly conductive glass–ceramic electrolytes. Electrochem Commun 5(8):701–705
Huang C et al (2014b) Manipulating surface reactions in lithium–Sulphur batteries using hybrid anode structures. Nat Commun 5(1):1–8
Huang J-Q et al (2014a) Ionic shield for polysulfides towards highly-stable lithium–sulfur batteries. Energy Environ Sci 7(1):347–353
Hwang JY et al (2016) High-energy, high-rate, lithium–sulfur batteries: synergetic effect of hollow TiO2-webbed carbon nanotubes and a dual functional carbon-paper interlayer. Adv Energy Mater 6(1):1501480
Jayaprakash N et al (2011) Porous hollow carbon@ sulfur composites for high-power lithium–sulfur batteries. Angew Chem Int Ed 50(26):5904–5908
Jeong YC et al (2016) Partially unzipped carbon nanotubes for high-rate and stable lithium–sulfur batteries. J Mater Chem A 4(3):819–826
Ji X, Nazar LF (2010) Advances in Li–S batteries. J Mater Chem 20(44):9821–9826
Ji X, Lee KT, Nazar LF (2009) A highly ordered nanostructured carbon–Sulphur cathode for lithium–Sulphur batteries. Nat Mater 8(6):500–506
Jin B, Kim J-U, Gu H-B (2003) Electrochemical properties of lithium–sulfur batteries. J Power Sources 117(1-2):148–152
Kim SJ et al (2019) Role and potential of metal sulfide catalysts in lithium-sulfur battery applications. ChemCatChem 11(10):2373–2387
Kim S-J et al (2020) Design considerations for lithium–sulfur batteries: mass transport of lithium polysulfides. Nanoscale 12(28):15466–15472
Lee YM et al (2003) Electrochemical performance of lithium/sulfur batteries with protected Li anodes. J Power Sources 119:964–972
Li Y, Guo S (2021) Material design and structure optimization for rechargeable lithium-sulfur batteries. Matter 4(4):1142–1188
Li Z et al (2014) A highly ordered meso@ microporous carbon-supported sulfur@ smaller sulfur core–shell structured cathode for Li–S batteries. ACS Nano 8(9):9295–9303
Li Z et al (2015a) Pie-like electrode design for high-energy density lithium–sulfur batteries. Nat Commun 6(1):1–8
Li Z, Zhang J, Lou XW (2015b) Hollow carbon nanofibers filled with MnO2 nanosheets as efficient sulfur hosts for lithium–sulfur batteries. Angew Chem 127(44):13078–13082
Liu S et al (2015) A 3D nanostructure of graphene interconnected with hollow carbon spheres for high performance lithium–sulfur batteries. J Mater Chem A 3(21):11395–11402
Liu S et al (2018) Hollow carbon spheres with nanoporous shells and tailored chemical interfaces as sulfur host for long cycle life of lithium sulfur batteries. Electrochim Acta 279:10–18
Lyu Z et al (2015) Hierarchical carbon nanocages confining high-loading sulfur for high-rate lithium–sulfur batteries. Nano Energy 12:657–665
Ma L et al (2015) Nanomaterials: science and applications in the lithium–sulfur battery. Nano Today 10(3):315–338
Manthiram A et al (2014) Rechargeable lithium–sulfur batteries. Chem Rev 114(23):11751–11787
Marom R et al (2011) A review of advanced and practical lithium battery materials. J Mater Chem 21(27):9938–9954
Martin RP et al (1973) Electrochemical reduction of sulfur in aprotic solvents. Inorg Chem 12(8): 1921–1925
Mikhaylik YV, Akridge JR (2004) Polysulfide shuttle study in the Li/S battery system. J Electrochem Soc 151(11):A1969
Nazar LF, Cuisinier M, Pang Q (2014) Lithium-sulfur batteries. MRS Bull 39(5):436–442
Palacin MR (2021) Battery materials design essentials. Account Mater Res 2(5):319–326

Pan H et al (2018) Addressing passivation in lithium–sulfur battery under lean electrolyte condition. Adv Funct Mater 28(38):1707234

Park J et al (2016) Graphene quantum dots: structural integrity and oxygen functional groups for high sulfur/sulfide utilization in lithium sulfur batteries. NPG Asia Materials 8(5):e272–e272

Park J et al (2017) Tungsten disulfide catalysts supported on a carbon cloth interlayer for high performance Li–S battery. Adv Energy Mater 7(11):1602567

Park J, Yu S-H, Sung Y-E (2018) Design of structural and functional nanomaterials for lithium-sulfur batteries. Nano Today 18:35–64

Peled E et al (1989a) Lithium-sulfur battery: evaluation of dioxolane-based electrolytes. J Electrochem Soc 136(6):1621

Peled E et al (1989b) Rechargeable lithium-sulfur battery. J Power Sources 26(3–4):269–271

Placke T et al (2017) Lithium ion, lithium metal, and alternative rechargeable battery technologies: the odyssey for high energy density. J Solid State Electrochem 21(7):1939–1964

Pu X, Yang G, Yu C (2014) Liquid-type cathode enabled by 3D sponge-like carbon nanotubes for high energy density and long cycling life of Li-S batteries. Adv Mater 26(44):7456–7461

Qu Y et al (2013) A simple SDS-assisted self-assembly method for the synthesis of hollow carbon nanospheres to encapsulate sulfur for advanced lithium–sulfur batteries. J Mater Chem A 1(45): 14306–14310

Rauh R et al (1979) A lithium/dissolved sulfur battery with an organic electrolyte. J Electrochem Soc 126(4):523

Su Y-S, Manthiram A (2012) A new approach to improve cycle performance of rechargeable lithium–sulfur batteries by inserting a free-standing MWCNT interlayer. Chem Commun 48(70):8817–8819

Sun L et al (2014) Sulfur nanocrystals confined in carbon nanotube network as a binder-free electrode for high-performance lithium sulfur batteries. Nano Lett 14(7):4044–4049

Sun XG et al (2012) Lithium–sulfur batteries based on nitrogen-doped carbon and an ionic-liquid electrolyte. ChemSusChem 5(10):2079–2085

Suo L et al (2013) A new class of solvent-in-salt electrolyte for high-energy rechargeable metallic lithium batteries. Nat Commun 4(1):1–9

Tao X et al (2014) Strong sulfur binding with conducting magnéli-phase Ti n O2 n–1 nanomaterials for improving lithium–sulfur batteries. Nano Lett 14(9):5288–5294

Tian Y et al (2019) Metal–organic framework derived yolk–shell NiS 2/carbon spheres for lithium–sulfur batteries with enhanced polysulfide redox kinetics. Chem Commun 55(22):3243–3246

Turcheniuk K et al (2018) Ten years left to redesign lithium-ion batteries, vol 559. Nature Publishing Group, p 467

Wang D-W et al (2013) Carbon–sulfur composites for Li–S batteries: status and prospects. J Mater Chem A 1(33):9382–9394

Wang J et al (2002a) A novel conductive polymer–sulfur composite cathode material for rechargeable lithium batteries. Adv Mater 14(13–14):963–965

Wang J et al (2002b) Sulfur–carbon nano-composite as cathode for rechargeable lithium battery based on gel electrolyte. Electrochem Commun 4(6):499–502

Wild M et al (2015) Lithium sulfur batteries, a mechanistic review. Energy Environ Sci 8(12): 3477–3494

Winter M, Barnett B, Xu K (2018) Before Li ion batteries. Chem Rev 118(23):11433–11456

Xiao Z et al (2019) Ultrafine Ti3C2 MXene nanodots-interspersed nanosheet for high-energy-density lithium–sulfur batteries. ACS Nano 13(3):3608–3617

Xin S et al (2012) Smaller sulfur molecules promise better lithium–sulfur batteries. J Am Chem Soc 134(45):18510–18513

Xu G et al (2017) Conductive graphene oxide-polyacrylic acid (GOPAA) binder for lithium-sulfur battery. Nano Energy 31:568–574

Xue W et al (2019) Intercalation-conversion hybrid cathodes enabling Li–S full-cell architectures with jointly superior gravimetric and volumetric energy densities. Nat Energy 4(5):374–382

Yang D-H et al (2019) Hollow N-doped carbon polyhedrons with hierarchically porous shell for confinement of polysulfides in lithium–sulfur batteries. Iscience 13:243–253
Yang X et al (2018a) Structural design of lithium–sulfur batteries: from fundamental research to practical application. Electrochem Energy Rev 1(3):239–293
Yang X et al (2018b) Structural design of lithium–sulfur batteries: from fundamental research to practical application. Electrochem Energ Rev 1:239–293
Yin YX et al (2013) Lithium–sulfur batteries: electrochemistry, materials, and prospects. Angew Chem Int Ed 52(50):13186–13200
Yu M et al (2016b) Atomic layer deposited TiO_2 on a nitrogen-doped graphene/sulfur electrode for high performance lithium–sulfur batteries. Energy Environ Sci 9(4):1495–1503
Yu S-H et al (2018) Direct visualization of sulfur cathodes: new insights into Li–S batteries via operando X-ray based methods. Energy Environ Sci 11(1):202–210
Yu X, Joseph J, Manthiram A (2016a) Suppression of the polysulfide-shuttle behavior in Li–S batteries through the development of a facile functional group on the polypropylene separator. Materials Horizons 3(4):314–319
Zang J et al (2015) Hollow-in-hollow carbon spheres with hollow foam-like cores for lithium–sulfur batteries. Nano Res 8(8):2663–2675
Zhang C et al (2012) Confining sulfur in double-shelled hollow carbon spheres for lithium–sulfur batteries. Angew Chem Int Ed 51(38):9592–9595
Zhang K et al (2013) Composite of sulfur impregnated in porous hollow carbon spheres as the cathode of Li-S batteries with high performance. Nano Res 6(1):38–46
Zhao M et al (2020) Lithium–sulfur batteries under lean electrolyte conditions: challenges and opportunities. Angew Chem Int Ed 59(31):12636–12652
Zhao X et al (2015) Nitrogen-doped mesoporous carbon: a top-down strategy to promote sulfur immobilization for lithium–sulfur batteries. ChemSusChem 8(19):3234–3241
Yu X et al (2004) Lithium storage in conductive sulfur-containing polymers. J Electroanal Chem 573(1):121–128

Chapter 6
Current Collectors for Li-S Batteries

Bilal Ahmad, Muhammad Sagir, Saima Nazir, and M. B. Tahir

Abstract The era of using fossil fuels as the main source of energy has been fading in our modern society. The advancement of several energy harvesting, converting, and storage technologies lessens the negative effects of human activity on the environment while also improving all facets of our life. Advanced large-scale energy storage technologies are essential for the efficient use of renewable resources and the maintenance of the future smart grid's dependability. For large-scale applications, there are typically two types of energy storage technologies: physical energy storage systems (such as pump hydro, compressed air, and flywheels) and electrochemical devices (i.e., rechargeable batteries). Rechargeable batteries stand out among them, thanks to their superior energy conversion efficiency, viability for distributed locations, and ease of maintenance.

Keywords Human activity · Energy conversion efficiency · Renewable resources

Our contemporary civilization is moving away from utilizing fossil fuels as an essential resource of energy. The development of several energy-gathering, energy-converting, and energy-storing technologies lowers the adverse consequences of human activities on the environment while simultaneously enhancing every aspect of our lives. Advanced large-scale energy storage technology is

B. Ahmad (✉) · M. B. Tahir
Institute of Physics, Khawaja Fareed University of Engineering and Information Technology, Rahim Yar Khan, Punjab, Pakistan

M. Sagir
Institute of Chemical and Environmental Engineering, Khawaja Fareed University of Engineering and Information Technology, Rahim Yar Khan, Punjab, Pakistan
e-mail: dr.msagir@kfueit.edu.pk

S. Nazir
University of Health Sciences, Lahore, Pakistan
e-mail: m.bilaltahir@kfueit.edu.pk

M. S. Tahir et al. (eds.), *Lithium-Sulfur Batteries: Key Parameters, Recent Advances, Challenges and Applications*, Springer Tracts in Electrical and Electronics Engineering, https://doi.org/10.1007/978-981-99-2796-8_6

essential for the efficient execution of the future smart grid and efficient use of renewable resources. Large-scale applications often make use of either electrochemical devices (like rechargeable batteries) or physical energy storage systems (such pump hydro, compressed air, and flywheels). Because of their superior energy conversion efficiency, adaptability to dispersed sites, and ease of maintenance, rechargeable batteries stand out from the crowd (Liu et al. 2019a). Battery systems for stationary storage applications must have components that are easily sourced on Earth, sufficient energy density, high efficiency, and low-maintenance requirements because battery systems can behave quite differently in terms of electrochemical properties (Gu et al. 2018; Liu et al. 2019a). Lithium-sulfur (Li-S) batteries are the best option because of their high theoretical energy density, low material prices, and little environmental impact.

Portable gadgets are now powered by lithium-ion batteries (LIBs), which have become a necessary component of daily life. However, lithium-ion batteries are unable to keep up with the rising needs of major markets, particularly in the area of electric cars, because they have a poor power and energy density. Research into lithium (Li) alternatives is motivated by the scarcity and expensive cost of lithium compounds (Liu et al. 2017; Zhu et al. 2019). With lithium, sulfur may have a multi-electron conversion-based electrochemistry process, producing a significant theoretical capability (1672 mAh g^{-1}). Sulfur (S) is a very plentiful and accessible substance (Li et al. 2019; Liu et al. 2020). As a cathode, S also provides the typical cell voltage of 2.15 V (in comparison with Li/Li^{+}). Accordingly, such a constructed theoretically, LSB may possess a high energy density (2600 Wh kg^{-1}) when coupled with a lithium metal anode (3860 mA h g^{-1}) (Cheon et al. 2003; Li et al. 2019).

Early Li-S battery research was started a few decades ago, but it wasn't until recent years' resurgence in interest in electric cars that this battery system came back into the spotlight. In 1962, Herbet and Ulam introduced the use of sulfur as a constituent of cathodes in electric dry cells and storage batteries, hence initiating the advancement of lithium-sulfur (Li-S) batteries (Duan et al. 2016). Later, in 1967, Argonne National Laboratory experimenters operated a high-temperature Li-S system (Birk and Steunenberg 1975). Nevertheless, the safety considerations surrounding high-temperature systems have posed significant challenges and have had little practical significance until the publication by E. Peled's research group on the electrochemical properties of sulfur in organic electrolytes at ambient temperature (Duan et al. 2016; Yamin et al. 1985). Since then, there has a surge in research toward improving the energy density, lifetime, and security of Li-S devices. Sulfur cathode (SC) has been the subject of extensive research and several chapters, all of which demonstrate the advancement of various schemes, material manipulations, reaction processes, and applied applications. The purpose of this chapter is to provide, maybe unexpectedly, a comprehensive evolutionary introduction to sulfur cathodes, complete including a historical analysis of the composition's evolution and revision.

Electrochemical Principles of the Sulfur Cathode

A typical SC is made up of polymeric binders, conductive, and electroactive materials (Yuan et al. 2018). For Li storage, elemental sulfur (S_8) and its solid reduction product, Li_2S, are the most often utilized electroactive materials. A typical lithium-sulfur battery has a sulfur cathode, which is positively charged; a lithium metal anode, which is negatively charged and linked to an outside circuit; and a permeable polymer separator immersed in a nonaqueous electrolyte with both electrodes, and which protects the two electrodes from coming into contact with one another (Xueliang Li et al. 2017; Rehman et al. 2017). In Fig. 6.1 we have a schematic representation of the whole operating system for Li-S batteries. During discharge, the separator performs the function of a barrier between the anode and the cathode. This makes it possible for Li^+ to diffuse while electrons travel via an external connection from the anode (A) to the cathode (C). It is feasible to condense all of the reactions that take place among the A and C into the following reaction, which is a good summary of everything that happens (Rehman et al. 2017).

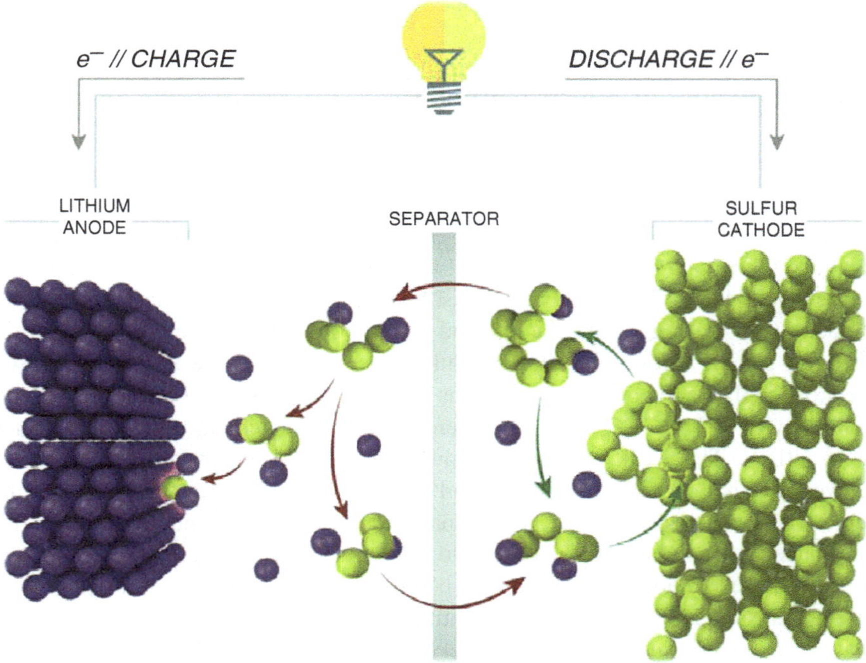

Fig. 6.1 Schematic layout of LSB operating system

Discharging Process
Lithium anode:

$$\mathrm{Li} \rightarrow \mathrm{Li}^{+} + \mathrm{e}^{-}$$

Sulfur cathode:

$$\mathrm{S}_8 + 16\mathrm{Li}^{+} + 16\mathrm{e}^{-} \rightarrow 8\mathrm{Li}_2\mathrm{S}$$

Charging Process
Lithium anode:

$$\mathrm{Li}^{+} + \mathrm{e}^{-} \rightarrow \mathrm{Li}$$

Sulfur cathode:

$$8\mathrm{Li}_2\mathrm{S} + \mathrm{S}_8 + 16\mathrm{Li}^{+} \rightarrow 16\mathrm{e}^{-}$$

The Challenges of Sulfur Cathode

Li-S batteries have shown great promise in terms of cheap cost and great energy density, yet they are still hampered by a plethora of cathode material difficulties related to real-world applications, severely limiting their potential for widespread commercialization (Chen and Shaw 2014). Sulfur concentrations in LSB cathodes range from 41% to 70%, with as low as 40% in some cases. Typically, a specific capacity (821 mAh g^{-1}) may be attained with a 70% sulfur content and utilization (Fang et al. 2017; Liu et al. 2019b). Sulfur-based cathodes are unable to receive electrons in the collector because sulfur is electrical and ionic insulator with a feeble conductivity of 5*1030 S cm^{-1} at ambient temperature (Rehman et al. 2022; Wang et al. 2017). The weak electrical conductivity of Li_2S makes the final step of converting Li_2S_2 to Li_2S challenging and complex. Due to inefficient use of sulfur, the theoretical capacity of the cathode material is very seldom met, even when the substance has a conductive agent added to it. The shuttle effect may also be observed in sulfur-based cathodes when they are being charged and discharged (Wang et al. 2020; Wu et al. 2018). The long-chain lithium polysulfides (LiPSs) Li_2Sn ($8 > n$ 4) that are created during this process quickly dissolve in the electrolyte. Once to the lithium anode, after they diffuse via the membrane, they interact with the lithium metal to create the insoluble PSs. All of this is possible because of the force of the electric field and the concentration gradient. The PSs then go to the SC, where they are converted back into long-chain lithium PS. The sulfur cathode undergoes a volumetric expansion/reduction of up to 80% throughout the electrochemical

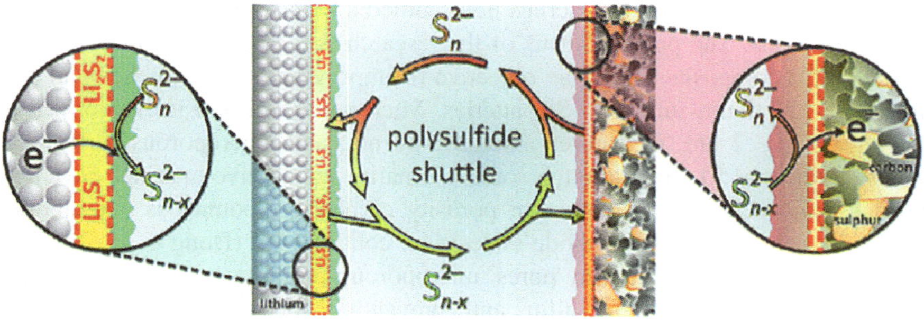

Fig. 6.2 The parasitic PS shuttle effect is shown diagrammatically in a Li/S 8 cell with a liquid electrolyte. The negative electrode is where long-chain PSs that are created by high discharge and charge plateaus diffuse before being further reduced to lower chain length PS. These PSs with shorter chains reoxidize to form PS with longer chains when they return to the positive electrode

process (Zhang et al. 2020) It will compromise the battery's safety and performance forever. The density of sulfur (S) is 2.03 g cm^3, but that of lithium sulfide (Li2S), a by-product of the discharge, is only 1.66 g cm^3 (Zhao et al. 2014). The following are features that should be present in high-capacity, cycling-stable sulfur cathodes: to have high conductivities, you need three things, (3) a flexible structure to allow for the volume expansion of sulfur and the preservation of the electrodes' morphology throughout the transit of lithium ions and electrons; (1) a closed structure that effectively encloses polysulfides (PSs), enhancing the interaction between the sulfur and the electrolyte; (2) high conductivities, such as those attained by combining conductive materials in the cathode; and (3) good sulfur-electrolyte contact (Fig. 6.2).

Different Matrices for Hosting Sulfur in LSBs

In order to boost discharge capacity, cyclability, and Coulombic efficiency, a number of methods were established to produce sulfur combinations with advantageous structures and characteristics. In addition, research is being done on new cell topologies with trapping interlayers, Li/dissolved polysulfide cells, and powerful electrolytes.

Sulfur-Carbon Composite Cathodes

Porous Carbon-Based Composites

Because of their strong electrical conductivity, comparatively constant physical and chemical characteristics, tunable pore structure, and simple-to-adjust surface chemical properties, carbon materials have earned a reputation as suitable hosts for active materials. The primary focus of the research is the development of carbon-based sulfur composites, with the objective of improving the electrochemical performance of lithium-sulfur (Li-S) batteries. Microporous carbon is categorized as having pores <2 nm, mesopores among 2–50 nm, and macroporous carbon has holes larger than 50 nm. Although some alternative conductive matrixes have also been examined, sulfur habitation in porosity carbon compounds is a preferable choice boosting the sulfur cathode's electrical conductivity (Dong et al. 2018; Li et al. 2018). Because of its tiny pores, microporous carbon acts as a potent sulfur immobilizer by containing sulfur and preventing intermediate products like polysulfides from dissolving into the electrolyte (Chen and Shaw 2014; Ma et al. 2015). Yoshida et al. have developed a technique of heating and mixing just microporous carbon materials and monomers in order to change microporous carbon with polyglycerol. When employing a sulfur cathode with microporous carbon that has been treated with polyglycerol, discharge capability at the 50th cycle as a percentage of discharge ability during the 2nd cycle increased from 57.6 to 75.2% (Yoshida et al. 2022). In relation to lithium-sulfur (Li-S) batteries and the well-known phenomenon of lithium polysulfide (LiPS) shuttle effect, Du et al. propose a design approach for a multifunctional covalent organic framework (COF) that effectively confines sulfur as a host material. This technique involves including an internal porous catalyst and an exterior layer of microporous COF. The integrated hollow catalyst serves the dual purpose of trapping and converting lithium polysulfides (LiPSs) while also providing lodging for sulfur and functioning as a catalytic center. The cell has the potential to attain a maximum rate performance of 5 °C, equivalent to 543.13 (mAh g^{-1}). After experiencing 50 cycles, the high area capacity reaches a value of 4.53 (mAh cm^{-2}), despite the presence of a weak electrolyte with an electrolyte-to-sulfur (E/S) ratio of 6.10 (L mg^{-1}) and a significant sulfur loading of 5.7 (mg cm^{-2}). Following a series of 50 cycles, the lithium-sulfur (Li-S) pouch cell demonstrates a specific capacity of 860.35 (mAh g^{-1}) when operated at a rate of 0.1 °C. The composite design that is being presented entails the confinement of COF (covalent organic framework), and it offers a feasible method for the production of Li-S (lithium-sulfur) batteries that exhibit improved performance characteristics (Du et al. 2022).

A novel method of fabricating one-dimensional ordered mesoporous carbon fiber was used, using the triblock copolymer Pluronic F127 as a template and the carbon supply Resol via the process of electrospinning. The encapsulation of sulfur inside ordered mesoporous carbon fibers is achieved using a simple heating process, as seen in Fig. 6.3. The mechanical stability of the interlaced fibrous threads' nanostructure enables effective transmission of sulfur and PS during cycling. The diffusion of long-chain PS is impeded inside the highly structured mesopores. The ordered mesoporous carbon fiber sulfur (OMCF-S) composite containing 63% sulfur exhibits notable characteristics, such as enhanced reversible capacity, improved capacity retention, and increased rate capacity, when used as the cathode in rechargeable lithium-sulfur batteries. After undergoing 300 cycles, the OMCF-S electrode

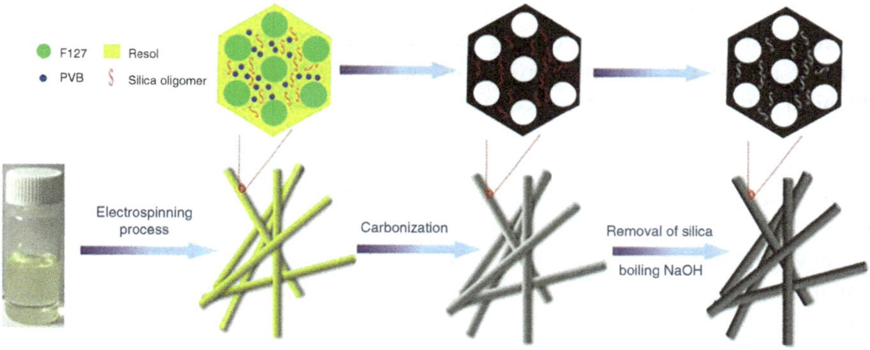

Fig. 6.3 Experimental technique and optimum mesoporous structure creation in OMCF are depicted schematically

exhibits stability and maintains a discharge capacity of 690 $mAhg^{-1}$ at a rate of 0.3 ° C (Wang et al. 2015).

The advancements in the manufacture of carbon hosts, namely, porous carbon, may be credited for the overall improvements seen in sulfur loading and electrochemical performance. In a general sense, carbon-based hosts characterized by large pore volumes may potentially benefit from increased sulfur loading. This is due to the potential enhancement of ion transference resulting from the impregnation of sulfur into the open pores. The rapid redox kinetics shown by lithium-sulfur batteries is advantageous for their overall performance. In order to optimize the transfer of lithium ions, accommodate volume fluctuations in the buffer, and facilitate effective impregnation of both sulfur and electrolyte, it is crucial to augment the overall accessible pore volume.

Binder-Free Composite Electrodes

Avoiding hazardous chemicals in the production of battery electrodes is tempting since green technologies have been established. The polymer binder and active chemicals used in producing Li-ion battery electrodes are typically dispersed in a solvent called N-methyl2-pyrrolidone (NMP). Meanwhile, NMP is harmful to both people and the planet (Yang et al. 2014). One option would be to develop electrodes that do not require the use of binders, thereby eliminating the need for NMP. The elimination of binders leads to an increase in the loading of active materials and the conductivity of electrodes (Kuang et al. 2019). The method of postpolymerization involves the self-assembly of polymers onto sulfide particles. This may be achieved by using either commercially available sulfur or sulfur generated by the thiosulfate reduction process in an aqueous solution with acid (Manthiram et al. 2013). Nevertheless, the restricted use of sulfur might be attributed to the presence of large particles, often in the micrometer size range. The incorporation of nanoscale sulfur particles into conductive polymer shells has the potential to enhance the utilization of active material. Nonetheless, the implementation of this approach necessitates the development of novel manufacturing methodologies.

Li et al. used a hybrid approach using in situ hydrothermal annealing and chemical vapor deposition to successfully fabricate clusters of nanofibers composed of NiCo2O4-decorated 3D graphene on a nickel foam substrate, referred to as NCO@3DGNF. After undergoing 480 cycles with a current density of 0.5 A g^{-1}, the discharge capacity of the NCO@3DGNF nanocomposite increased to 707.7 mAh g^{-1}, indicating a high level of reversibility (with a Coulombic efficiency of 98–99%) and efficient performance at different rates (Li et al. 2022a). In order to manufacture Mn_3O_4/graphene hybrid electrodes without the need of a binder and to address both issues at once, Yan et al. presented a unique electrophoretic deposition technique. The nanoparticles of Mn_3O_4 are contained in a conductive graphene network, thanks to the structure's well-thought-out construction. Based on this perspective, the Mn_3O_4/graphene hybrid electrode has notable characteristics, including a substantial reversible specific capacity of 880 mAh g-1 at 0.1 A g-1, a commendable rate capability of 678 mAh g^{-1} at 5 A g^{-1}, and an amazing cycling durability with little capacity drop even after 500 cycles at 1 A g^{-1} (Yan et al. 2022).

As a potential high-performance anode for LIBs, a composite of silicon, carbon nanofiber, and graphene was recently investigated for its interfacial properties. Composite films made from silicon nanoparticles (SiNPs), carbon nanofibers (CNFs), and reduced graphene oxides (rGO) were produced using a simple physical filtering procedure and an environmentally friendly thermal reduction method. For LIBs that don't need a binder, the films acted as a high-performance anode component. The anodes in lithium-ion batteries (LIBs) were composed of Si/CNF/rGO composites. These composites demonstrated an initial specific capacity of 1894.54 mAh/g when exposed to a current density of 0.1 Ag^{-1}. The reversible specific capacity remained at 964.68 mAh/g after 100 cycles, whereas the Coulombic effectiveness at the same current density increased to 93.8%. In addition to increasing electrode conductivity and providing a swift and efficient conduit for electron and ion transport, the Si/CNF/rGO composite films provide a stable solid-electrolyte interface by efficiently accommodating and buffering changes in silicon nanoparticle volume (Cong et al. 2021) (Fig. 6.4).

Hybrid Sulfur Composite Cathodes

The cyclability of sulfur-carbon composites has been proven to be quite high; however, the electrode is mostly inert carbon and binder. In order to increase energy density, the loading of active materials in the cathode must be increased. One potential tactic is to substitute a lithium insertion oxide or sulfide for some of the carbon and binder. To add capabilities, these materials must operate in a voltage range equal to that of sulfur (31.5 V) (Kulkarni et al. 2017).

By integrating a sulfur-loaded Al_2O_3-modified carbon nanotube cathode (S@ACNTs) with a malleable hybrid solid electrolyte (HSE), Hoang et al. presented an outstanding performance solid-state LiSB system. The polycationic binder, poly

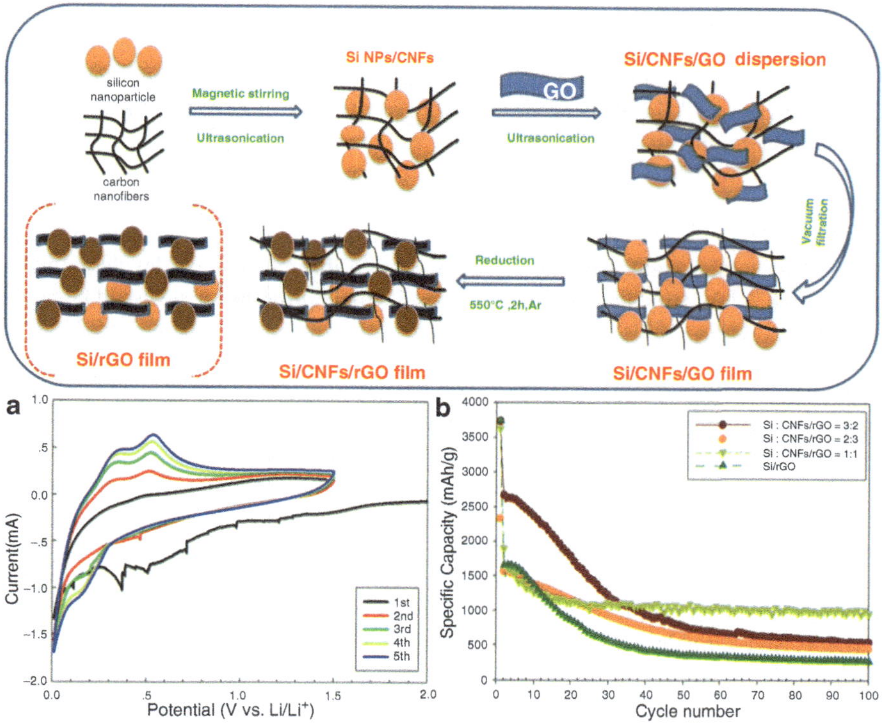

Fig. 6.4 The investigation is focused on the production of composite films that include silicon (Si), carbon nanofibers (CNF), and reduced graphene oxide (rGO), as well as films formed solely of Si and rGO. The cyclic voltammetry data for the Si/CNF/rGO = 1:1 composite, which has been synthesized, is shown for the first five cycles. The cycling performance of composite electrodes consisting of synthetic Si/rGO and Si/CNF/rGO in various ratios (1:1, 2:3, and 3:2) was assessed at a current density of 0.1 A/g (Li 2012)

(diallyldimethylammonium bis(trifluoromethylsulfonyl)imide) (PDATFSI), has exceptional thermal stability of up to 450 °C when used in combination with S@ACNTs. Additionally, it demonstrates exceptional bonding strength with aluminum current collectors, reaching up to 24 MPa. Moreover, PDATFSI offers prolonged nonflammability and desirable flexibility. The cathode layer composed of S@ACNTs/PDATFSI-60IL demonstrates excellent compatibility (with a compatibility index of 65) and effectively traps polysulfide species. As a result, it exhibits a high initial discharge capacity of 1203 $mAhg^{-1}$ at a rate of 0.2 °C during the first cycle. Furthermore, this cathode layer maintains long-term stability, retaining 91.69% of its discharge capacity after undergoing 200 charge/discharge cycles (Hoang and Kim 2022). The specific capacity at a rate of 1 °C for the 3D N-doped carbon/sulfur (S/CNx) composite developed by Li et al. is measured to be 830 $mA\ h\ g^{-1}$ after undergoing 100 cycles. This indicates a high retention rate of 98.8% in comparison with the initial capacity. The used synthetic process in this study exhibits a high degree of innovation, hence offering significant promise for engineers engaged in the advancement of lithium-sulfur (Li-S) batteries with enhanced capacity (Li et al. 2022b).

The electrode exhibits a significant capacity, as shown by the measurement of 572 mAh $g_{cathode}^{-1}$ (in terms of the overall mass of the electrode) or 1866 mAh gs^{-1} (in relation to the mass of sulfur). This enhanced capacity may be due to the involvement of both S and $Li_4Ti_5O_{12}$ in redox reactions. The aforementioned measurements were acquired at a rate of 0.1 °C, with 1 °C corresponding to a current of 1675 mAh gs^{-1}. The electrode has a substantial sulfur composition, namely, up to 90% of its total weight. The battery exhibits two discrete voltage levels, namely, 2.35 V and 2.1 V, which are produced by the S cells. Furthermore, it should be noted that the $Li_4Ti_5O_{12}$ cells generate a distinct base with a voltage of 1.55 V. It is possible to get a cycle lifespan beyond 100 cycles, accompanied by a notable rate capability of 566 mAh $g_{cathode}^{-1}$ at a rate of 0.25C and 376 mAh $g_{cathode}^{-1}$ at a rate of 1C. The findings of a study including Raman and electron microscope analysis of graphite/$Li_4Ti_5O_{12}$ layers during operation indicate that the presence of polysulfides is effectively hindered from dissolving or migrating. This phenomenon leads to enhanced utilization of sulfur and a reduced rate of capacity decline. The high-tech hybrid battery presented in this study employs a multilayer design that combines voltage plateaus from several electrode materials. This innovative approach provides a means to achieve customizable applications that need energy devices with high-capacity and variable-voltage platforms (Ming et al. 2016) (Fig. 6.5).

Sulfur-Polymer Hybrid Cathode Materials

Polymers play a crucial role in augmenting charge transfer and minimizing the shuttle effect in lithium-sulfur (Li-S) batteries with high energy density. This is primarily attributed to the ability of polymers to chemically and covalently bind elemental sulfur to their frameworks. Additionally, the remarkable flexibility and inherent ability to self-heal shown by conductive polymers render them very promising for utilization in sulfur cathodes with substantial loadings. Numerous studies have been conducted to explore various sulfur-polymer hybrid materials with the aim of enhancing the energy density of Li-S batteries. Conductive polymers provide a promising option for their ability to facilitate electron transport. However, their suitability as a substitute for the sulfur electrode is limited owing to their comparatively lower electrical conductivity when compared to carbon materials. Various techniques are available for the synthesis of conductive polymers, enabling more versatility in terms of manipulating morphology and designing materials.

In their investigation, Yeşilot et al. used a novel polythiophene derivative that included an acryloyloxy moiety inside its side chain as a functional comonomer. Scientists conducted the synthesis of a polymer called poly(3-[(2-acryloyloxy) methyl] thiophene in order to enhance the incorporation of elemental sulfur (S8) into the inverse vulcanization process. The technique of inverse vulcanization was used to change the polyacrylamide-thiophene material by introducing varying

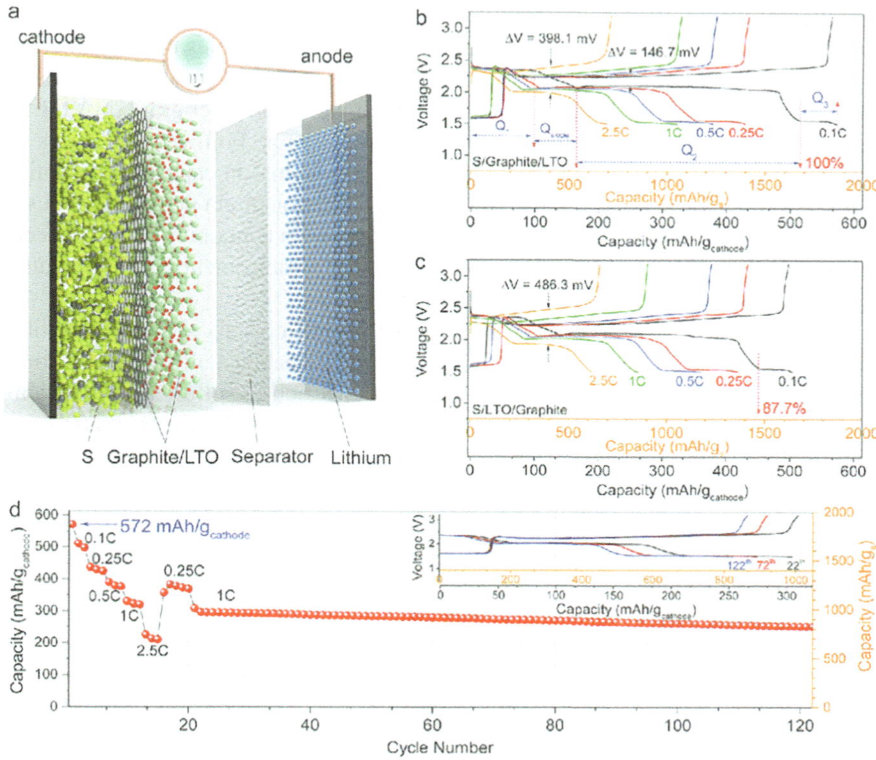

Fig. 6.5 The schematic representation illustrates the electrochemical characteristics of a lithium-sulfur battery. The electrode in the hybrid lithium-sulfur battery is composed of 90% sulfur by mass and 10% Super P composite, as seen in figure (**a**). The voltage versus capacity curves for hybrid batteries using cathodes with different compositions, namely, (**b**) S/graphite/LTO and (**c**) S/LTO/graphite, are examined at various C rates. The hybrid battery's capacity rate and cycle performance are evaluated in relation to its S/graphite/LTO cathode. The figure (**d**) illustrates the voltage vs capacity curves for a current of 1C

amounts of molten sulfur. Consequently, graft materials with elevated levels of sulfur were developed, known as S-poly(3-[(2-acryloyloxy) methyl) thiophenes (52 and 74%). The electrochemical characteristics of S-PAMT, with a concentration of 74%, were examined to assess its viability as a cathode material for lithium-sulfur batteries. After performing trials at a current density of C/5, it was revealed that the cells consistently exhibited a capacity of around 400 $mAh{\bullet}g^{-1}$ after enduring 100 cycles. The present study used electrochemical impedance spectroscopy (EIS) to examine the underlying mechanism involved in the storage of lithium in S-PAMT (74%) both before to and during cycling (Yeşilot et al. 2022). The development of a cathode material for lithium-sulfur (Li-S) batteries was undertaken by Xiang Li and his research team. The copolymerization of ring-opening S and triallyl isocyanurate is successfully accomplished by a straightforward two-step heating protocol. The unique two-phase structure of the organosulfur polymer composite, known as

S-triallyl isocyanurate (STI), is attributed to the integration of ultrafine monoclinic S particles into the matrix of the organosulfur polymer. This enables the attainment of a weight concentration of up to 90% for S. The cathode composed of STI has a significant initial reversible capacity of 904 milliampere-hours per gram (mA h g^{-1}) when operated at a temperature of 0.5 Coulombs (C). Following a series of 350 cycles at a rate of 1 °C, the cathode has had a little loss in capacity, amounting to just 0.017%. The results obtained from this investigation might potentially be used to improve the structural attributes of composite cathodes containing a high concentration of sulfur. Consequently, this would aid in the advancement of lithium-sulfur batteries with enhanced durability (Li et al. 2020).

The freshly synthesized porous trithiocyanuric acid crystals serve as a flexible template for the ring-opening polymerization of elemental sulfur. This method enables the three-dimensional connectivity of phases rich in sulfur. After exposing the lithium-sulfur cells to 100 cycles at a discharge rate of 0.2 C, an observation was made that the cells exhibited a discharge capacity of 945 (mAh g^{-1}). In addition, the cells had a lifespan of 450 cycles. The observed capacity retention rate of 92% is worthy of commendation. The incorporation of structured amine groups into the crystalline framework leads to an augmentation in the rate of Li^{+} ion transfer, hence yielding rate capabilities of 1210 mAh g^{-1} at 0.1 °C and 730 mAh g^{-1} at 5 °C (Kim et al. 2015) (Fig. 6.6).

Schematics describing the synthesis of sulfur-rich polymers with tunable morphology show the steps involved in preparing TTCA co-crystals, removing the solvent through heat treatment to produce porous TTCA frameworks, impregnating the frameworks with sulfur, and polymerizing elemental sulfur at the thiol surfaces of the porous TTCA template.

Novel Li-S Cell Configurations

Conductive and Porous Carbon Interlayers

Materials made of highly conductive carbon might greatly reduce interface resistance on the sulfur cathode's side, allowing for more efficient use of the element's active form. In this scenario, LiPS could be prevented from dissolving and spreading due to the wide pore structure of carbon materials. Consequently, carbon interlayers have become a widely used functional interlayer in the context of lithium-sulfur batteries. Kong et al. introduced a novel approach for fabricating porphyrin-derived graphene-based nanosheets (PNG) by depositing a thin layer of 2D porphyrin organic framework (POF) directly onto graphene, followed by pyrolysis to form a nitrogen-enriched skin rich in pyrrolic and pyridinic nitrogen functionalities. The PNG nanosheets were recommended as an interlayer material (Wang et al. 2022). A novel design for Li-S batteries has been created in order to enhance their reusability. The design incorporates carbon interlayers that are strategically placed between the separator and the typical sulfur electrodes. Initial inquiries indicate that the

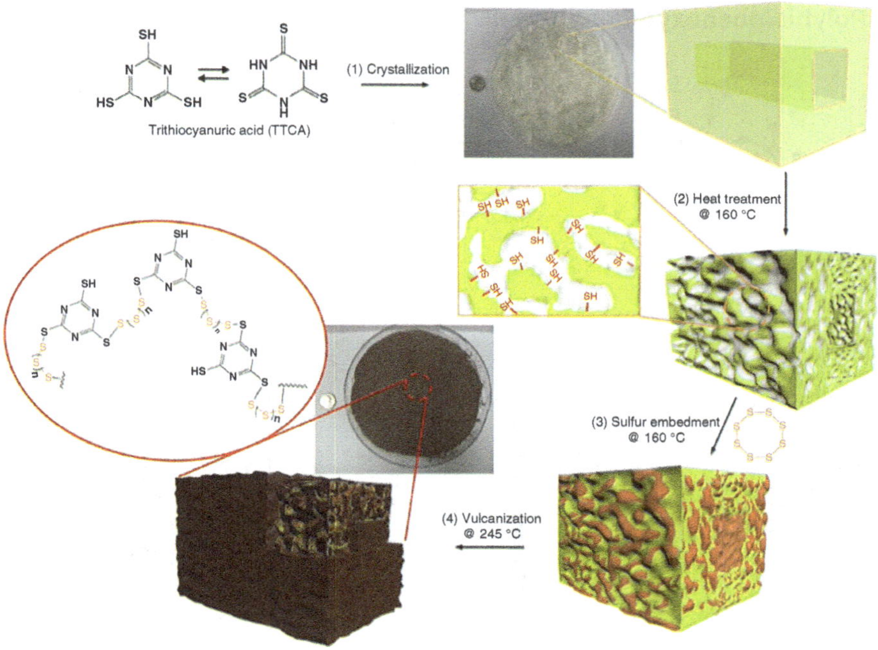

Fig. 6.6 The schematics illustrate the process of synthesizing sulfur-rich polymers with adjustable morphology. The steps involved include the preparation of TTCA co-crystals, the removal of solvent through heating to generate porous TTCA frameworks, the impregnation of sulfur into these porous frameworks, and the ring-opening polymerization of elemental sulfur at the thiol surfaces of the porous TTCA template (Zhu et al. 2019)

incorporation of conductive multi-walled carbon nanotube (MWCNT) paper into a lithium-sulfur (LiS) cell results in exceptional cyclability. This enhanced performance is characterized by a capacity surpassing 800 (mA h g^{-1}) over the course of 100 cycles. In order to promote the passage of active materials during the cycling process, it is necessary for the interlayer structure to include porosity or many interspaces. The achievement of a composite material with outstanding stability, high Coulombic efficiency, exceptional cyclability, and a longer cycle life is realized by the chemical bonding of dissolved polysulfides with the interlayer subsequent to the charging and discharging process.

To fabricate a hierarchical porous carbon (HPC) material with enhanced polysulfide capture capabilities, the experimental procedure included the carbonization of bis(phenoxy)phosphazene, followed by the incorporation of triple heteroatoms into the carbon lattice. The carbon material demonstrates chemisorption of polysulfides, physical confinement, and effective ion transport due to its integration of a hierarchical porous structure, a significant specific area, and several dopants including nitrogen, oxygen, and phosphorus. The use of the high-performance computing (HPC) interlayer has yielded substantial enhancements in the electrochemical

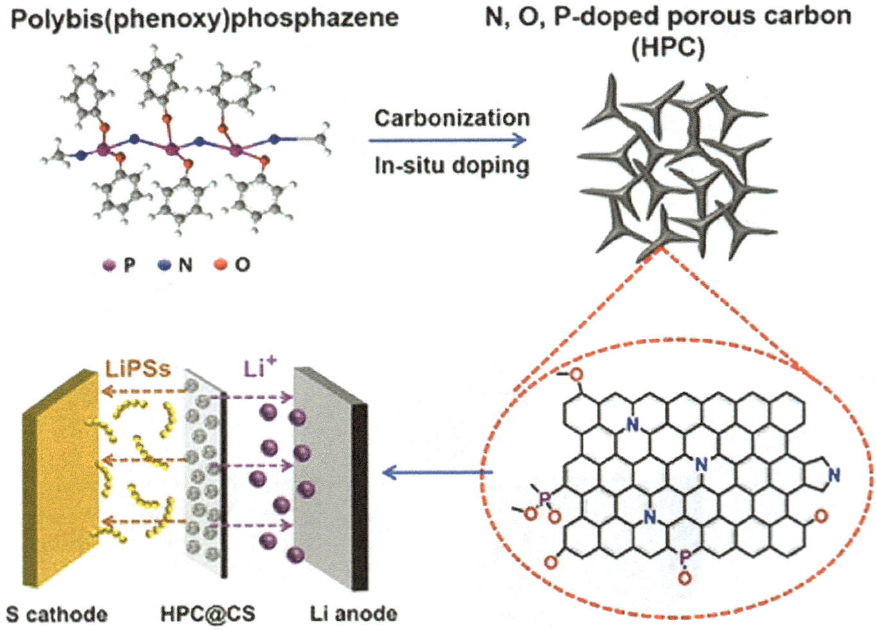

Fig. 6.7 Diagrams that show the processes for creating a polysulfide-trapping interlayer with a controlled morphology for use in Li-S batteries

efficacy of lithium-sulfur (Li-S) batteries, leading to notable discharge capacities of 1509 mA h/g at a rate of 0.06 C. In addition, the batteries demonstrated a capacity retention of 83.7% after 250 cycles at a discharge rate of 0.3 C (Qiu et al. 2021) (Fig. 6.7).

Li/Dissolved Polysulfide Cells

The inclusion of electrolytes is an essential attribute shared by traditional redox flow batteries and the newly found cathode-flow Li-ion batteries. Both battery types use catholytes, which are dissolved active materials, to enhance the reaction kinetics inside the electrodes. One of the organic solvents, tetrahydrofuran (THF), has a high solubility for lithium polysulfides. THF can produce up to 10 M polysulfide, and a lithium/dissolved sulfide rechargeable battery was successfully demonstrated in 1979 (Rauh et al. 1979). Despite minimal cyclability being accomplished, the first discharge demonstrated a capacity of 1532 $mAhg^{-1}$, which is comparable to 1.83 electrons per sulfur, with a depth of discharge (DoD) as little as 0.5 electrons per sulfur, resulting in 418 $mAhg^{-1}$ (Chen et al. 2015).

Encapsulation techniques are frequently employed to reduce polysulfide dissolution and diffusion; however, they frequently suffer irreparable structural breakdown

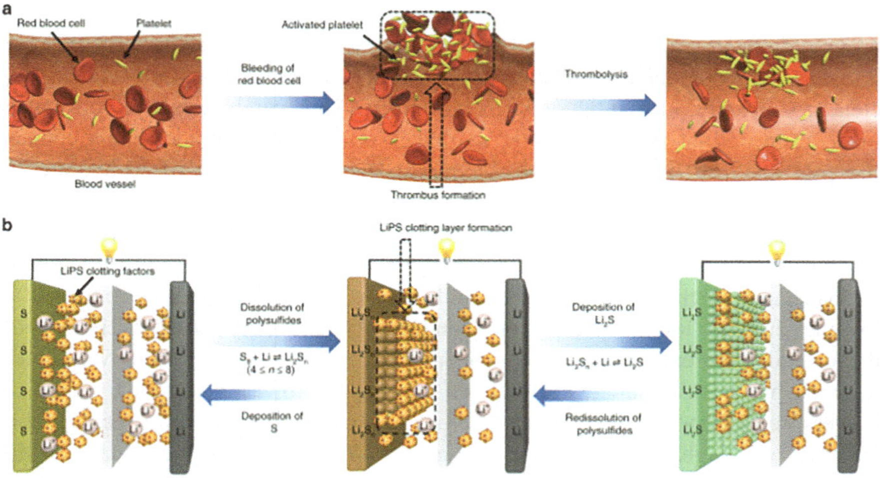

Fig. 6.8 (**a**) In animals, when a blood artery is broken, platelets activate and form a thrombus to halt the bleeding. Thrombolysis occurs spontaneously once the wound has healed. (**b**) During the discharge/charge process, dissolved LiPS interacts with LiPS clotting factors to produce a protective clotting layer at the electrode-electrolyte interface, thereby blocking the diffusion of LiPS and allowing the battery to be recharged

because lithium-sulfur batteries undergo large volume changes every time they are cycled. Here, we present a technique for creating a polysulfide protective layer that may be electrochemically recovered with the addition of electrolytes. The nitrogen-doped carbon dots included in the additive exhibit solubility in the electrolyte, even under conditions of maximum charge. Furthermore, a portion of these dots function as active sites for the formation of lithium sulfide at maximum discharge. Nitrogen-doped carbon dots exhibit a pronounced reactivity toward polysulfides, resulting in the formation of a solid and retrievable polysulfide encapsulation layer. This phenomenon occurs specifically at the interface between sulfur and lithium sulfide. The prevention of polysulfide shuttling is achieved without affecting the structural integrity, thereby demonstrating a remarkable design. The sulfur cathodes exhibit remarkable cycling stability for duration of 1000 cycles at a rate of 2 °C, accompanied by a Coulombic efficiency of 99.5%. Additionally, these cathodes demonstrate a substantial reversible capacity of 891 $mAhg^{-1}$ when operated at a rate of 0.5 °C (Fu et al. 2020) (Fig. 6.8).

Conclusion and Perspectives

Among the most intriguing approaches for energy storage in recent years has been the manufacturing of Li-S batteries, which display impressive theoretically particular energy densities. However, the commercialization of Li-S batteries has been hampered by issues with low active material utilization, limited cycle life, and risks brought on by volumetric growth. The main focus of research to address these issues is cathode substantial modification to boost Li-S batteries' electrochemical efficiency. Currently, elemental sulfur is combined using conductors and substances made from carbon polymer matrix to make sulfur-based cathodes. This chapter presents the most recent findings on modifying the cathodes of Li-S batteries utilizing such matrix materials which have been described. The most popular type of material used to increase sulfur-cathode conductivity is carbon-based. We have covered a wide range of methods for achieving high sulfur loading in Li-S batteries in this chapter, including changing the sulfur hosts. In an effort to enhance the sulfur loading, enhance ionic and electronic conductivity, and attach soluble LiPS inside the cathode, current efforts have concentrated on these areas. Moreover, much research has been conducted on several components of Li-S batteries, including binders, separators, and interlayers. There exists a significant need for rechargeable batteries with high capacity, namely, over 600 W h kg^{-1}, in order to cater to the requirements of transportation and large-scale energy storage on the electrical grid.

References

Birk J, Steunenberg R (1975) Chemical investigations of lithium-sulfur cells. In: ACS Publications

Chen L, Shaw LL (2014) Recent advances in lithium–sulfur batteries. J Power Sources 267:770–783

Chen H, Wang C, Dong W, Lu W, Du Z, Chen L (2015) Monodispersed sulfur nanoparticles for lithium–sulfur batteries with theoretical performance. Nano Lett 15(1):798–802

Cheon S-E, Ko K-S, Cho J-H, Kim S-W, Chin E-Y, Kim H-T (2003) Rechargeable lithium sulfur battery: I. Structural change of sulfur cathode during discharge and charge. J Electrochem Soc 150(6):A796

Cong R, Choi J-Y, Song J-B, Jo M, Lee H, Lee C-S (2021) Characteristics and electrochemical performances of silicon/carbon nanofiber/graphene composite films as anode materials for binder-free lithium-ion batteries. Sci Rep 11(1):1–11

Dong C, Gao W, Jin B, Jiang Q (2018) Advances in cathode materials for high-performance lithium-sulfur batteries. IScience 6:151–198

Du B, Luo Y, Yang Y, Xue W, Liu G, Li J (2022) COFs-confined multifunctional sulfur-host design towards high-performance lithium-sulfur batteries. Chem Eng J 442:135823

Duan L, Zhang F, Wang L (2016) Cathode materials for lithium sulfur batteries: design, synthesis, and electrochemical performance. In: Alkali-ion Batteries. IntechOpen, London

Fang R, Zhao S, Sun Z, Wang DW, Cheng HM, Li F (2017) More reliable lithium-sulfur batteries: status, solutions and prospects. Adv Mater 29(48):1606823

Fu Y, Wu Z, Yuan Y, Chen P, Yu L, Yuan L, …, Kan E (2020) Switchable encapsulation of polysulfides in the transition between sulfur and lithium sulfide. Nat Commun 11(1):845

Gu S, Sun C, Xu D, Lu Y, Jin J, Wen Z (2018) Recent progress in liquid electrolyte-based li–s batteries: shuttle problem and solutions. Electrochem Energ Rev 1(4):599–624

Hoang HA, Kim D (2022) High performance solid-state lithium-sulfur battery enabled by multifunctional cathode and flexible hybrid solid electrolyte. Small, 2202963

Kim H, Lee J, Ahn H, Kim O, Park MJ (2015) Synthesis of three-dimensionally interconnected sulfur-rich polymers for cathode materials of high-rate lithium–sulfur batteries. Nat Commun 6(1):1–10

Kuang Y, Chen C, Kirsch D, Hu L (2019) Thick electrode batteries: principles, opportunities, and challenges. Adv Energy Mater 9(33):1901457

Kulkarni P, Nataraj S, Balakrishna RG, Nagaraju D, Reddy M (2017) Nanostructured binary and ternary metal sulfides: synthesis methods and their application in energy conversion and storage devices. J Mater Chem A 5(42):22040–22094

Li X, Ding Z, Zhang L, Tang R, He Y (2017) Enhanced performance of lithium sulfur batteries with sulfur embedded in Sm_2O_3-doped carbon aerogel as cathode material. Electrochim Acta 241: 197–207

Li G, Lei W, Luo D, Deng Y, Deng Z, Wang D, …, Chen Z (2018) Stringed "tube on cube" nanohybrids as compact cathode matrix for high-loading and lean-electrolyte lithium–sulfur batteries. Energy Environ Sci 11(9):2372–2381

Li F, Liu Q, Hu J, Feng Y, He P, Ma J (2019) Recent advances in cathode materials for rechargeable lithium–sulfur batteries. Nanoscale 11(33):15418–15439

Li X, Yuan L, Liu D, Li Z, Chen J, Yuan K, …, Huang Y (2020) High sulfur-containing organosulfur polymer composite cathode embedded by monoclinic S for lithium sulfur batteries. Energy Storage Materials 26:570–576

Li S, Guo, X, Zhao L, Chen F, Wang X, Chu Y, …, Zhu Y (2022a) $NiCo_2O_4$ nanofiber cluster assisted by NH4F growing on 3D graphene as a binder-free electrode for lithium-ion batteries. Energy Fuel 36(2):1072–1080

Li Y, Xu J, Mi L, Huo K (2022b) 3D interconnected N-doped carbon/sulfur derived from organic-inorganic hybrid ZnS Superlattice Nanorods for high-performance lithium-sulfur batteries. Chem Lett 51(4):386–390

Liu Y, Wang L, Cao L, Shang C, Wang Z, Wang H, …, Li J (2017) Understanding and suppressing side reactions in Li–air batteries. Mater Chem Front 1(12):2495–2510

Liu D, Li Z, Li X, Cheng Z, Yuan L, Huang Y (2019a) Recent advances in cathode materials for room-temperature sodium– sulfur batteries. ChemPhysChem 20(23):3164–3176

Liu M, Zhang C, Su J, Chen X, Ma T, Huang T, Yu A (2019b) Propelling polysulfide conversion by defect-rich MoS2 nanosheets for high-performance lithium–sulfur batteries. ACS Appl Mater Interfaces 11(23):20788–20795

Liu T, Hu H, Ding X, Yuan H, Jin C, Nai J, …, Tao X (2020) 12 years roadmap of the sulfur cathode for lithium sulfur batteries (2009–2020). Energy Storage Materials 30:346–366

Ma L, Hendrickson KE, Wei S, Archer LA (2015) Nanomaterials: science and applications in the lithium–sulfur battery. Nano Today 10(3):315–338

Manthiram A, Fu Y, Su Y-S (2013) Challenges and prospects of lithium–sulfur batteries. Acc Chem Res 46(5):1125–1134

Ming J, Li M, Kumar P, Li L-J (2016) Multilayer approach for advanced hybrid lithium battery. ACS Nano 10(6):6037–6044

Qiu M, Fu X, Yang F, Qi S, Wu Z, Zhong WH (2021) In-situ synthesis of N, O, P-doped hierarchical porous carbon from poly-bis (phenoxy) phosphazene for polysulfide-trapping interlayer in lithium-sulfur batteries. Chemistry–A Eur J 27(38):9876–9884

Rauh R, Abraham K, Pearson G, Surprenant J, Brummer S (1979) A lithium/dissolved sulfur battery with an organic electrolyte. J Electrochem Soc 126(4):523

Rehman S, Khan K, Zhao Y, Hou Y (2017) Nanostructured cathode materials for lithium–sulfur batteries: progress, challenges and perspectives. J Mater Chem A 5(7):3014–3038

Rehman S, Pope M, Tao S, McCalla E (2022) Evaluating the effectiveness of in situ characterization techniques in overcoming mechanistic limitations in lithium–sulfur batteries. Energy Environ Sci 15(4):1423–1460

Wang H, Zhang C, Chen Z, Liu HK, Guo Z (2015) Large-scale synthesis of ordered mesoporous carbon fiber and its application as cathode material for lithium–sulfur batteries. Carbon 81:782–787

Wang X, Khachai H, Khenata R, Yuan H, Wang L, Wang W, …, Guo R (2017) Structural, electronic, magnetic, half-metallic, mechanical, and thermodynamic properties of the quaternary Heusler compound FeCrRuSi: a first-principles study. Sci Rep 7(1):1–13

Wang, B., Jin, F., Xie, Y., Luo, H., Wang, F., Ruan, T., . . . Dou, S. (2020). Holey graphene modified LiFePO4 hollow microsphere as an efficient binary sulfur host for high-performance lithium-sulfur batteries. Energy Storage Materials 26:433–442

Wang M, Bai Z, Yang T, Nie C, Xu X, Wang Y, …, Wang N (2022) Advances in high sulfur loading cathodes for practical lithium-sulfur batteries. Adv Energy Mater 2201585

Wu F, Chen S, Srot V, Huang Y, Sinha SK, van Aken PA, …, Yu Y (2018) A sulfur–limonene-based electrode for lithium–sulfur batteries: high-performance by self-protection. Adv Mater 30(13):1706643

Yamin H, Penciner J, Gorenshtain A, Elam M, Peled E (1985) The electrochemical behavior of polysulfides in tetrahydrofuran. J Power Sources 14(1–3):129–134

Yan B-L, Jun D, Wang J, Yang T, Mao X-H (2022) A simplified electrophoretic deposition route for sandwiched structure-based Mn_3O_4/G composite electrodes as high-capacity anodes for lithium-ion batteries. J Alloys Compd 905:164121

Yang D, Liao X-Z, Shen J, He Y-S, Ma Z-F (2014) A flexible and binder-free reduced graphene oxide/Na 2/3 [Ni 1/3 Mn 2/3] O_2 composite electrode for high-performance sodium ion batteries. J Mater Chem A 2(19):6723–6726

Yeşilot S, Küçükköylü S, Mutlu T, Demir E, Demir-Cakan R (2022) Highly sulfur-rich polymeric cathode materials via inverse vulcanization of sulfur for lithium–sulfur batteries. Mater Chem Phys 285:126168

Yoshida L, Hakari T, Matsui Y, Ishikawa M (2022) Polyglycerol-functionalized microporous carbon/sulfur cathode for Li-S battery. Electrochim Acta 429:141000

Yuan H, Huang JQ, Peng HJ, Titirici MM, Xiang R, Chen R, …, Zhang Q (2018) A review of functional binders in lithium–sulfur batteries. Adv Energy Mater 8(31):1802107

Zhang S, Zhong N, Zhou X, Zhang M, Huang X, Yang X, …, Liang X (2020) Comprehensive design of the high-sulfur-loading Li–S battery based on MXene nanosheets. Nano-Micro Lett 12(1):1–13

Zhao Y, Wu W, Li J, Xu Z, Guan L (2014) Encapsulating MWNTs into hollow porous carbon nanotubes: a tube-in-tube carbon nanostructure for high-performance lithium-sulfur batteries. Adv Mater 26(30):5113–5118

Zhu J, Zou J, Cheng H, Gu Y, Lu Z (2019) High energy batteries based on sulfur cathode. Green Energy Environ 4(4):345–359

Chapter 7
Sulphur Cathodes: An Introduction

Tausif Ahmad, Muhammad Sagir, and M. B. Tahir

Abstract As fossil fuels are burned more and more for energy, the world's environmental problems get worsen. Therefore, the research and development of renewable, clean energy sources is an urgent priority. Since 2009, there has been a surge in interest in studying the sulphur cathode used in lithium sulphur batteries (LSBs) due to high theoretical capacity and likely low manufacturing cost. However, electrical isolation of S and Li_2S, the strong shuttle effect of lithium polysulphides, and the major volume change from S to Li_2S upon lithiation are the key issues, which prompted scientists and engineers to work tirelessly for finding scientific and technological solutions. We have aimed to provide an all-encompassing fresh insight into the sulphur cathode's future direction, specifically, the carbon architecture design of the sulphur host, advanced characterization techniques for in-depth mechanism understanding, and full-cell evaluation for a truly viable LSB. Volumetric expansion during charging and discharging also has a negative impact on LSB electrochemical performance, including rate performance, cycle life, and coulombic efficiency. Consequently, different sulphur composites containing carbon materials, metallic oxides, and conductive polymers, and different composite cathode materials recently developed for application to LSBs were discussed. Hence, it's very important to learn more about sulphur cathode to get better understandings of the material

T. Ahmad (✉)
Institute of Chemical and Environmental Engineering, Khwaja Fareed University of Engineering and Information Technology, Rahim Yar Khan, Punjab, Pakistan

Institute of Energy & Environmental Engineering, University of the Punjab, Lahore, Pakistan
e-mail: tausif.ieee@pu.edu.pk

M. Sagir
Institute of Chemical and Environmental Engineering, Khwaja Fareed University of Engineering and Information Technology, Rahim Yar Khan, Punjab, Pakistan
e-mail: dr.msagir@kfueit.edu.pk

M. B. Tahir
Institute of Physics, Khwaja Fareed University of Engineering and Information Technology, Rahim Yar Khan, Punjab, Pakistan
e-mail: m.bilaltahir@kfueit.edu.pk

M. S. Tahir et al. (eds.), *Lithium-Sulfur Batteries: Key Parameters, Recent Advances, Challenges and Applications*, Springer Tracts in Electrical and Electronics Engineering, https://doi.org/10.1007/978-981-99-2796-8_7

chemistries, interfaces, and engineering of LSBs, which will open up a wealth of new possibilities for battery scientists.

Keywords Fossil fuels · Clean energy · Coulombic efficiency

Introduction

Electric power provides more advantages than fossil fuels, including availability to renewable energy sources like solar and wind, higher efficiency, lower building costs, and simpler transmission. The conflict between rising worldwide fossil fuel usage and progressively stricter greenhouse gas emissions could easily be resolved through widespread electrification (Li et al. 2018a). The widespread use of electrification in synthesis and related usage may be directly attributed to the many global projects on state-of-the-art energy storage and alteration technologies (Obama 2017; Dresselhaus and Thomas 2001; Chu and Majumdar 2012; Liu et al. 2019; Zapata Dominguez et al. 2022). The battery industry has paid a lot of attention to lithium sulphur battery (LSB) because of their high energy density and low material cost (Zapata Dominguez et al. 2022; Cano et al. 2018; Bruce et al. 2012; Lou et al. 2018). Due to its theoretically higher capacity of 1672 $mAhg^{-1}$, S may be preferable to lithium in multielectron reaction-based conversion electrochemistry (Van Noorden 2014; Ogoke et al. 2017; Liu et al. 2017a). In contrast to Li/Li+, the average cell voltage provided by sulphur is 2.15 V. Combining the finished LSB with a lithium metal anode (3860 mAh g-1), theoretic energy density might approach 2600 Wh kg^{-1} (Liu et al. 2017a). The LSB system was initially introduced in 1960s, and its popularity kept on evolving till 2009. To encapsulate sulphur, Nazar and colleagues developed a nanostructured mesoporous carbon (CMK-3) host, and by coating the composite's exterior with polymer, the composite's specific capacity and cycle durability were significantly improved (Ji et al. 2009). Since then, a lot of money and time have been poured into research on sulphur cathode. As can be shown in Fig. 7.1, the number of research articles addressing hosts, interlayers, binders, and organosulfides as potential solutions for sulphur cathode has increased dramatically in recent years.

Figure 1b shows that between 2009 and 2020, there will be a total of 5260 papers discussing sulphur cathode, which is about 96% of the entire amount of literature. There are still problems with sulphur cathode that prevent it from being used in lithium-sulphur battery systems (LSBs) that can be sold commercially (Fang et al. 2017; Wang et al. 2018a; Gu and Lai 2019; Manthiram et al. 2014), electrically separated sulphur and its solid reduction product, Li_2S; electrode cracking and its pulverization because of substantial change in volume during lithiation/de-lithiation procedures; and (ii) the severe shuttle effect of soluble LiPSs in electrolyte. Consequently, it is of great interest to battery researchers to learn more about sulphur cathode in terms of energy density and cycle durability (Dunn et al. 2011; Sun et al. 2016). Sulphur cathode in LSBs has been around for quite some time, but now is the time for a bold new vision for its development.

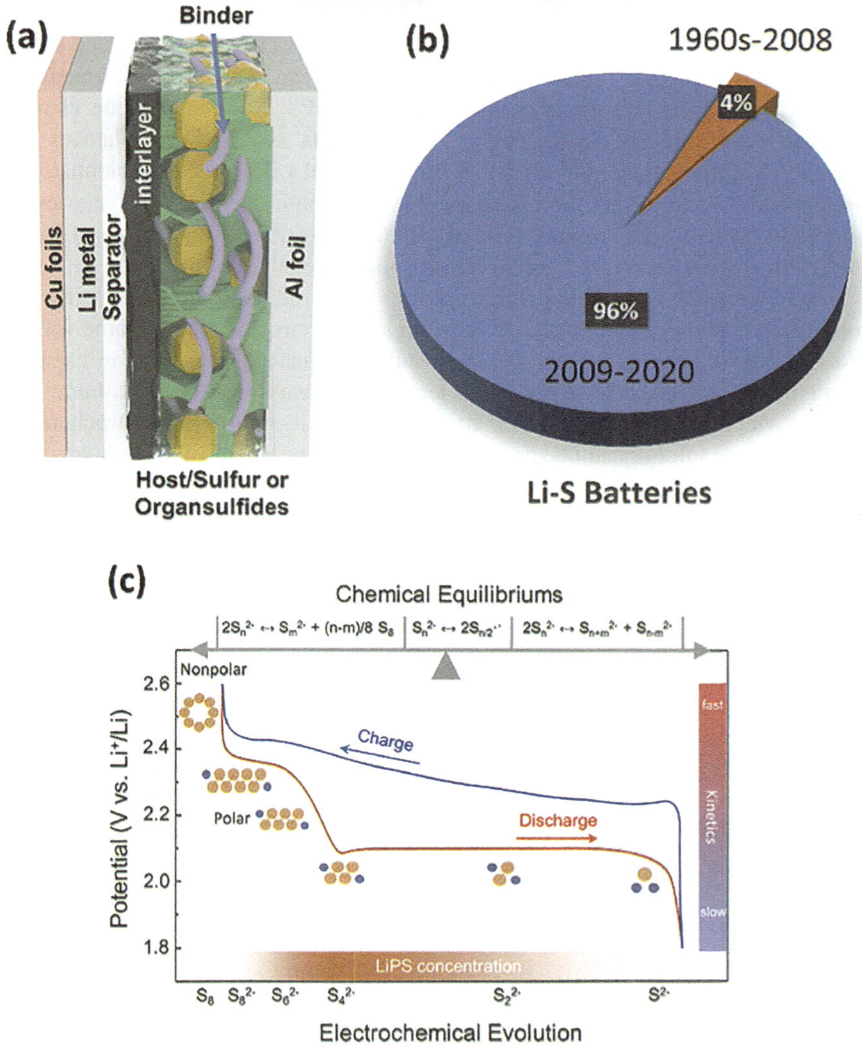

Fig. 7.1 (**a**) Methods to create a better sulphur cathode (**b**) a literature review covering the years 1960–2020 and (**c**) characteristics of the sulphur cathode (Li et al. 2018b)

Electrochemical Fundamentals of the Sulphur Cathode

Electroactive compounds, conductive agents, and polymeric binders may all be found in a common sulphur cathode (Yuan et al. 2018a). Elemental sulphur (S_8), in the form of its solid reduction product Li_2S, is the electroactive material chosen for lithium storage. The overall electrochemical reaction between them is defined as follows: $16\ Li + S_8 \rightarrow 8\ Li_2S$ (Bresser et al. 2013). However, because of their multi-electron reaction and poor electrical conductivity (Peng et al. 2017), it is difficult to directly convert sulphur into Li2S via solid-state chemistry. The actual redox processes of the sulphur cathode are enabled via a chain of intermediates (Li_2S_8, Li_2S_6, Li_2S_4, etc.) known collectively as LiPS. Sulphur cathode charge-discharge waveforms are shown in Fig. 1c. It exhibits two discharge plateaus and multi-step electrochemical behaviour, forming several LiPS species. Scientists are working towards a mechanistic understanding of sulphur chemistry in batteries by controlling the reduction and oxidation of sulphur alongside different types of LiPSs during reversible conversion of solid-liquid-solid (Li et al. 2018b).

Since elemental S is nonpolar, it is often believed that the polar property, mobility, and high solubility of LiPSs in organic electrolyte are to blame for the losses of electroactive materials and the result in enhanced deterioration capacity (Wang et al. 2018b). Electroactive materials may vary greatly in volume and kinetics, which further complicate matters. As a result, improved electrochemical behaviour at the sulphur cathode necessitates a shift in the way hosts, interlayers, and binders are conceptualized.

Cathode sulphur is studied systematically in terms of its different initial states (Fang et al. 2017). Sulphur molecules and sulfurized polymers contain the short-chain sulphur. A single discharge plateau indicates that these sulphur-containing compounds do not undergo a transition from S_8 to S_4. Upon formation of LiPSs, they undergo a solid-solid phase transition from the S phase to the Li_2S phase. Catholyte for the cathode applications of soluble LiPS has also been developed (Fu et al. 2013). Since they are more evenly distributed throughout the host and have a greater reactive activity by nature, their electrochemical performance exceeds that of the aforementioned solid-solid mixing of sulphur materials. In contrast to employing sulphur allotropes or LiPS as the cathode, Li_2S undergoes a straight solid-solid conversion following de-lithiation to sulphur (Yan et al. 2019; Zhang et al. 2017). There must be a greater termination voltage for isolated Li_2S (about 3.8 V) due to this resistance. When this is done, the reactions that follow are identical to those seen in cathodic reactions involving elemental sulphur.

Focusing on the Host Substances

Sulphur cathodes have substantially higher electrochemical and chemical stability than ceramic cathodes due to the "host-less" electrochemical reaction that occurs between the two materials. These investigations aim to enhance the sulphur cathode's stability and electrochemical behaviour and hence propose a host framework to communicate this goal. By providing a confinement framework for the sulphur cathode, in particular, the electrical conductivity of sulphur and Li_2S is increased, the intermediate LiPS is tightly trapped inside the electrode, and the conversion process between sulphur and Li_2S is accelerated (Pei et al. 2016; Hart et al. 2015; Li et al. 2016a). There are many competing theories for improving sulphur cathodes.

Despite the low strength of the bonds between sulphur and carbon, the nonpolar S8 may be adsorbed onto the carbon surface and held there by van der Waals attraction (Wang et al. 2013). Sulphur elemental must be efficiently infused into host materials and evenly distributed across their pores; it is crucial to advocate for its use. Cathode discharging via the lithiation process results in the formation of polar LiPSs with varying chain lengths, which then slowly dissolve in the electrolyte. The powerful interaction between LiPSs and host is highlighted by polar adsorption (Tao et al. 2016). LiPSs can be retained on the host longer if they are chemically anchored to the surface (this works particularly well for heteroatom-doped carbon and metal-based hosts). Given the abundance of LiPSs, monolayer chemisorption cannot keep up with demand; therefore, catalytic conversion is required to expedite the reduction of LiPSs (Li_2S_8, Li_2S_6, Li_2S_4, Li_2S_3, and Li_2S_2) to Li_2S as an end product (Yuan et al. 2016).

Although most host materials presumably already possess these features, they have been proposed in succession throughout time as an indication of deep knowledge of the sulphur cathode reaction. Host material design and production for high performance sulphur cathode therefore incorporate these mechanism-oriented methodologies. As is the case with most scientific and technical domains, recent advances in nanotechnology and characterization had a significant impact on the synthesis.

Selection of Host Substances Based on Physisorption

Prior to the now-famous discovery by Nazar and colleagues, it was stated that carbonaceous materials were used in sulphur cathode. Wang et al., in one of the first research of a S/C cathode, proposed that sulphur cathodes are hosted by active carbon (Wang et al. 2002). By heating a combination of sulphur powder and active carbon to 200 °C in an Ar environment, the elementary sulphur will melt and permeate into the carbon pores. It is possible that an additional heat treatment at 300 °C will eliminate the sulphur from the active carbon surface. The reversible capacity of such a sulphur/carbon nanocomposite is around 440 mAh g^{-1}. Although this is only a proof of concept, it serves as a useful reminder that a sulphur-carbon composite may be made by melting sulphur and diffusing it into porous carbon.

Studies of sulphur and porous carbon cathodes have garnered considerable interest from the battery community since the early 2000s (Ji et al. 2009; Lai et al. 2009; Yuan et al. 2009; Liang et al. 2009). Here are a few key features of a typical Nazar illustration (Fig. 7.2a): (I) For a composite containing as much as 70% sulphur, the porous carbon with an SSA of up to 1976 m^2 g^{-1} makes the ephemeral material readily accessible to the carbon framework. (II) The electrochemical behaviour of carbon/sulphur composites may be enhanced by subjecting them to a heat treatment (at least 155 °C) during synthesis. The cycling stability is enhanced by adding polyethylene glycol (PEG) to diminish the quantity of LiPSs that leach into the electrolyte. In subsequent studies of sulphur hosts (Seh et al. 2016; Manthiram

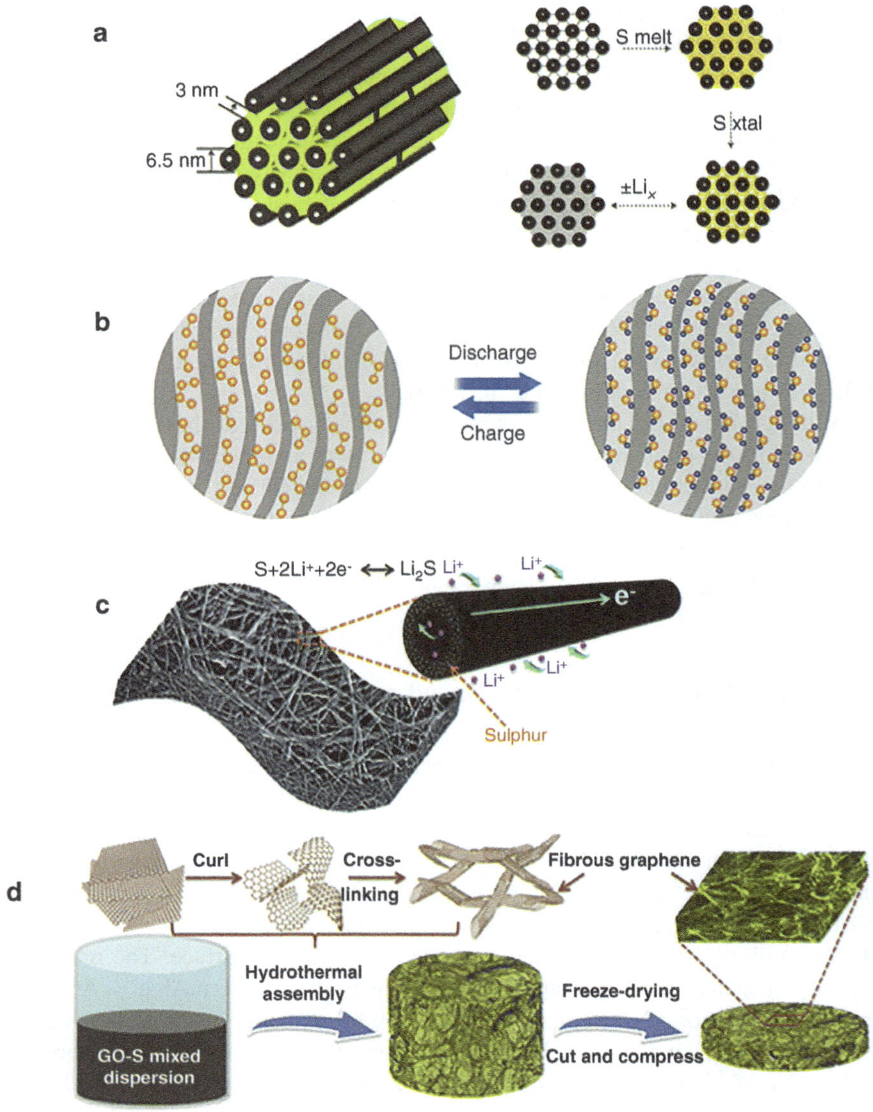

Fig. 7.2 (**a**) A schematic of the manufacturing method and a diagram of the graphene/sulphur hybrid are shown below. (**a**) Sulphur contained in the pore structure of CMK-3; (**b**) smaller sulphur molecules in a carbon host; (**c**) template-directed production of sulphur/CNTs; and (**d**) production of the graphene/sulphur hybrid (Ji et al. 2009; Xin et al. 2012; Zhou et al. 2012, 2013)

et al. 2013), these experimental methods and parameters have proven to be highly useful references.

Reversible capacities of above 1000 mAh g^{-1} at tolerable rates with great cycle steadiness have been shown for sulphur cathodes based on porous carbon hosts; the next step is to research large-scale carbon host materials for sulphur cathode (Zhou et al. 2015a; Xu et al. 2015a; Strubel et al. 2015; Jung et al. 2014; Sun et al. 2014). Since the LiPSs are enclosed by the electrode, the shuttling process is sped up (Ogoke et al. 2017). This is the widely accepted explanation for the increased capacity and lifetime of sulphur cathodes. It has also been important to select carbon hosts with a high SSA, as this provides sulphur with adequate access to the carbon framework, ensuring the sulphur atoms are evenly dispersed in the carbon host, preventing them from clumping together, and thereby improving the electric connection. These ideas have had a huge influence on the porous carbon design (host materials) for sulphur cathode and are now widely used in the industry.

In light of these results, a thorough investigation of the role of pore volume and pore size of the carbon host in sulphur storage is done. Sulphur cathode should be supported in mesoporous carbon, according to Liu's group (Li et al. 2011), as a result of its high pore volume and incompletely filled sulphur. The former permits high sulphur loading, whereas the later takes into consideration the volume fluctuation of sulphur. On the other hand, some groups have zeroed in on carbon hosts with micropores small enough to contain metastable small sulphur ($S_{2–4}$) (Xin et al. 2012; Li et al. 2014a; Chen et al. 2015). Small $S_{2–4}$ materials exhibit very diverse electrochemical responses compared to conventional S_8 materials. Solvent molecules cannot pass through ultra-micropores, and $S_{2–4}$ materials can be dissolved in them.

Figure 3b shows that under these circumstances, shuttle suppression occurs when the lithiation/de-lithiation of tiny S_2–$_4$ progresses as a quasi-solid-solid process. Due to this, sulphur exhibits a monotonic electrochemical behaviour, where its cathode's capacity retention is vastly enhanced. As we now see it, different pore size types each have their own unique characteristics (Li et al. 2014b; Xu et al. 2015b; Wang et al. 2019a; Tian et al. 2020). Carbon hosts with macropores (>50 nm) give ample space on the inside, allowing for high sulphur loading, buffering volume expansion upon lithiation, and rapid ion diffusion with high-rate performance. However, LiPSs are difficult to contain within the cathode because of their open pore structure and lower adsorption. Carbon hosts for sulphur cathodes often have mesopores (2–50 nm) in them. This greatly raises the sulphur-to-carbon atomic ratio (SSA), enabling a larger sulphur loading, uniform sulphur distribution in the carbon host, stronger the composite's electric conductivity, a shorter ionic diffusion route, and a higher SSA than would be feasible with macropores. Higher adsorption energy is provided by a micropore structure with a pore size of 2 nm when physically embedding LiPS into the electrode. In addition, studies have demonstrated that LiPSs are impenetrable to electrolyte solvent molecules (0.6–0.8 nm) and that short-chain sulphur allotrope may be stored in ultra-micropores (0.6 nm) (S_2–$_4$) (Zhu et al. 2017; Hu et al. 2017).

The precursors are indispensable to the porous carbon host. Tao's group suggested a mesoporous carbon foam (MCF) with very tiny pores (Tao et al. 2013) as the carbon host derived from biomass. MCF in its as-prepared form has an excellent pore size distribution for sulphur impregnation and a high specific surface area (SSA) of 1478.55 $m^2\ g^{-1}$ with a primary discharge of 1285 $mAh\ g^{-1}$, after 50 cycles and a retention of 878 mA h g^{-1}, the MCF/sulphur nanocomposite containing 57.22 wt percent sulphur shows the excellent performance. Biomass is advantageous due to its plentiful supply, low price, convenience of availability, and low impact on the natural world. The ability to receive the inherent morphologies of biomass, such as its pore size and volume, is an appealing feature that facilitates the manufacture of the carbon host without requiring a complicated process. As a result, researchers are looking far and wide for a carbon host that can be successfully utilized in the production of sulphur cathode (Tao et al. 2013; Liu et al. 2020a, b; Jin et al. 2018; Shi et al. 2019a).

Selection of Host Substances Based on Chemisorption

Sulphur may be enclosed inside the carbon host, which improves the sulphur cathode's electrochemical behaviour. However, researchers are finding that nonpolar carbon hosts provide challenges when attempting to halt polar LiPS for longer-range cycling (Yin et al. 2013; Xu et al. 2014). Individual LiPSs, which establish rather strong connections with their altered carbon host, are capable of providing efficient confinement inside the cathode. In this regard, heteroatom-doped modifications are generally in favour of carbon host materials (Pei et al. 2016; Hou et al. 2016; Jin et al. 2017; Krüner et al. 2018). Ji et al, (Ji et al. 2011) found that graphene oxide's epoxy and hydroxyl groups make sulphur more securely attached to the host (Fig. 7.3a). In particular, the Ling-led team presented N supplemented mesoporous carbons having adjustable nitrogen concentration (Sun et al. 2013), showing unambiguously that boosted exterior contact between N modifications and polysulfide structures aids to reduce the shuttle of LiPSs (Fig. 7.3b).

Researchers' interest in the chemical standing changes of carbon host surfaces was piqued by the aforementioned works. This carbon host has P, O, B, and N introduced on purpose (Jin et al. 2017; Kong et al. 2018a; Zou et al. 2019). Having a lone pair of electrons, higher electronegativity than carbon, delocalized bonds in the conjugated system, a tiny radius that matches Li's, and a steady link to the carbon surface plane are all desirable qualities in a doping atom for a carbon scaffold (Hou et al. 2016). Take the instance of N-doped graphene as mentioned in Fig. 4c, used in sulphur cathode as an example to show how the N dopant becomes competitive to toughen the contact of the carbon hosts and the LiPSs (Wang et al. 2014a, b; Qu et al. 2015; Qiu et al. 2014). Additionally, Zhang and coworkers (Wang et al. 2014b) explored two distinct sheets made of N-doped graphene and served as a sulphur cathode. When comparing sulphur/graphene cathodes, the one rich in pyridinic-N outperforms the other in terms of electrochemical performance. In addition, Tao's squad exhibited the highest capture capacity of pyridinic N-doped carbon among LiPSs with an aid of the lowest dissociation energy of Li_2S (Yuan et al. 2018b). The host is also testing materials with a high nitrogen concentration. Nazar's group, for instance, proposed a graphitic carbon nitride (g-C3N4)-based lightweight polar host

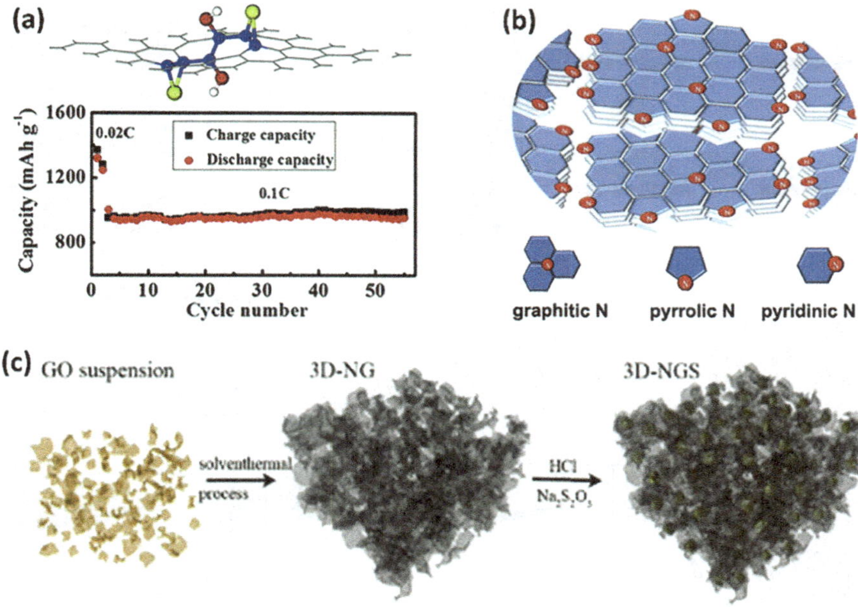

Fig. 7.3 (**a**) Schematic representation of the three nitrogen species (**b**), design of GO immobilization of S and pedalling presentation (**c**), and construction of the 3D N-doped graphene complex (Ji et al. 2011; Sun et al. 2013; Wang et al. 2014a)

(Pang 2016). When compared to N-doped carbons, the g-promising C3N4's chemical interaction with the LiPSs promises to significantly increase cycling stability.

Carbon host design has recently benefited from the incorporation of metal-organic frameworks (MOFs), which have enabled the simultaneous attainment of well-defined porosity structure, nonmetal-rich doping, SSA, and high electric conductivity. A sulphur battery, for instance, might be made from nitrogen-doped porous carbon anchored on graphene sheets (NPC/G), which has a high specific capacity of 1372 mAh g^{-1} and great cycling stability beyond 300 cycles (Pang 2016; Zhang et al. 2018a). To a lesser extent, similar efforts may be made with developing materials. With its low material density and readily available asset, carbon matrix remains an ideal host material. Carbon compounds are also crucial components of other hosts, providing properties like as adsorption and conductivity (Rehman et al. 2016).

Sulphur cathodes have been studied extensively, and it has been shown that metals, due to their large quantity of oxygen, nitrogen, and sulphur atoms, have a significant affinity for LiPSs and hence prevent LiPSs from dissolving into the electrolyte (Liu et al. 2017b). When the morphology project is added in (which accounts for things like shape, porosity, size, and so on), the resulting host has an architecture that is well-suited to both the physical and chemical binding of LiPSs. The inorganic adsorption of inorganic host materials for the LiPS was initially suggested by Ahn et al. (Song et al. 2004) in 2004, utilizing nanosized

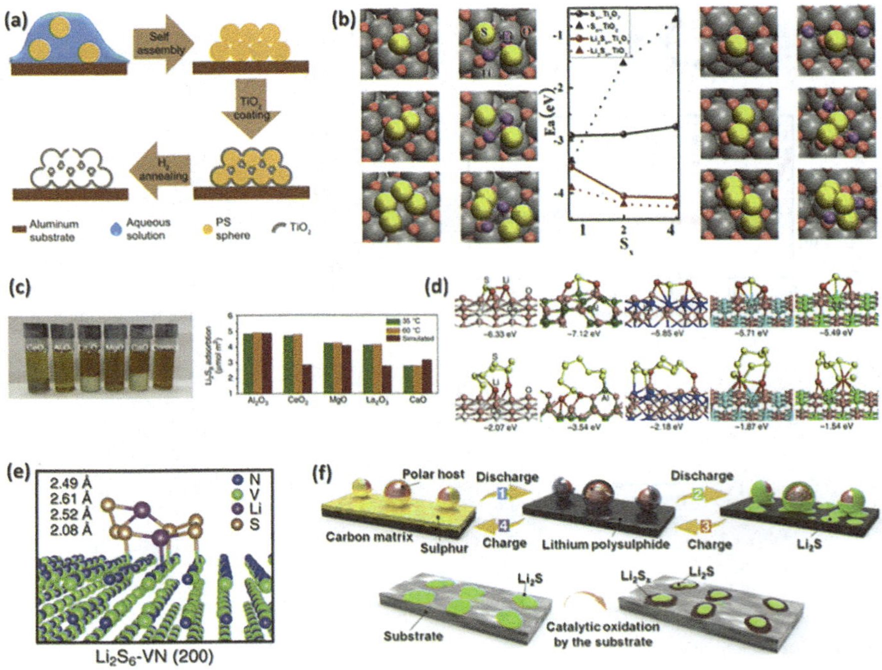

Fig. 7.4 (**a**) Hydrogen-reduced titanium dioxide fabrication procedures (**b**) A density functional theory (DFT) analysis of the adsorption of LiPSs species on Ti_4O_7 and TiO_2 surfaces (**c**) Deposition on various metal oxides (**d**) Comparative structures of sulphide systems adsorption on metal oxide surfaces (**e**) A Li_2S_6 molecule adsorption on the VN (200) surface (**f**) An illustration of the sulphur conversion process and Li_2S catalytic oxidation

Mg0.6Ni0.4O. It's worth noting that the last time inorganic materials were given serious consideration was in 2014. Sulphur cathode development is given new hope by the use of inorganic host materials. The work done by Cui and colleagues is remarkable, since it demonstrates the ability metal-derived structures to oblige as a host, uptaking of LiPS and the enhanced pedalling performance of sulphur cathode. Figure 7.4a demonstrates that the solubility of LiPSs may be reduced by chemically tuning TiO_2 species (Liang et al. 2014). After 200 cycles at a rate of 0.2 C, the TiO2/ sulphur cathode retained 81% of its initial capacity, making it a good candidate for use in rechargeable batteries. The cycling was carried out while maintaining a Coulombic efficiency (CE) of around 99.5%. Further investigation revealed that Ti4O7 formed an even more robust bond with the LiPSs (Tao et al. 2014; Pang et al. 2014; Liang et al. 2016). Density functional theory (DFT) calculations showed that compared to TiO_2, the adsorption of LiPSs to Ti_4O_7's low-coordinated Ti sites is much stronger, as shown in Fig. 7.4b.

Sulphur's high capacity retention of 99% is made possible by the Ti_4O_7 host (Tao et al. 2014). In addition, it has also shown the deposition of LiPSs onto tin-doped indium oxide, as opposed to glassy carbon, during electrochemical charge/discharge

(Yao et al. 2014). Sulphur cathode research has been energized by a resurgence of interest in controlling the morphology of various metal oxides (MnO_2, Ti_O, Fe_2O_3, V_2O_5, etc.) (Xu et al. 2018; Chen et al. 2018; Park et al. 2018; Carter et al. 2017; Wang et al. 2016). Most notably, in Fig. 4c and d, Tao and coworkers (Tao et al. 2016) equated the cycling abilities of cathodes, made of sulphur cathodes MgO and La_2O_3. The authors hypothesized that the capture of LiPS by metal oxides occurs via a mechanism that is a synergy between monolayer chemisorption and surface diffusion. Motivated by the success of metal oxide in anchoring LiPSs, a plethora of researchers have tested other host materials for sulphur capping groups such as metal hydroxide (Zhang et al. 2018b; Jiang et al. 2015; Peng et al. 2016a), metal sulphide (Yuan et al. 2016; Cheng et al. 2018; Lei et al. 2017), metal carbides (Peng et al. 2016b) (Gao et al. 2018; Zeng et al. 2018), and metal phosphides (Zeng et al. 2018). The theoretical calculations and experimental characterizations provided by the authors deliver strong support for the proposed ideas and strategies.

The transition metals hold significant importance in forming S-binding configurations containing improved attaching which was highlighted by Zhang's group (Chen et al. 2017), while dipole-dipole interactions have a major role in inducing Li-binding.

Selection of Host Substances Based on Electrocatalysis

Researchers are shifting their attention away from chemisorption enhancement in host material design and instead studying the electrocatalytic activities of metal-based materials in LiPSs' redox processes (Yuan et al. 2019a). In 2014, researchers discovered that these hosts can act as a catalyst to speed up the redox reaction between S and Li_2S (Pang et al. 2014; Liang et al. 2015). CoS_2 drives redox reaction of LiPSs (Yuan et al. 2016), which significantly strengthens this idea. LiPSs' accelerated redox reactions, like those of conductive enhancers and strong adsorbents, significantly lessen the likelihood of their diffusing and dissolving into organic electrolyte. A high initial capacity of 1368 mAh g^{-1} at 0.5 C and a low decay rate of 0.034 percent/cycle were observed in the experiments after 2000 cycles at 2.0 C.

Then, Dong et al. (Li et al. 2016b) described Co-N-GC that accelerates S redox and entraps LiPSs in a perfect electric atmosphere. The performance of electrode made by S@Co-N-GC was significantly enhanced via suggested catalytic function method.

Therefore, amplified rate at which sulphur is converted into Li_2S is likely to help in the effort to lower the total amount of LiPSs in solution. The cathode of an LSB can also be a conductive porous VN/G structure, as reported by Li et al. (Sun et al. 2017) as shown in (Fig. 7.4e). Strong chemical anchoring of LiPSs is demonstrated by the VN/G composite, which also speeds up the kinetics of the redox reactions within them, resulting in excellent rate and cycling performances. In addition, as shown in Fig. 7.4f (Zhou et al. 2017), Cui's team discovered that Li_2S, the discharge product, can be oxidized using metal sulphide as an activation catalyst, resulting in sulphur, once again being the charge product.

Recently, over the course of the past few years, all of the three ideas including physisorption, chemisorption, and electrocatalysis are taken into account to factor into the design of host materials. To achieve the "All-in-One" goal for the hosts, many different types of materials are incorporated into the host. There has been extensive research into integrated designs, including those using SnS_2 (Gao et al. 2019), V associated (Wang et al. 2019b), perovskite (Kong et al. 2018b), and Co made systems (Li et al. 2016b; Ma et al. 2018; Yuan et al. 2019b). Similarly hosts without metals are revitalized to rapid Li_2S renovation (Yuan et al. 2018b). Lin's team has reported a high N-carbon/graphene sheet integration that exhibits conductivity, immobility, and catalytic abilities (Xu et al. 2020).

A rationally designed catalytic host for the sulphur cathode enables this in situ anchoring-conversion of LiPSs, which yields higher capacity, great cycling steadiness without the use of $LiNO_3$ in the electrolyte. As an added bonus, Koratkar et al. (Xu et al. 2020) reported a CNF host for Li_2S_5 catholyte, which enabled for 500 cycles of charge-discharge.

Sulphur usage has been vastly enhanced in the search of a more effective hosts that has been addressed through creating the awareness of single material-based catalyst into host structures (Shi et al. 2019b; Du et al. 2019; Liu et al. 2018). In order to expedite the surface-mediated reaction of LiPSs (Du et al. 2019), Wan's group developed a composite of monodisperse cobalt atoms embedded in nitrogen-doped graphene (Co-N/G). As an electrocatalyst with dual functional, the recently built Co-N-C coordination centre forms and decomposes Li_2S during discharge and charge operations. At 0.2 C and 100 cycles, the S@Co-N/G composite's capacity is maintained at 1210 mAh g^{-1} gravimetrically with a measurements declining degree of 0.029 percent each rotation, thanks to electrode settings including a 90 wt percent S sulphur mass ratio and frame load of 6.0 mg cm^{-2}.

Focusing Interlayer Substances

One problem with LSBs is that their capacity degrades with time because LiPSs diffuses from the sulphur cathode to the lithium metal anode. To further prevent LiPSs from diffusing from the sulphur cathode to the anode, an interlayer is inserted between the cathode and the separator, in addition to the host materials already stated. In 2012, Manthiram's group presented a unique proposal for increasing the LSBs' cycle stability by injecting a microporous carbon interlayer among the sulphur cathode and the extractor (Su and Manthiram 2012; Chung and Manthiram 2014), as depicted in Fig. 7.5a. This self-contained interlayer had two distinct but complementary functions: (a) it helped keep LiPSs localized on the sulphur cathode side, decreasing the likelihood that they would diffuse into anode, made of lithium; (b) it served as a higher receiver of current, optimizing the efficiency with which active materials were utilized. By reducing the sulphur cathode's charge transfer

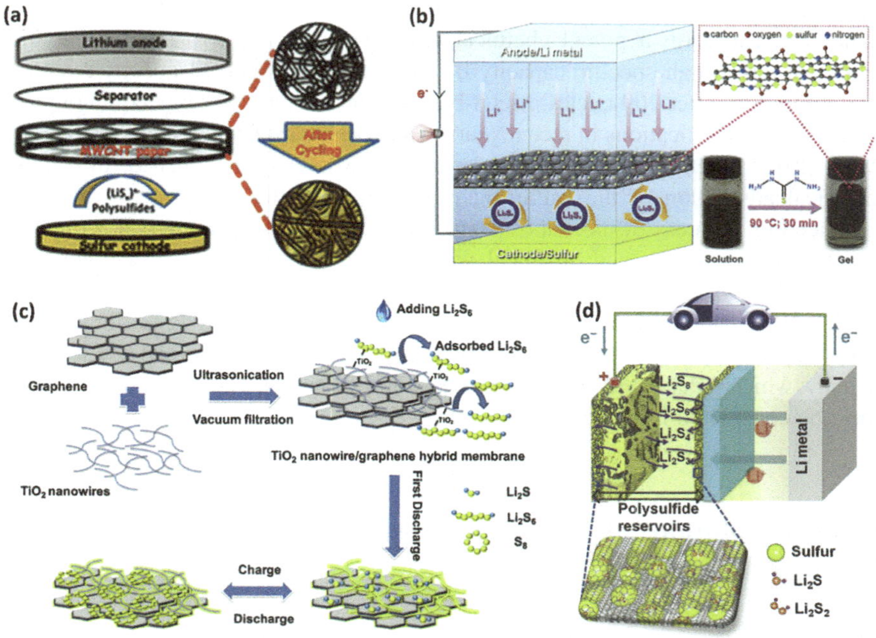

Fig. 7.5 (**a**) Interlayer of multi-walled carbon nanotubes (MWCNT) configuration (**b**) microstructure representation and constructed LSBs (**c**) TiO_2 nanowire-embedded graphene (**d**) A simplified diagram of an LSB electrode setup (Xing et al. 2016; Li et al. 2017; Zhou et al. 2015b; Zhou 2017)

impedance, the conductive interlayer improved the reaction's kinetics. After being subjected to 100 cycles at 1 C, the battery's release operational limit was alleviated at over 1000 mAh g^{-1}. The battery industry has quickly embraced the idea of an interlayer. The cycling stability of LSBs can also be enhanced by utilizing interlayers, made of a variety of materials between the sulphur cathode and the separator.

Due to the simplicity of 2D fabrication, carbon materials like CNT (Singhal et al. 2015; Huang et al. 2016; Hua et al. 2017) and graphene (Zhao et al. 2014; Huang et al. 2015) have the most potential as a forerunner to permeable interlayers. In addition, N and S are doped into these carbon materials before the interlayer is fabricated to simulate the effects of heteroatom changes on the efficient uptake of LiPSs. Kumar et al. (Xing et al. 2016) reported on the sponge-like graphene with N, S doping and its interlayer function, as shown in Fig. 7.5b.

The volume of 2193.2 mAh g^{-1} is attained by sulphur cathodes at a current density of 0.2 C. Metal-associated materials similarly have an interlayer embellishment added to them during production (Zhou et al. 2015b; Xiao et al. 2015; Hwang et al. 2016). An early example of it was presented by Manthiram's group of a hybrid membrane, made of TiO_2 nanowires and graphene that can stand on its own (Zhou et al. 2015b). Sulphur cathode internal resistance was effectively lowered by using the high-conductivity graphene material as a current collector to physically immobilize the dissolved LiPSs, as shown in Fig. 7.5c. Membrane-introduced TiO_2 nanowires can form a potent chemical bond with LiPSs. At 0.2 C, this hybrid electrode has a high specific capacity of 1327 mAh g^{-1} with a capacity retention of around 80% after 200 cycles. Metal-based materials have become commonplace in interlayer modifications in recent years (Tan et al. 2018; Hwang et al. 2018; Song et al. 2018). Metal-based components, as we've already established, aid in the chemical adsorption of the LiPSs interlayer. Further, the novel interlayers have enhanced in the areas of mechanical steadiness, SSA, and especially LiPS-trapping capability by physical and chemical adsorption, as a result of the emergence of concepts like MOF (Bai et al. 2016; Mao et al. 2017) and carbon quantum dot (Pang et al. 2018). The interlayer serves only as a barrier and does not contribute to the cell's capacity in any way. Metal-based materials for the interlayer have a higher density, which reduces the overall cathode's energy density. Therefore, the interlayer must be considered in any cell expansion.

Possible Solutions of Challenges Associated with Cathode Sulphur

The obstacles to using sulphur as a cathode for LSBs have been discussed along with their promising solutions including hosts, interlayers, and binders. The subsequent experimental and characterization methods have confirmed the viability of these new ideas and approaches. It is clear that S, LiPS, and Li_2S may be rapidly converted inside the electrode, thanks to the development of the host design and synthesis from physisorption to chemisorption to electrocatalysis. In the cathode form, a conductive interlayer is suggested and put between the separator and the cathode as a host, where it plays an essential function in controlling the diffusion of LiPS and the rate at which it is converted. Polymer materials, including coating and binder components, are also studied extensively in LiPS environments. As a result of its ability to simultaneously provide both strong mechanical adhesion and operative LiPS trapping, the latter is gaining in significance. Organosulfides are a capable new sulphur-based cathode material that can replace traditional sulphur and host combinations.

Conclusion

Sulphur cathode of LSB has lately attracted a lot of attention and investment from the battery industry because of its high energy density and low raw material cost. Scientists working on batteries should focus on improving sulphur cathode, which has a number of drawbacks including low electric conductivity and poor cycle performance. Possible causes of these flaws include the presence of isolated sulphur and Li_2S, the LiPS shuttle effect, and the blend of the aforementioned factors. Based on our analysis of the evolution of sulphur cathode, researchers have shifted their attention towards more holistic assessment of real-world needs for achieving better electrochemical performances including a larger sulphur content, high mass loading, and a low E/S ratio. The complexity of sulphur cathode has undoubtedly showed the devastating effect on electrode integrity and performance. Carbon resources have proved themselves as ideal hosts for sulphur cathode, owing to their portability, adaptability, inexpensiveness, and abundance of renewable resources. In order to realize effective LiPS confinement and high sulphur loading, certain design principles, such as SSA (more than 1000 $m^2\ g^{-1}$), pore size, and electronic conductivity, should be kept in mind. The melting of sulphur into pores shows that the carbon host's high SSA facilitates better performance of sulphur-based structures to the carbon systems through van der Waals connections. Carbon graphitization and heteroatom-doped alterations also have substantial effects on electronic conductivity. Among the carbon materials, the electronic conductivity of graphene and CNT is the highest due to their superior carbon graphitization and heteroatom doping. To realize the desired efficient energy density and longer life LSBs, practical carbon host designs should be centralized on a fusion plan in the permeability of carbon host. The carbon host's design principles call for high sulphur loading in the inner meso/macropores, uniform sulphur distribution, and strong geometrical confinement in the outer micropores, which prevent the LiPS from accessing the electrolyte by creating a quasi-solid environment. One idea is to combine biomass as carbon sources with practical alkaline treatments. Additionally, carbon-associated materials from metal-organic frameworks (MOFs) will become increasingly popular for this type of carbon-based approach in the near future because of their large SSA, tunability of porosity through modification of metal species at molecular-level, and organic linkers and ionic conductivities from 3D-based carbon structures. As an added bonus, the electrochemical performances of LSB might be enhanced by elements (such Co, V, Fe, and Ti) naturally present in the carbon hosts that are left as adsorption spots and LiPS catalyst.

References

Bai S, Liu X, Zhu K, Wu S, Zhou H (2016) Metal–organic framework-based separator for lithium–sulfur batteries. Nat Energy, 1(7):1–6

Bresser D, Passerini S, Scrosati B (2013) Recent progress and remaining challenges in sulfur-based lithium secondary batteries–a review. Chem. Commun., 49(90):10545–10562

Bruce PG, Freunberger SA, Hardwick LJ, Tarascon J-M (2012) Li–O2 and Li–S batteries with high energy storage. Nat. Mater., 11(1):19–29

Cano ZP et al (2018) Batteries and fuel cells for emerging electric vehicle markets. Nat Energy 3(4): 279–289

Carter R et al (2017) Polysulfide anchoring mechanism revealed by atomic layer deposition of V2O5 and sulfur-filled carbon nanotubes for lithium–sulfur batteries. ACS Appl. Mater. Interfaces, 9(8):7185–7192

Chen H, Wang C, Dong W, Lu W, Du Z, Chen L (2015) Monodispersed sulfur nanoparticles for lithium–sulfur batteries with theoretical performance. Nano Lett, 15(1):798–802

Chen M et al (2018) MnO2 nanosheets grown on the internal/external surface of N-doped hollow porous carbon nanospheres as the sulfur host of advanced lithium-sulfur batteries. Chemical Engineering Journal, 335:831–842

Chen X et al (2017) An analogous periodic law for strong anchoring of polysulfides on polar hosts in lithium sulfur batteries: S-or Li-binding on first-row transition-metal sulfides? ACS Energy Lett., 2(4):795–801

Cheng Z, Xiao Z, Pan H, Wang S, Wang R (2018) Elastic sandwich-type rGO–VS2/S composites with high tap density: structural and chemical cooperativity enabling lithium–sulfur batteries with high energy density. Advanced Energy Materials, 8(10):1702337

Chu S, Majumdar A (2012) Opportunities and challenges for a sustainable energy future. Nature, 488(7411):294–303

Chung SH, Manthiram A (2014) Carbonized eggshell membrane as a natural polysulfide reservoir for highly reversible Li-S batteries. Advanced Materials, 26(9):1360–1365

Dresselhaus MS, Thomas I (2001) Alternative energy technologies. Nature414, (6861):332–337

Du Z et al (2019) Cobalt in nitrogen-doped graphene as single-atom catalyst for high-sulfur content lithium–sulfur batteries. J. Am. Chem. Soc., 141(9):3977–3985

Dunn B, Kamath H, Tarascon J-M (2011) Electrical energy storage for the grid: a battery of choices. SCIENCE ,334(6058):928–935

Fang R, Zhao S, Sun Z, Wang DW, Cheng HM, Li F (2017) More reliable lithium-sulfur batteries: status, solutions and prospects. Advanced Materials, 29(48):1606823

Fu Y, Su YS, Manthiram A (2013) Highly reversible lithium/dissolved polysulfide batteries with carbon nanotube electrodes. Angewandte Chemie International Edition, 52(27):6930–6935

Gao X et al (2019) Cobalt-Doped SnS2 with dual active centers of synergistic absorption-catalysis effect for high-S loading Li-S batteries. Advanced Functional Materials, 29(8):1806724

Gao Z, Schwab Y, Zhang Y, Song N, Li X (2018) Ferromagnetic nanoparticle–assisted polysulfide trapping for enhanced lithium–sulfur batteries. Advanced Functional Materials, 28(20):1800563

Gu X, Lai C (2019) One dimensional nanostructures contribute better Li–S and Li–Se batteries: progress, challenges and perspectives. Energy Storage Materials, 23:190–224

Hart CJ, Cuisinier M, Liang X, Kundu D, Garsuch A, Nazar L (2015) Rational design of sulphur host materials for Li–S batteries: correlating lithium polysulphide adsorptivity and self-discharge capacity loss. Chem. Commun., 51(12):2308–2311

Hou TZ et al (2016) Design principles for heteroatom-doped nanocarbon to achieve strong anchoring of polysulfides for lithium–sulfur batteries. Small, 12(24):3283–3291

Hu L et al (2017) Optimization of microporous carbon structures for lithium–sulfur battery applications in carbonate-based electrolyte. Small, 13(11):1603533

Hua W et al (2017) Polysulfide-scission reagents for the suppression of the shuttle effect in lithium–sulfur batteries. ACS Nano, 11(2):2209–2218

Huang J-Q, Zhuang T-Z, Zhang Q, Peng H-J, Chen C-M, Wei F (2015) Permselective graphene oxide membrane for highly stable and anti-self-discharge lithium–sulfur batteries ACS Nano, 9(3):3002–3011

Huang J-Q, Xu Z-L, Abouali S, Garakani MA, Kim J-K (2016) Porous graphene oxide/carbon nanotube hybrid films as interlayer for lithium-sulfur batteries. Carbon, 99:624–632

Hwang JY et al (2016) High-energy, high-rate, lithium–sulfur batteries: synergetic effect of hollow TiO2-webbed carbon nanotubes and a dual functional carbon-paper interlayer. Advanced Energy Materials,6(1):1501480

Hwang JY, Kim HM, Shin S, Sun YK (2018) Designing a high-performance lithium–sulfur batteries based on layered double hydroxides–carbon nanotubes composite cathode and a dual-functional graphene–polypropylene–Al2O3 separator. Advanced Functional Materials, 28(3):1704294

Ji L et al (2011) Graphene oxide as a sulfur immobilizer in high performance lithium/sulfur cells. J. Am. Chem. Soc.,133(46):18522–18525

Ji X, Lee KT, Nazar L (2009) A highly ordered nanostructured carbon–sulphur cathode for lithium–sulphur batteries. Nature Materials, 8(6):500–506

Jiang J et al (2015) Encapsulation of sulfur with thin-layered nickel-based hydroxides for long-cyclic lithium–sulfur cells. Nature Communications, 6(1):1–9

Jin C et al (2017) Enhanced sulfide chemisorption using boron and oxygen dually doped multi-walled carbon nanotubes for advanced lithium–sulfur batteries. Journal of Materials Chemistry, 5(2):632–640

Jin C et al (2018) Sustainable, inexpensive, naturally multi-functionalized biomass carbon for both Li metal anode and sulfur cathode. Energy Storage Materials,15:218–225

Jung DS et al (2014) Hierarchical porous carbon by ultrasonic spray pyrolysis yields stable cycling in lithium–sulfur battery. Nano Lett.,14(8):4418–4425

Kong L et al (2018a) Porphyrin-derived graphene-based nanosheets enabling strong polysulfide chemisorption and rapid kinetics in lithium–sulfur batteries. Advanced Materials, 8(20): 1800849

Kong L et al (2018b) A bifunctional perovskite promoter for polysulfide regulation toward stable lithium–sulfur batteries. Advanced Energy Materials,30(2):1705219

Krüner B et al (2018) Gyroidal Porous Carbon Activated with NH3 or CO2 as Lithium– Sulfur Battery Cathodes. Batteries & Supercaps, 1(2):83–94

Lai C, Gao X, Zhang B, Yan T, Zhou Z (2009) Synthesis and electrochemical performance of sulfur/highly porous carbon composites. J. Phys. Chem., C113(11):4712–4716

Lei T et al (2017) Multi-functional layered WS2 nanosheets for enhancing the performance of lithium–sulfur batteries. Advanced Energy Materials,7(4):1601843

Li G, Wang S, Zhang Y, Li M, Chen Z, Lu J (2018b) Revisiting the role of polysulfides in lithium–sulfur batteries. Advanced Materials,30(22):1705590

Li L et al (2017) Phosphorene as a polysulfide immobilizer and catalyst in high-performance lithium–sulfur batteries. Advanced Materials, 29(2):1602734

Li M, Lu J, Chen Z, Amine K (2018a) 30 years of lithium-ion batteries. Advanced Materials, 30(33):1800561

Li X et al (2011) Optimization of mesoporous carbon structures for lithium–sulfur battery applications. Journal of Materials Chemistry, 21(41):16603–16610

Li Y-J, Fan J-M, Zheng M-S, Dong Q-F (2016b) A novel synergistic composite with multi-functional effects for high-performance Li–S batteries. Energy & Environmental Science, 9(6):1998–2004

Li Z et al (2014a) Insight into the electrode mechanism in lithium-sulfur batteries with ordered microporous carbon confined sulfur as the cathode. Advanced Energy Materials, 4(7):1301473

Li Z et al (2014b) A highly ordered meso@ microporous carbon-supported sulfur@ smaller sulfur core–shell structured cathode for Li–S batteries. ACS Nano, 8(9):9295–9303

Li Z, Wu HB, Lou XWD (2016a) Rational designs and engineering of hollow micro–/ nanostructures as sulfur hosts for advanced lithium–sulfur batteries. Energy Environ. Sci., 9(10):3061–3070

Liang C, Dudney NJ, Howe J (2009) Hierarchically structured sulfur/carbon nanocomposite material for high-energy lithium battery. Chem. Mater., 21(19):4724–4730

Liang X, Hart C, Pang Q, Garsuch A, Weiss T, Nazar L (2015) A highly efficient polysulfide mediator for lithium–sulfur batteries. Nature Communications, 6(1):1–8

Liang X et al (2016) Tuning transition metal oxide–sulfur interactions for long life lithium sulfur batteries: the "Goldilocks" principle. Advanced Energy Materials, 6(6):1501636

Liang Z et al (2014) Sulfur cathodes with hydrogen reduced titanium dioxide inverse opal structure. ACS Nano, 8(5):5249–5256

Liu J et al (2017a) Exploiting a robust biopolymer network binder for an ultrahigh-areal-capacity Li–S battery. Energy & Environmental Science, 10(3):750–755

Liu J et al (2020b) Recent progress on biomass-derived ecomaterials toward advanced rechargeable lithium batteries. EcoMat, 2(1):e12019

Liu T et al (2019) Exploring competitive features of stationary sodium ion batteries for electrochemical energy storage. Energy & Environmental Science, 12(5):1512–1533

Liu T et al (2020a) 12 years roadmap of the sulfur cathode for lithium sulfur batteries (2009–2020). Energy Storage Materials, 30:346–366
Liu X, Huang JQ, Zhang Q, Mai L (2017b) Nanostructured metal oxides and sulfides for lithium–sulfur batteries. Advanced materials, 29(20):1601759
Liu Z et al (2018) Atomic iron catalysis of polysulfide conversion in lithium–sulfur batteries. ACS Appl. Mater. Interfaces, 10(23):19311–19317
Lou S et al (2018) Pseudocapacitive Li+ intercalation in porous Ti2Nb10O29 nanospheres enables ultra-fast lithium storage. Energy Storage Materials, 11:57–66
Ma L et al (2018) Strong capillarity, chemisorption, and electrocatalytic capability of crisscrossed nanostraws enabled flexible, high-rate, and long-cycling lithium–sulfur batteries. ACS Nano, 12(5):4868–4876
Manthiram A, Fu Y, Su Y (2013) Challenges and prospects of lithium–sulfur batteries. Acc. Chem. Res., 46(5):1125–1134
Manthiram A, Fu Y, Chung S-H, Zu C, Su Y-S (2014) Rechargeable lithium–sulfur batteries. Chem. Rev., 114(23):11751–11787
Mao Y et al (2017) Foldable interpenetrated metal-organic frameworks/carbon nanotubes thin film for lithium–sulfur batteries. Nature communications, 8(1):1–8
Obama B (2017) The irreversible momentum of clean energy. Science, 355(6321):126–129
Ogoke O et al (2017) Effective strategies for stabilizing sulfur for advanced lithium–sulfur batteries. J. Mater. Chem. A, 5(2):448–469
Pang Q, Nazar LF (2016) Long-life and high-areal-capacity Li–S batteries enabled by a light-weight polar host with intrinsic polysulfide adsorption. ACS nano, 10(4):4111–4118
Pang Q, Kundu D, Cuisinier M, Nazar L (2014) Surface-enhanced redox chemistry of polysulphides on a metallic and polar host for lithium-sulphur batteries. Nature communications, 5(1):1–8
Pang Y, Wei J, Wang Y, Xia Y (2018) Synergetic protective effect of the ultralight MWCNTs/NCQDs modified separator for highly stable lithium–sulfur batteries. Advanced energy materials, 8(10):1702288
Park GD, Lee J, Piao Y, Kang YC (2018) Mesoporous graphitic carbon-TiO2 composite microspheres produced by a pilot-scale spray-drying process as an efficient sulfur host material for Li-S batteries. Chemical Engineering Journal, 335:600–611
Pei F et al (2016) From hollow carbon spheres to N-doped hollow porous carbon bowls: rational design of hollow carbon host for Li-S batteries. Advanced Energy Materials, 6(8):1502539
Peng HJ et al (2016a) A cooperative interface for highly efficient lithium–sulfur batteries. Advanced Materials, 28(43):9551–9558
Peng HJ et al (2016b) Enhanced electrochemical kinetics on conductive polar mediators for lithium–sulfur batteries. Angewandte Chemie International Edition, 55(42):12990–12995
Peng HJ, Huang JQ, Cheng XB, Zhang Q (2017) Review on high-loading and high-energy lithium–sulfur batteries. Advanced Energy Materials,7(24):1700260
Qiu Y et al (2014) High-rate, ultralong cycle-life lithium/sulfur batteries enabled by nitrogen-doped graphene. Nano Lett., 14(8):4821–4827
Qu Y et al (2015) Highly ordered nitrogen-rich mesoporous carbon derived from biomass waste for high-performance lithium–sulfur batteries. Carbon, 84:399–408
Rehman S, Guo S, Hou Y (2016) Rational design of Si/SiO2@ hierarchical porous carbon spheres as efficient polysulfide reservoirs for high-performance Li–S battery. Advanced materials, 28(16):3167–3172
Seh ZW, Sun Y, Zhang Q, Cui Y (2016) Designing high-energy lithium–sulfur batteries. Chem. Soc. Rev., 45(20):5605–5634
Shi Y, Liu G, Jin R, Xu H, Wang Q, Gao S (2019a) Carbon materials from melamine sponges for supercapacitors and lithium battery electrode materials: a review. Carbon Energy, 1(2):253–275
Shi Z et al (2019b) A soluble single atom catalyst promotes lithium polysulfide conversion in lithium sulfur batteries. Chem. Commun., 55(80):12056–12059

Singhal R, Chung S-H, Manthiram A, Kalra V (2015) A free-standing carbon nanofiber interlayer for high-performance lithium–sulfur batteries. J. Mater. Chem. A, 3(8):4530–4538

Song M-S et al (2004) Effects of nanosized adsorbing material on electrochemical properties of sulfur cathodes for Li/S secondary batteries. J. Electrochem. Soc. 151(6):A791

Song X et al (2018) Fe-N-doped carbon nanofiber and graphene modified separator for lithium-sulfur batteries. Chemical Engineering Journal, 333:564–571

Strubel P et al (2015) ZnO hard templating for synthesis of hierarchical porous carbons with tailored porosity and high performance in lithium-sulfur battery. Advanced Functional Materials, 25(2): 287–297

Su Y-S, Manthiram A (2012) A new approach to improve cycle performance of rechargeable lithium–sulfur batteries by inserting a free-standing MWCNT interlayer. Chem. Commun., 48(70):8817–8819

Sun F et al (2013) High efficiency immobilization of sulfur on nitrogen-enriched mesoporous carbons for Li–S batteries. ACS Appl. Mater. Interfaces, 5(12):5630–5638

Sun L et al (2014) Sulfur nanocrystals confined in carbon nanotube network as a binder-free electrode for high-performance lithium sulfur batteries. Nano Lett., 14(7):4044–4049

Sun Y, Liu N, Cui Y (2016) Promises and challenges of nanomaterials for lithium-based rechargeable batteries. Nature Energy,1(7):1–12

Sun Z et al (2017) Conductive porous vanadium nitride/graphene composite as chemical anchor of polysulfides for lithium-sulfur batteries. Nature communications, 8(1):1–8

Tan L, Li X, Wang Z, Guo H, Wang J (2018) Lightweight reduced graphene oxide@ MoS2 interlayer as polysulfide barrier for high-performance lithium–sulfur batteries. ACS Appl. Mater. Interfaces, 10(4):3707–3713

Tao X et al (2013) Highly mesoporous carbon foams synthesized by a facile, cost-effective and template-free Pechini method for advanced lithium–sulfur batteries. J. Mater. Chem. A,1(10): 3295–3301

Tao X et al (2014) Strong sulfur binding with conducting magnéli-phase Ti n O2 n–1 nanomaterials for improving lithium–sulfur batteries. Nano Lett.14(9):5288–5294

Tao X et al (2016) Balancing surface adsorption and diffusion of lithium-polysulfides on nonconductive oxides for lithium–sulfur battery design. Nature communications, 7(1):1–9

Tian W, Zhang H, Duan X, Sun H, Shao G, Wang S (2020) Porous carbons: Structure-oriented design and versatile applications. Advanced Functional Materials, 30(17):1909265

Van Noorden RJN (2014) A better battery. Nature,507(7490):26

Wang C et al (2014a) High sulfur loading composite wrapped by 3D nitrogen-doped graphene as a cathode material for lithium–sulfur batteries. J. Mater. Chem. A, 2(14):5018–5023

Wang D-W et al (2013) Carbon–sulfur composites for Li–S batteries: status and prospects. J. Mater. Chem. A, 1(33):9382–9394

Wang H, Zhang W, Xu J, Guo Z (2018b) Advances in polar materials for lithium–sulfur batteries Advanced Functional Materials, 28(38):1707520

Wang J, Yang J, Xie J, Xu N, Li Y (2002) Sulfur–carbon nano-composite as cathode for rechargeable lithium battery based on gel electrolyte. Electrochemistry Communications4(6): 499–502

Wang M et al (2019a) Porous carbon hosts for lithium–sulfur batteries. Chemistry–A European Journal, 25(15):3710–3725

Wang S et al (2019b) Designing a highly efficient polysulfide conversion catalyst with paramontroseite for high-performance and long-life lithium-sulfur batteries., Nano Energy, 57:230–240

Wang X, Zhang Z, Qu Y, Lai Y, Li J (2014b) Nitrogen-doped graphene/sulfur composite as cathode material for high capacity lithium–sulfur batteries. Journal of Power Sources, 256:361–368

Wang X et al (2016) Structural and chemical synergistic encapsulation of polysulfides enables ultralong-life lithium–sulfur batteries. Energy Environ. Sci., 9(8):2533–2538

Wang Y, Huang X, Zhang S, Hou Y (2018a) Sulfur hosts against the shuttle effect. Small Methods, 2(6):1700345

Xiao Z et al (2015) A lightweight TiO2/graphene interlayer, applied as a highly effective polysulfide absorbent for fast, long-life lithium–sulfur batteries. Adv. Mater, 27(18):2891–2898
Xin S et al (2012) Smaller sulfur molecules promise better lithium–sulfur batteries. J. Am. Chem. Soc. 134(45):18510–18513
Xing L-B et al (2016) Nitrogen, sulfur-codoped graphene sponge as electroactive carbon interlayer for high-energy and-power lithium–sulfur batteries. Journal of Power Sources, 303:22–28
Xu C et al (2015a) Sulfur/three-dimensional graphene composite for high performance lithium–sulfur batteries. Journal of Power Sources, 275:22–25
Xu G, Ding B, Pan J, Nie P, Shen L, Zhang X (2014) High performance lithium–sulfur batteries: advances and challenges. J. Mater. Chem. A, 2(32):12662–12676
Xu H, Jiang Q, Zhang B, Chen C, Lin Z (2020) Integrating conductivity, immobility, and catalytic ability into high-N carbon/graphene sheets as an effective sulfur host. Advanced Materials, 32(7):1906357
Xu J, Zhang W, Chen Y, Fan H, Su D, Wang G (2018) MOF-derived porous N–Co 3 O 4@ N–C nanododecahedra wrapped with reduced graphene oxide as a high capacity cathode for lithium–sulfur batteries. J. Mater. Chem. A,6(6):2797–2807
Xu Y et al (2015b) Confined sulfur in microporous carbon renders superior cycling stability in Li/S batteries. Advanced Functional Materials, 25(27):4312–4320
Yan Y, Cheng C, Zhang L, Li Y, Lu J (2019) Deciphering the reaction mechanism of lithium–sulfur batteries by in situ/operando synchrotron-based characterization techniques. Advanced Energy Materials, 9(18):1900148
Yao, H., Zheng, G., Hsu, PC. et al. Improving lithium–sulphur batteries through spatial control of sulphur species deposition on a hybrid electrode surface. Nat Commun 5, 3943 (2014).
Yin YX, Xin S, Guo YG, Wan LJ (2013) Lithium–sulfur batteries: electrochemistry, materials, and prospects. Angewandte Chemie International Edition, 52(50):13186–13200
Yuan H et al (2018a) A review of functional binders in lithium–sulfur batteries. Advanced Energy Materials, 8(31):1802107
Yuan H et al (2018b) Facilitation of sulfur evolution reaction by pyridinic nitrogen doped carbon nanoflakes for highly-stable lithium-sulfur batteries. Energy Storage Materials, 10:1–9
Yuan H, Peng HJ, Huang JQ, Zhang Q (2019a) Sulfur redox reactions at working interfaces in lithium–sulfur batteries: a perspective. Advanced Materials Interfaces, 6(4):1802046
Yuan H et al (2019b) Conductive and catalytic triple-phase interfaces enabling uniform nucleation in high-rate lithium–sulfur batteries. Advanced Energy Materials, 9(1):1802768
Yuan L, Yuan H, Qiu X, Chen L, Zhu W (2009) Improvement of cycle property of sulfur-coated multi-walled carbon nanotubes composite cathode for lithium/sulfur batteries. Journal of Power Source, 189(2):1141–1146
Yuan Z et al (2016) Powering lithium–sulfur battery performance by propelling polysulfide redox at sulfiphilic hosts. Nano Lett., 16(1):519–527
Zapata Dominguez D et al (2022) (De) Lithiation and strain mechanism in crystalline Ge nanoparticles, ACS Nano, vol 16, p 9819
Zeng X et al (2018) Conductive molybdenum carbide as the polysulfide reservoir for lithium–sulfur batteries. J. Mater. Chem. A,6(35):17142–17147
Zhang J, Huang M, Xi B, Mi K, Yuan A, Xiong S (2018a) Systematic study of effect on enhancing specific capacity and electrochemical behaviors of lithium–sulfur batteries. Advanced Energy Materials, 8(2):1701330
Zhang J, Li Z, Chen Y, Gao S, Lou XW (2018b) Nickel–iron layered double hydroxide hollow polyhedrons as a superior sulfur host for lithium–sulfur batteries. Angewandte Chemie,130(34): 11110–11114
Zhang L, Sun D, Feng J, Cairns EJ, Guo J (2017) Revealing the electrochemical charging mechanism of nanosized Li2S by in situ and operando X-ray absorption spectroscopy. Nano Lett., 17(8):5084–5091
Zhao M-Q et al (2014) Unstacked double-layer templated graphene for high-rate lithium–sulphur batteries. Nature communications, 5(1):1–8

Zhou G (2017) Graphene–pure sulfur sandwich structure for ultrafast, long-life lithium-sulfur batteries. In: Design, fabrication and electrochemical performance of nanostructured carbon based materials for high-energy lithium–sulfur batteries. Springer, Springer Theses. Springer, Singapore. https://doi.org/10.1007/978-981-10-3406-0_5

Zhou G et al (2012) A flexible nanostructured sulphur–carbon nanotube cathode with high rate performance for Li-S batteries. 5(10):8901–8906

Zhou G et al (2013) Fibrous hybrid of graphene and sulfur nanocrystals for high-performance lithium–sulfur batteries. 7(6):5367–5375

Zhou G, Zhao Y, Manthiram A (2015a) Dual-confined flexible sulfur cathodes encapsulated in nitrogen-doped double-shelled hollow carbon spheres and wrapped with graphene for Li–S batteries. 5(9):1402263

Zhou G, Zhao Y, Zu C, Manthiram A (2015b) Free-standing TiO2 nanowire-embedded graphene hybrid membrane for advanced Li/dissolved polysulfide batteries. 12:240–249

Zhou G et al (2017) Catalytic oxidation of Li2S on the surface of metal sulfides for Li− S batteries. 114(5):840–845

Zhu Q, Zhao Q, An Y, Anasori B, Wang H, Xu B (2017) Ultra-microporous carbons encapsulate small sulfur molecules for high performance lithium-sulfur battery. 33:402–409

Zou G, Hou H, Hu J, Ji X (2019) General synthesis of heteroatom-doped hierarchical carbon toward excellent electrochemical energy storage. 2(8):712–722

Chapter 8
Nanomaterial-Based Sulfur Composite Cathodes

Muhammad Arif, Muhammad Sagir, Saima Nazir, and M. B. Tahir

Abstract Lithium-sulfur (Li-S) batteries fascinated the excessive intension of research community due to higher energy density, cost-effective, and greater efficiency as compared to lithium-ion batteries. Though under the huge benefits and countless achievements, there are some limitations such as rapid capacity declining and low utilization with sulfur cathode because of shuttle effect and uncontrollable high polysulfide deposition which restricted to practical usage of such effective energy storage device. Under such circumstances, we discussed nanomaterial-based sulfur cathode composites in this chapter specifically which provide great support in better electrochemical pathways. However, design/fabrication of different carbon-sulfur or other polymer-sulfur composite cathodes have been discussed with addressed shuttle effect and solution of this problem by these composites. Furthermore, porous, hollow/core-shell, nanosheets/nanorods and hierarchical nanomaterials have also been discussed with their size effects on Li^+/electron transfer and their morphological impact on electrochemical performance as well.

M. Arif (✉)
Institute of Chemical and Environmental Engineering, Khwaja Fareed University of Engineering and Information Technology, Rahim Yar Khan, Punjab, Pakistan

Centre for Thermal and Renewable Energy Research, Khwaja Fareed University of Engineering and Information Technology, Rahim Yar Khan, Punjab, Pakistan
e-mail: muhammad.arif@kfueit.edu.pk

M. Sagir
Institute of Chemical and Environmental Engineering, Khwaja Fareed University of Engineering and Information Technology, Rahim Yar Khan, Punjab, Pakistan
e-mail: dr.msagir@kfueit.edu.pk

S. Nazir
Nawaz Sharif medical College, University of Gujrat, Gujrat, Pakistan

University of Health Sciences, Lahore, Pakistan

M. B. Tahir
Institute of Physics, Khwaja Fareed University of Engineering and Information Technology, Rahim Yar Khan, Punjab, Pakistan
e-mail: m.bilaltahir@kfueit.edu.pk

M. S. Tahir et al. (eds.), *Lithium-Sulfur Batteries: Key Parameters, Recent Advances, Challenges and Applications*, Springer Tracts in Electrical and Electronics Engineering, https://doi.org/10.1007/978-981-99-2796-8_8

We summarized the different porous carbon hosts with their steady incorporation ways of carbon-sulfur cathodes. At the end, we summarized sufficient regarding challenges with sulfur cathode which restricted the applications in electronic field at commercial level. Additionally, our discussion concluded with their developing trend in the future about porous nanomaterial-based sulfur composite of sulfur cathode.

Keywords Nanomaterials · Carbon-sulfur cathodes · High polysulfide deposition

Introduction

There has been required to develop clean, efficient, economical, and highly significant rechargeable energy storage devices in front of insufficient, hazardous energy sources and unsafe to environmental climate changes. Concerns over an inadequate global energy supply have prompted the development of possible electrode materials for energy conversion and storage. Because of its huge volume and high gravimetric energy density, rechargeable lithium batteries are a prevailing and dominating power source for a wide range of portable electronic gadgets and public transportation vehicles. Now, significant improvement has been required in rechargeable lithium batteries to overcome CO_2 emission and fossil fuel consumption. Consequently, lower specific capacities of cathode components which are about ~170 mAh/g for $LiFePO_4$ and ~150 mAh/g for layered oxides as compared to anode materials which is about 4200 mAh/g for Si and 370 mAh/g for graphite have been warning features to the energy density of rechargeable batteries. However, theoretical specific capacity of sulfur as promising cathode material is 1672 mAh/g which ~5 times higher than transition metal oxides/phosphates-based cathode material as well as low cost and eco-friendly. The poor electrical conductivity of sulfur, volume expansion after discharge, and polysulfides in the electrolyte, however, have been identified as constraints and challenges, and they have hindered the development of practical Li-S battery applications (Wang et al. 2011, 2014; Oakes et al. 2016; Zhang et al. 2017; Wu et al. 2020). (i) Due to shielding nature of sulfur, low utilization efficiency of active material occurred during its discharging products (Li_2S/Li_2S_2) which restricted charge transmission in the cathode material; and (ii) shuttle effect has been occurring; and soluble polysulfides (LiPSs:Li_2S_x, $4 \leq x \leq 8$) created during cycling process. Because of this, coulombic efficiency and capacity were severely impacted. It's been observed that lithium-sulfur batteries have a low specific capacity, short cycle life, and low energy proficiency. For better Li^+ transportation, good conductivity and specific capacity (more than 1000 mAh/g) have been achieved by various nanocomposites like nanotubes (Li et al. 2017), carbon or sulfur composites utilizing active carbon/mesoporous carbon (Ye et al. 2017), MXenes (Pang et al. 2019), layered metal oxides (Wang et al. 2020), and fluorinated carbon (Liu et al. 2019) but still challenging to maintain specific capacity of sulfur after 100 cycles (Wu et al. 2021). Furthermore, these advanced developments create interest of research community to synthesize more efficient cathode material that maintains greater gravimetric capacity of sulfur cathode, and high spatial capacity and high areal load are obtained. These fundamental changes required thick electrode architectures for commercial Li^+ battery cathodes (generally between 2 and 4 mAh/cm^2) (Manthiram et al. 2014).

In this chapter, we elaborated recent advancements and associated problems in easiest way for better understandings of sulfur cathode nanomaterials in lithium-sulfur batteries. We mainly focused on the nanomaterials for sulfur cathode in Li-sulfur batteries where we explore it in the different dimensions. The catholyte technique also includes the advantages of many S-based nanocomposites cathode materials. As a host material, we demonstrated different types of nanomaterials (carbon-based, polymeric, metal oxides, and metal sulfide composites) as well as versatility and surface modifications in Li-S batteries as cathode to reduce the shuttle effects and density declining issue associated during the cyclic process.

Detailed Mechanism of Catalyst and Shuttling Effect in Li-S Batteries

The charging/discharging mechanism in Li-S batteries is a mutistep redox reaction which has been escorted through polysulfide-ions conversion Sn^{2-} ($1 \leq n \leq 8$) with different valences. Furthermore, at the discharging time, S_8 which is the most constant system counters with Li^+ produced by Li-anode. However, chain of numerous polysulfides has been started, and insoluble Li_2S_2 and Li_2S are produced. The following reactions are explained below (Barchasz et al. 2012a, b; Manthiram et al. 2015; Cheon et al. 2003; Mikhaylik and Akridge 2004; Su et al. 2013):

$$S_8 + 2Li^+ + 2e^- \rightarrow Li_2S_8 \tag{8.1}$$

$$3Li_2S_8 + 2Li^+ + 2e^- \rightarrow 4Li_2S_6 \tag{8.2}$$

$$2Li_2S_6 + 2Li^+ + 2e^- \rightarrow 3Li_2S_4 \tag{8.3}$$

$$Li_2S_4 + 2Li^+ + 2e^- \rightarrow 2Li_2S_2 \tag{8.4}$$

$$Li_2S_2 + 2Li^+ + 2e^- \rightarrow 4Li_2S \tag{8.5}$$

$$S_8 + 16Li \rightarrow 8Li_2S$$

Under the observations of abovementioned reaction mechanism (also schematically representation in Fig. 8.1), carbon matrixes are very important for lithium batteries. For this reason, the following factors are mentioned below to improve electrical conductivity:

1. Increasing the conductivity: Due to low electrical conductivity of final product (Li_2S) and sulfur in abovementioned mechanism, must be combined with carbon to attain better electronic pathway to emphasize electrical conductivity (Fig. 8.2).

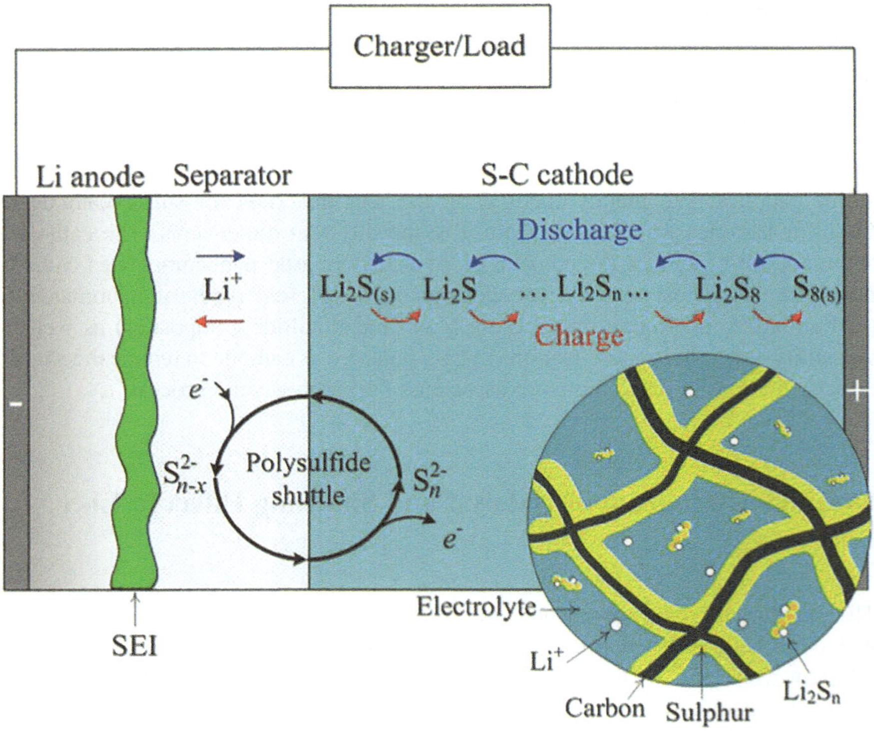

Fig. 8.1 Schematic illustration of shuttling effect and electrical hindrance due to insoluble polysulfides during charge/discharge process in in Li-S batteries (Wild et al. 2015)

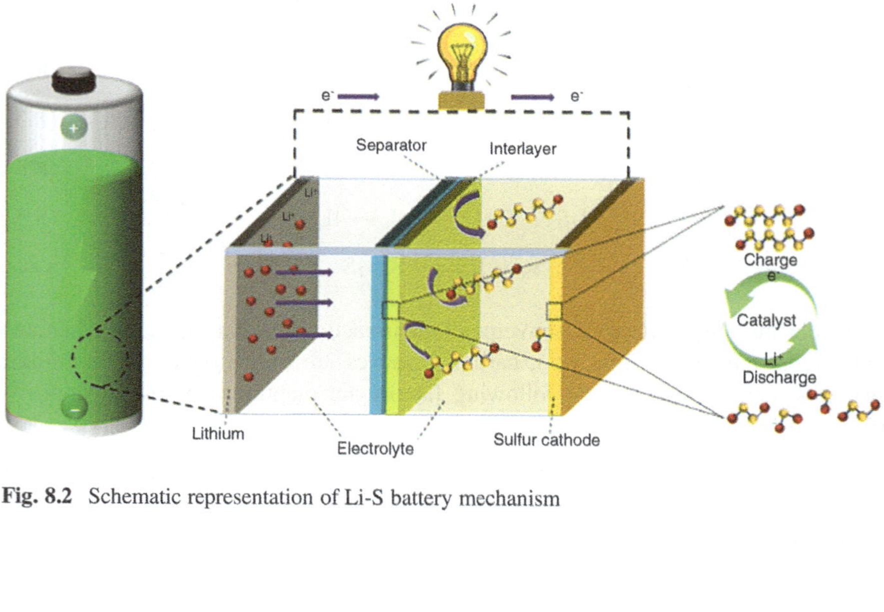

Fig. 8.2 Schematic representation of Li-S battery mechanism

2. Fastening effect: The carbon/sulfur cathode requires sulfur to be incorporated in carbon matrixes and media.
3. Hindering effect: Polysulfides chain is dissolved into an electrolyte and moving toward cathode to anode. Hence, carbon matrixes can deliver chemical or physical support to reducing/blocking effect and strengthen the electrochemical performance by suppressing the shuttle effect. However, large surface area, porosity of the compound, and distribution are the major requirements for carbon matrixes/hosts to improve electrical conductivity.

Morphological Investigations of S-Based Nanocomposites

Porous carbon not only increase the electrical conductivity due to porous structure in sulfur/carbon cathode composite by creating short ionic as well as electronic pathways but also stop the polysulfides transitions and increased the dramatically volume variations during the discharge/charge procedures of Li-S batteries. Hence, it created great interest among the scientists nowadays to solve these complications of sulfur cathodes as matrix. It is difficult to synthesized nanostructures with equally pore size dispersal to unitize it with full advantages in LSBs cathode (Wang et al. 2002; Li et al. 2015) (Fig. 8.3).

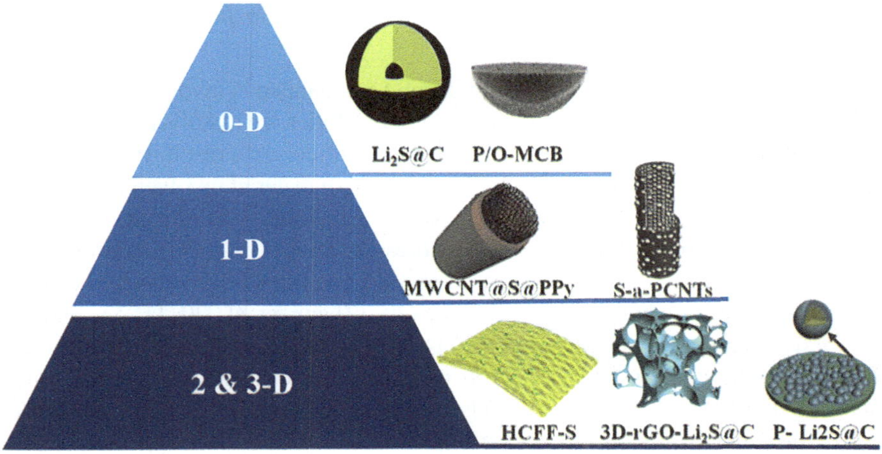

Fig. 8.3 Schematic representations of sulfur-containing porous carbon nanocomposites (Liang et al. 2016)

0-D Nanomaterials-Based Sulfur Composites Cathode

Various nanosulfur-containing composites attracted great interactions of researchers about zero-dimensional (0D) sulfur nanomaterials in the recent years including sulfur quantum and sulfur nanoparticles. The importance of 0D nanomaterials and the coherent investigation of advanced new and potential low-dimensional nanomaterials and derivatives associated with capable cathode materials in LSBs. To increase the electrochemical performance of sulfur-containing cathode composites, doping of sulfur has been required to increased functional sites and volume expansion. When the nanosulfur combined with any other polar inorganic framework and carbonaceous polymers, multifunctional products have been significantly increased the trapping and conductivity in cathode material and exhibited excellent reversible capacity, original capacity, and extra efficient capability with long-term cycling durability (Bai et al. 2016). Chen et al. have fabricated encapsulated and dual confined graphene-wrapped sulfur nanoparticles in hollow TiO_2 spheres (G-TiO_2@S) through hydrolyzing method with acid engraving. This unique structure with great potential not only facilitated the effectively trapped polysulfide matrix but also provide physical support due to hollow spheres as well as chemical support through chemical bonding between the TiO_2 and polysulfides. Moreover, it is responsible for dramatically volume change during the discharge/charge procedure as revealed in Fig. 8.4 (Fan et al. 2016).

1-D Nanomaterials-Based Sulfur Composites Cathode

In the reaction cell of LSBs, solid–liquid–solid mechanism is involved between the lithium and sulfur reaction which react, and polysulfides are dissolved in electrolyte and deposited on conductive material as lithium sulfide. As a result, lithium sulfide pores/precipitation are formed which lower the active material involvement and thus lowering the cycling proceeds. To overcome these problems, one-dimensional (1-D) carbon nanomaterials play key role to addressing these issues. Furthermore, 1-D carbon composites potentially addressed above mentioned difficulties and provide better electrical conduction in sulfur cathode (Zheng et al. 2011). Yu and his coworkers successfully synthesized self-assembled 1-D nanotubes wrapped in few layered of vertically oriented 1T-MoS_2 on porous graphene/MoS_2 hybrid through spray-drying method. Due to its highly oriented nanostructure, a large number of active sites and greater surface area of highly conductive graphene structure facilitated electron transfer, Li-ion transfer, and efficient electrolyte infiltration. It showed well-adsorption of polysulfides and highly efficient electrocatalyst for long-term cyclic ability with very low capacity decayed (0.039% per cycle) over the 500 cycles at 1 C as mentioned in Fig. 8.5 (Zheng et al. 2006).

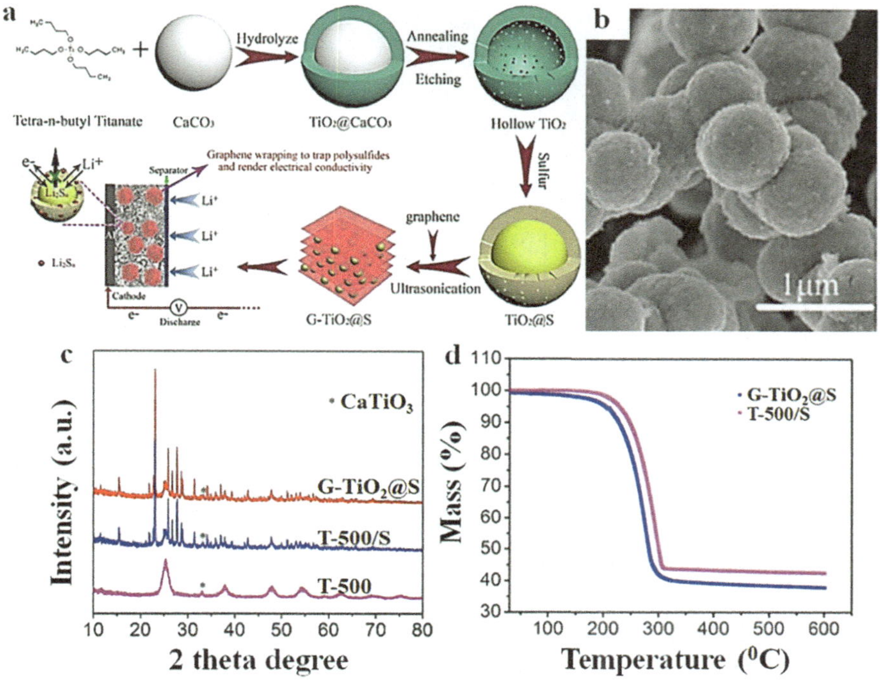

Fig. 8.4 Illustrations of nanomaterial-based sulfur cathode: (**a**) G-TiO_2@S. (**b**) SEM. (**c**) XRD. (**d**) TG curves

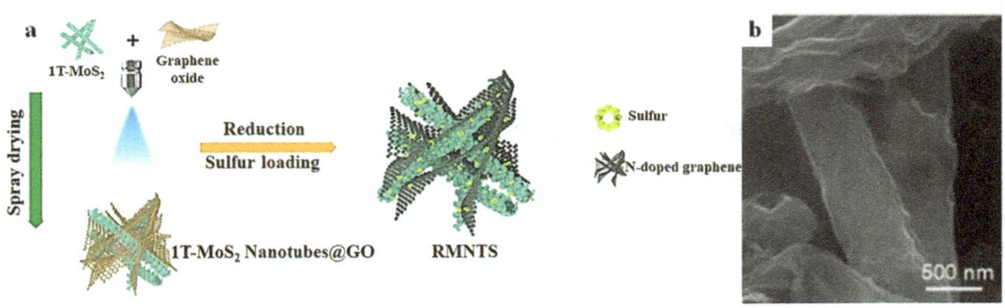

Fig. 8.5 Schematic illustration of 1-D nanotubes-based sulfur cathode: (**a**) GMNTS. (**b**) SEM (Zheng et al. 2006)

2-D and 3-D Nanomaterials-Based Sulfur Composites Cathode

As zero-dimensional and one-dimensional nanostructures, 2-D and 3-D also have their own morphological preferences in the catalytic performance of sulfur cathode nanomaterials. However, 3-dimensional N-PC/Ti_3C_2 nanosheets have been successfully synthesized by Zhang and his teammates via facile self-sacrificing template process for sulfur cathode composite in lithium sulfur batteries. N-doped porous

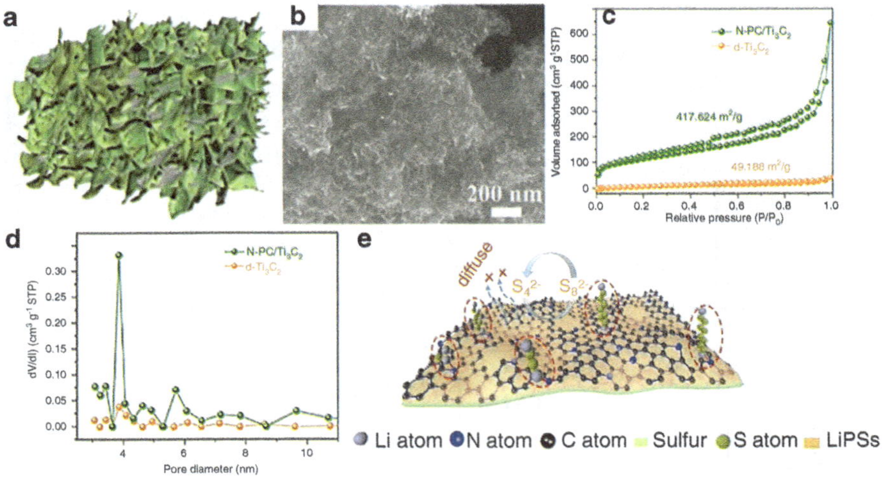

Fig. 8.6 Diagrammatical representations of nanomaterial-based 3-D sulfur cathode composite: (**a**) 3-D (N-PC/Ti_3C_2). (**b**) SEM. (**c**) Shuttle effect of LiPSs. (**d**) Adsorption/desorption isotherm and (**e**) pore size distribution (Zhang et al. 2019a)

carbon and MXene nanosheets coexisted have enhanced surface area of sulfur cathode which well-adsorbed the polysulfides composites that are produced during the charge/discharge mechanism and created hindrance in ion/electron transportation. Additionally, more than 80% of the sulfur content is present in N-PC/Ti_3C_2 composite that exhibited excellent electrochemical performance with lower fading rate (0.04% per cycle) up to 800 cycles as shown in Fig. 8.6 (Zhang et al. 2019a).

Host Materials for Sulfur Composites

At room temperature, sulfur shows extremely low electrical conductivity in most of the solvents. So, it required productive support of any other material to extent its performance in LSBs as cathode. Hence, to overcome these restrictions, researchers have need to add some other constituents like carbon, polymeric, and inorganic host material. Nanocomposites (Ketjen Black/Mg0.6Ni0.4O and G/M@CNT) have also been created in order to increase the electrochemical performance of separators as interlayer. The carbonaceous materials provide conductive support in framework, and polymers provide trapping and stability of polysulfides (Li et al. 2015; Tang et al. 2017; Bhargav et al. 2019).

Carbonaceous Host Materials

To advance the electric conduction of sulfur fused cathode, carbon is the major constituent of electrode as conductive material in Li-S batteries and makes a medium to trap dissolved polysulfides. Porous carbon materials have been shown to be the most operative matrix for sulfur loading and to enhance Li+/electron motions in the cathode, out of the wide range of carbon host nanomaterials that have been investigated. Mostly, carbon categories into three range according to size: (i) microporous carbon (<2 nm), (ii) mesoporous carbon (2–50 nm), and (iii) macro-porous carbon (>50 nm). However, microporous carbon acts like container, and others (meso and macro) act as electrolytes channels. Thereby, carbon improves great capability of LSBs (Liu et al. 2020). Inside the structure, nanopores can also facilitate the volume expansion. Mesoporous carbon materials are a good carrier for sulfur loading as microporous carbon material (<40%). Porosity is the most important factor for sulfur loading of carbonaceous materials to enhance electrochemical activity of LSBs. Nanocarbon-sulfur composites with different dimensions shown ultrahigh (up to 90%) sulfur loading. To increase electrochemical performance, advancements are not limited to size distribution but also improved by different other techniques like doping of various heteroatoms to create effective polar sites and electron pathways (Zeng et al. 2017). Gao et al. fabricated 3-D graphene nanosheet–carbon nanotube (GN–CNT) matrix by one-pot pyrolysis method. Carbon nanotubes grow on the surface of graphene nanosheets with covalent interactions between them. There were several heteroatoms scattered on graphene sheets in which cobalt nanoparticles existed on top of CNTs in wrapped form and N, O, are scattered. Additionally, heteroatoms can help in the immobilization of sulfur species. As a result, electrochemical performance has been increased due to favorable electrolyte infiltration and rapid charge transfer by S/GN–CNT material as active matrix as shown in Fig. 8.7 (Zhang et al. 2017).

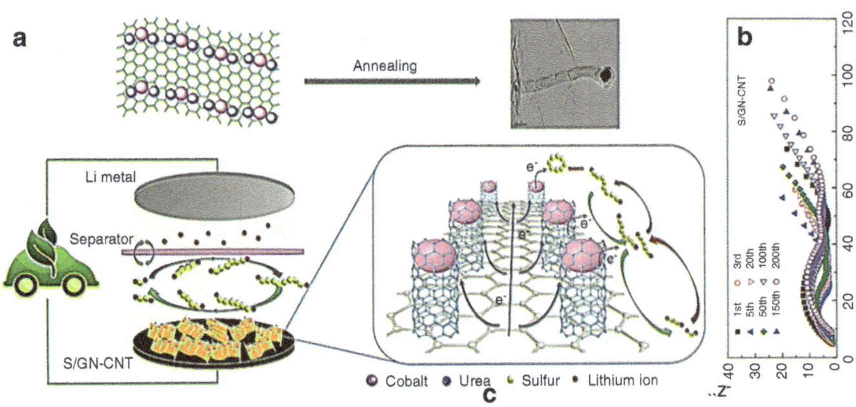

Fig. 8.7 Illustrations of nanocomposite-based sulfur cathode material: (**a**) 3-D (GN–CNT) matrix with electron transfer. (**b**) EIS after fully charged for different cycles (Zhang et al. 2017)

Polymeric Host Materials

Polymers were discovered at 1970s as first conductive polymers (polyacetylene) in the past several decades. Due to their large number of advantages, conductive polymers were observed most promising and efficient host materials for sulfur loading to enhanced sulfur volume and hindering polysulfide dissolution. Furthermore, chain-like assemblies and rich functional groups of them provided great conduction of complex cathode in Li-S batteries. Additionally, polymers are also used as binder, separators, defensive coverings, and electroactive ingredients. In the Li-S redox reactions at cathode, polymers had been used as protective coatings to restrict reaction and reduce the soluble lithium polysulfides diffusion effectively. Since 2002, polymer-sulfur composite was initially described for LSBs, particularly to enhance the cathode's catalytic activity. Moreover, hybrid carbon-polymer buildings had been especially designed to reduce polysulfides diffusion which hindered in catalytic performance of electrode. As a result, poor energy density and cyclic performance have been reported (Dirlam et al. 2017; Yang et al. 2013).

Host Materials: Metal Oxide and Metal Sulfide

As other host materials, metal oxides, sulfides, and carbides were also exposed surfaces with different structures to further trap polysulfides. Due to rich polar positions for the absorption of polysulfides, polysulfides trap easily, and electrochemical pathways have been efficiently conductive of the electrons or lithium ions. As a result, good electrochemical performance is exhibited in LSBs. It is concluded when metals incorporated carbon, reaction between the metals and sulfur stabilizes the porous active material, improves cyclic activity and coulombic efficiency, and provides physical/chemical absorption barriers to polysulfides (Liu et al. 2020; Zeng et al. 2017). He and his coworkers fabricated MOF-derived cobalt sulfide arrays on 3-D graphene sheets (Co_9S_8-3DGF) as active host substantial for highly efficient cathode in Li-S batteries. Due to its unique hierarchical nanoarchitecture, it may also be employed as a free-standing cathode without a binder. In synthesized composite, Co_9S_8 entrapped the polysulfides by strong chemical bindings and exhibited pronounced electrochemical activity because of easy and fast transportation. As a result, Co_9S_8-3DGF cathode has been achieved a greater areal capacity (10.9 mA h cm^{-2}) with so high sulfur loading of 10.4 mg cm^{-2} in which 86.9 wt% sulfur amount was present as exposed in Fig. 8.8 (He et al. 2018).

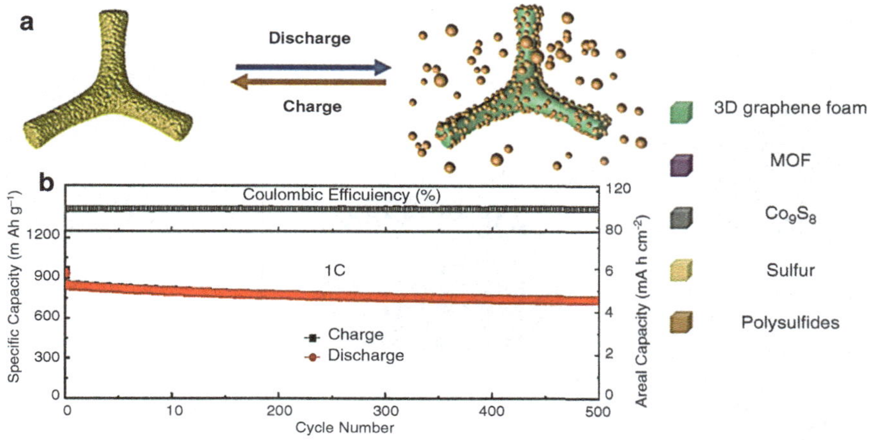

Fig. 8.8 Discharge/charge process in sulfur-based nanomaterial with graphical representation. (**a**) Co_9S_8-3DGF/S cathode. (**b**) Remarkably high areal capacity (He et al. 2018)

Versatility and Surface Modifications in S-Based Nanocomposites

Elemental sulfur is mixed with any other conductive material like active carbon/ conductive polymer and heated to get required composite structure in most of the fabricating processes of sulfur electrode. Particle of sulfur commonly remains within the electrode framework in this type of matrix, without encapsulation of pores. For this reason, through the charge/discharge procedure of LSBs, these morphological variations play a significant role between the active material and current collector to reduce sudden capacity loss or any electrical contact between them (He et al. 2009; Elazari et al. 2010).

Porous Nanocomposites

Porous carbon compounds, such as those containing sulfur (Zheng et al. 2011; Cheng et al. 2014), conductive polymer-based composites (Liang et al. 2015; Li et al. 2014), and inorganic materials, had been presented into the cathodes to efficiently restrict polysulfide distribution and advance the Li-S batteries (Zhang et al. 2010). Due to the increased conductivity of the sulfur cathodes and the sulfur being embedded inside the pores, it is known that sulfur host materials with great aspect ratios and hollow structures might give improved electrochemical properties (Liang et al. 2011). Noel Dez et al. showed that a porous nanocomposite sample made from polypyrrole nanoparticles and sodium thiosulfate may yield porous

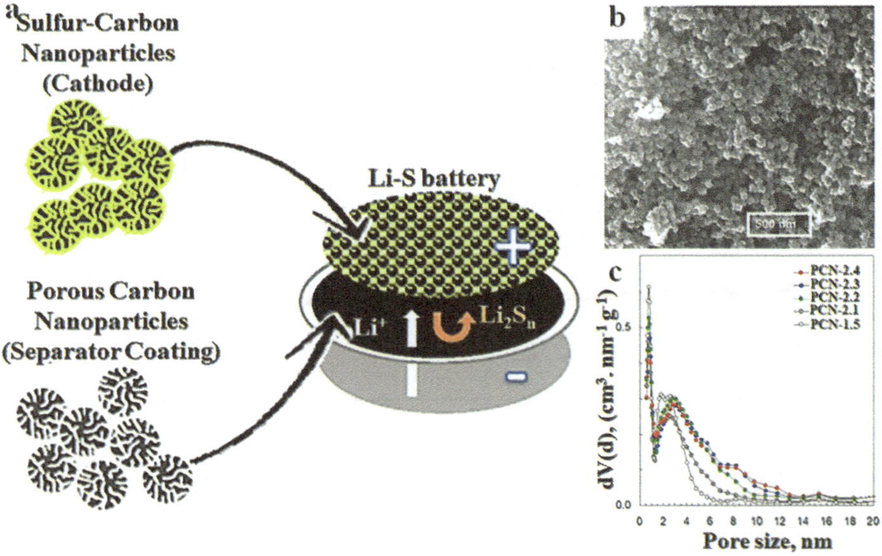

Fig. 8.9 Illustrations and graphical representation of porous nanoparticles-based sulfur cathode nanomaterial; (**a**) S/PCN composite. (**b**) SEM. (**c**) Pore size distribution (Díez et al. 2020)

carbon nanoparticles (PCN) co-doped with nitrogen and sulfur using a simple template-free technique.

PCN and S/PCN (nanocomposites containing elemental sulfur penetrated into the PCN) may be generated by making slight adjustments to the synthesis technique. This provides a great deal of adaptability. Two materials of a lithium-sulfur cell have been used for testing these materials.

The cathode is made of S/PCN nanocomposite, and the separator is coated with a thin coating of PCN to limit the polysulfide's mobility. A capacity loss of just 0.089% per cycle was observed throughout 500 charge-discharge cycles at 2 °C for the cell (Fig. 8.9) (Díez et al. 2020).

As a sulfur carrier for Li-S cathodes, micro/mesoporous SnO_2 has been shown to provide higher electrochemical performance as a sulfur carrier for S-M-O chemical bonding. However, the poor conductivity of the cathodes would restrict the specific capacity and rate capabilities of the Li-S batteries (Wang et al. 2019). Sulfur-restrained activated porous carbon nanotubes were synthesized by Arumugam Manthiram and coworkers and used as a cathode material in Li–S batteries using a simple technique of dual nozzle co-electrospinning and a chemical redox process. For example, the micro/mesopores may bring nanoscale sulfur ornamentation and integrate up to 75% sulfur into carbon structure. As trap sites for hydroxyl groups, the hydroxyl groups help in micropores and electrolyte diffusion into the channel structure, which results in an outstanding cycle stability (917 mAh g^{-1} after 300 cycles) and rate capability (859 mAh g^{-1} at 5 C). This research demonstrates a successful technique for addressing the most pressing matters in practical Li–S battery systems (Lee et al. 2017).

Hollow at Core-Shell Nanomaterials

Materials with multi-structures and catalytic components have been combined by many researchers now to increase the electrochemical performance of LSB (Zhang et al. 2019b; Li et al. 2020). Porosity in the porous core–shell structure may hold active material while keeping polysulfides contained, and CoS_2 has a high conductivity of electrons, making it an ideal catalyst component. Combining them is a good technique to boost LSBs' electrochemical efficiency (Li et al. 2019a). Coaxial electrospinning and a selective vulcanization process were used by Zheng et al. to create CoS_2-TiO_2 at carbon core–shell fibers. Because of their ability to effectively host sulfur and restrict polysulfides while also speeding up the conversion of intermediates, core–shell fiber systems are ideal. After 300 cycles, this electrode has a high coulombic efficiency of over 99.5% and a primary specific capacity of 1181.1 mAh g^{-1} (capacity decay of 0.06% per cycle) (Li et al. 2022). Xu et al. have created 3D hollow nickel hydroxide polyhedrons as cathode material by using a template-sacrificed approach. It had excellent electrochemical properties and decayed at a very slow rate (0.067% per cycle). It is SiO_2 that provides the most structural stability and cycle durability, as can be shown in Fig. 8.10 (Liu et al. 2021).

It is possible to create lithium-sulfur battery heterostructures using this technique of selective vulcanization and isolating interactant.

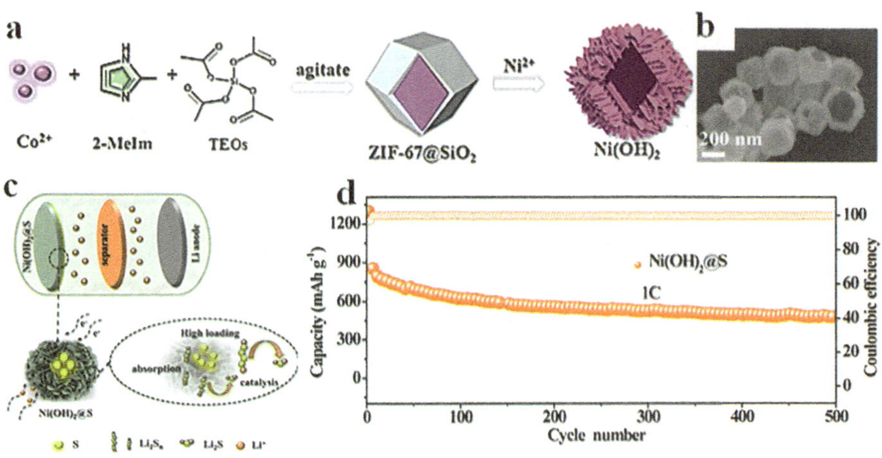

Fig. 8.10 Nanomaterial-based coreshell/hollow sulfur cathode material: (**a**) $Ni(OH)_2$ composite. (**b**) SEM. (**c**) Illustrations of $Ni(OH)_2$@S used as cathode. (**d**) Long-term cyclic performance (Liu et al. 2021)

Nanosheets at Nanotubes Composites

SnO_2 nanoparticles anchored to tubular polypyrrole (T-PPy) as the sulfur carrier with a porous basis for loading sulfur in LiS cathodes have been developed as another strategy to fabricate hierarchically porous S/(TPPy) at SnO_2 ternary nanocomposite cathodes for long-life Li-S batteries (Fig. 8.11b). This strategy inhibits polysulfide diffusion by using SnO_2 nanoparticle (Wei et al. 2020). For the purpose of decreasing the shuttling effect during discharging and recharging of lithium ion batteries (LSBs), Shao et al. synthesized fissile mesoporous Co, N-doped honeycomb-like carbon nanosheets (MC-NS) by the fissile technique. MC-NS, on the other hand, showed a remarkable capacity to adsorb polysulfides and a remarkable electrochemical performance, as shown in Fig. 8.11a, making it a viable catalyst (Li et al. 2019b).

Hierarchical Nanostructures

As a vast family of materials, transition metal dichalcogenides (TMDs) have received increasing interest in the field of electrochemical energy storage in recent years. They have special layered structures and have strong chemical connections with LiPS and thus capably depress the shuttle effect (Yu et al. 2017; Cai et al.

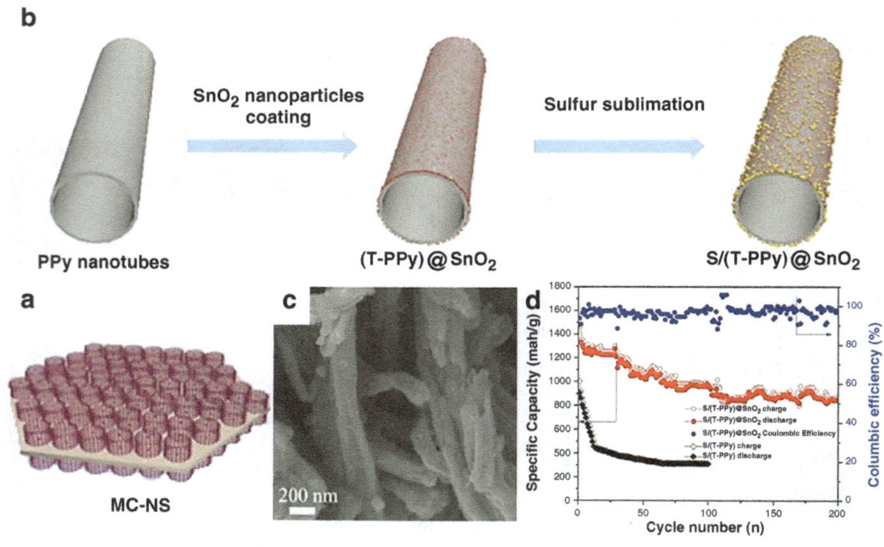

Fig. 8.11 Schematic illustration of the nanosheets/nanotubes-based sulfur composites for cathode material; (**a**) MC-NS (Li et al. 2019b). (**b**) S/(T-PPy)@SnO_2. (**c**) SEM image of S/(T-PPy)@SnO_2 (**d**) effect on electrochemical performance by S/(T-PPy)@SnO_2 nanotubes architecture (Wei et al. 2020)

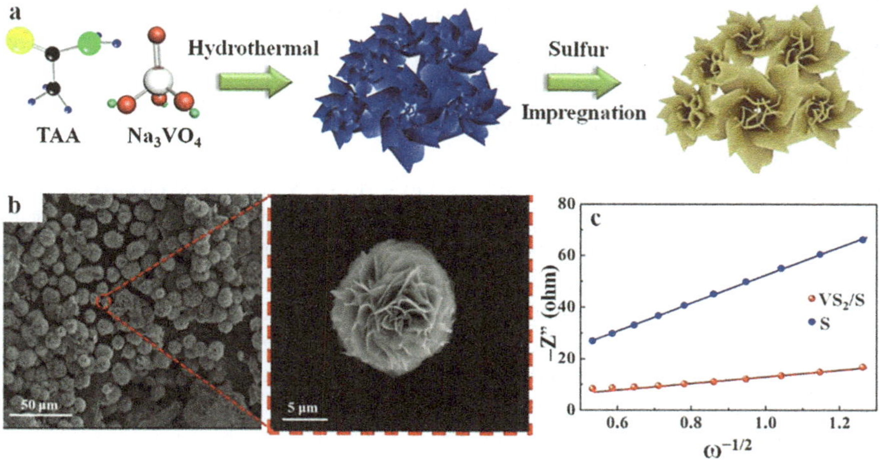

Fig. 8.12 Illustrations of hierarchical nanocomposite-based sulfur cathode: (**a**) VS_2/S composite. (**b**) SEM. (**c**) Tafel slope (Wu et al. 2019)

2018). VS_2 as an important member of the TMDs intimate possesses numerous excellent chemical and physical characteristics. Theoretic research predicts the VS_2 possesses metallic properties and has faster electron diffusion rate compared with MoS_2 and graphite (Zhou et al. 2017). The metallic feature of VS_2 makes sure that VS_2 becomes a multipurpose stand for many implications as an example supercapacitor (Feng et al. 2011), nanoelectronics (Feng et al. 2011; Ji et al. 2017), and hydrogen evolution (Yuan et al. 2015). Potential in electrochemical energy loading has not yet been completely realized, unfortunately. And it's also worth mentioning that the VS_2 has not been used as the sulfur host independently for the cathode in these pioneer reports (Zhu et al. 2018). Xiaoqin Yan et al. used a facile hydrothermal method to make VS_2 NFs comprised of aligned ultrathin nanosheets and synthesized the VS_2/S composites by one-step method. The sulfur can be evenly distributed on the surface of VS_2 nanosheets, which provide plentiful adsorption and catalytic sites for the polysulfides during the charge-discharge process. It will ensure unhindered charge transport to indorse polysulfide redox kinetics and depress the shuttle effect. Tafel slope indicated (Fig. 8.12c) electrocatalytic efficiency toward cathode material in lithium-sulfur batteries. When utilized as a cathode for LSBs, VS2/S composite achieved excellent rate capability and extended cycle life as shown in Figure 8.12 (Wu et al. 2019).

Electrochemical Measurements of Nanomaterials as Sulfur Cathode

A typical sulfur cathode material consists of electroactive matrix, polymer binder, and conductive material. However, in this section, we demonstrated most commonly used sulfur element (S_8) in Li-S batteries and form Li_2S. Complete electrochemical reaction described as $16Li + S_8 \rightarrow 8Li_2S$. Researchers attempted many times to modify the redox reaction process of sulfur by different species of LiPSs. Elemental sulfur shows nonpolar properties, while subsequent LiPSs show polar characteristics, high solubility, and mobility in organic electrolyte (Qin et al. 2019; Meng et al. 2020). Pang et al. successfully developed 3D conductive network of intercalated bentonite N-doped carbon-coated architecture (Bent@C) with interlamellar ion path to improve electrochemical performance through feasible ion/electron transportation in cathode for Li-S batteries. However, modified and unique architecture of S/Bent@C cathode exhibits high reversible capacity of 618.1 m Ah g^{-1} at 2 C charge and also shows 1361 mA h g^{-1} of initial capacity at 0.2 C over 500 cycles with 2 mg cm^{-2} of sulfur loading. Furthermore, this cathode material still exhibited 560.2 mA h g^{-1} of reversible capacity at 0.1 C with sulfur loading (3.0 mg cm^{-2}) over 100 cycles as revealed in Fig. 8.13 (Wu et al. 2021).

Yan and his teammates successfully fabricated urchin-like Co-Mo bimetallic nitride (Co_3Mo_3N) without additional nitrogen by one-step mechanism to reduce the shuttle effect in the sulfur cathode of LSBs. In above-discussed work it is claimed that, successfully catalytic conversion of polysulfides lowers the hindrance in ions or electron movement. Thus, electrochemical reaction has been accelerated and exhibited highest electrochemical activity of 705 mAh g^{-1} at 3 C charge rate over the 600 repeating cycles. Furthermore, it shows lowest capacity declining (about 0.08% per cycle) even at high loading of sulfur (5.4 mg cm^{-2}) as demonstrated in Fig. 8.14 (Yan et al. 2021).

Future Expects

In the energy storage applications, sulfur achieved great importance because of profusion, low price, and high catalytic activity in the latest research, but some limitations have been reported. Research community tries to overcome these limitations by different techniques and modifications in the composites structures to get efficient electrochemical performance in lithium-sulfur batteries. In this chapter, we integrated the structure by different point of views in easiest way for better understandings. However, properties of various sulfur-containing structures (elemental sulfur, polymeric sulfur, porous, nanosheets, hollow, coreshell composites) have also been explained. We explained recent challenges like effects of elemental sulfur, polymeric sulfur, and others (nanosulfur composites) like size on the electrochemical properties of cathode performance. Simply, this chapter provided a systematic

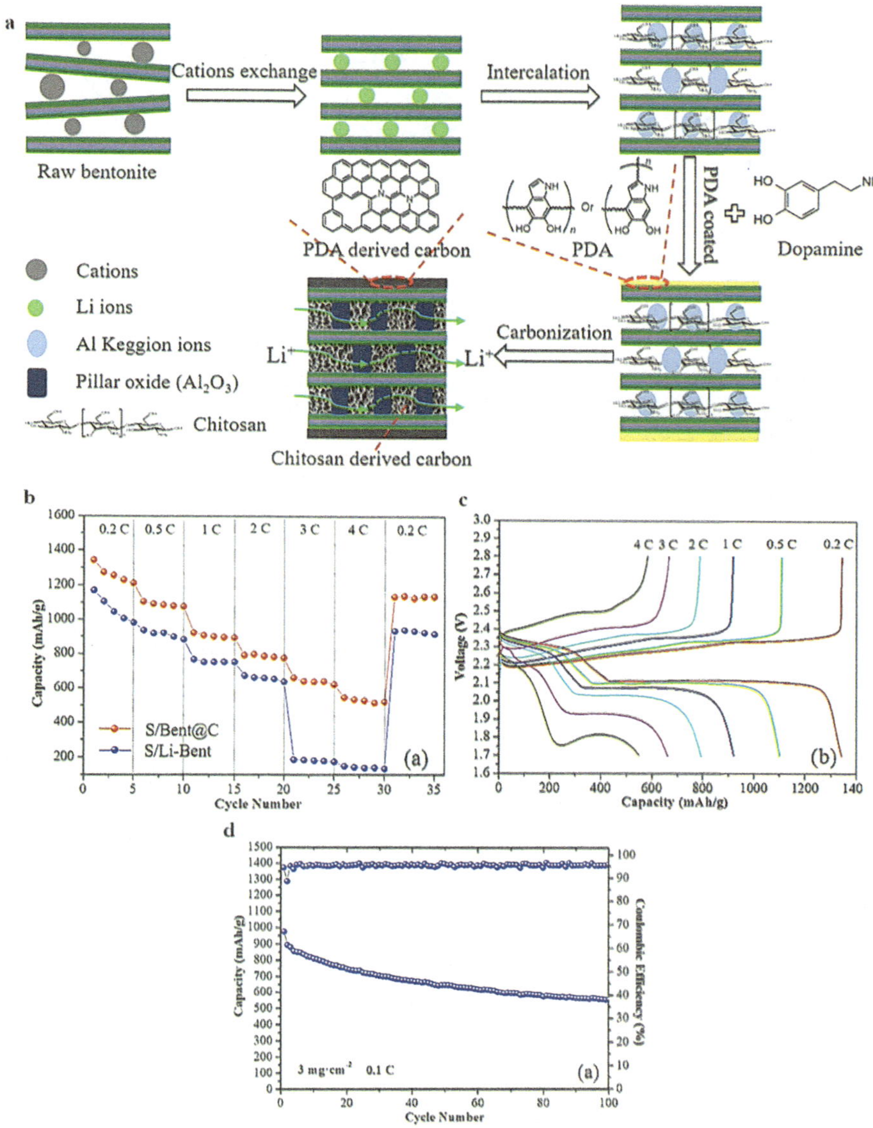

Fig. 8.13 (**a**) Illustration of the host materials for sulfur cathodes: (**a**) Bent@C. (**b**) Rate performances with sulfur loading. (**c**) Galvanostatic charge/discharge profiles and (**d**) cyclic performance (Wu et al. 2021)

information about sulfur matrix in different sizes and forms to obtain high electronic performance, up till now, while sulfur studies have been discussed in possible aspects. Long-acting LSBs might benefit from the use of sulfur in diverse host materials to enhance/modify their activity or catalytic efficacy. To achieve the

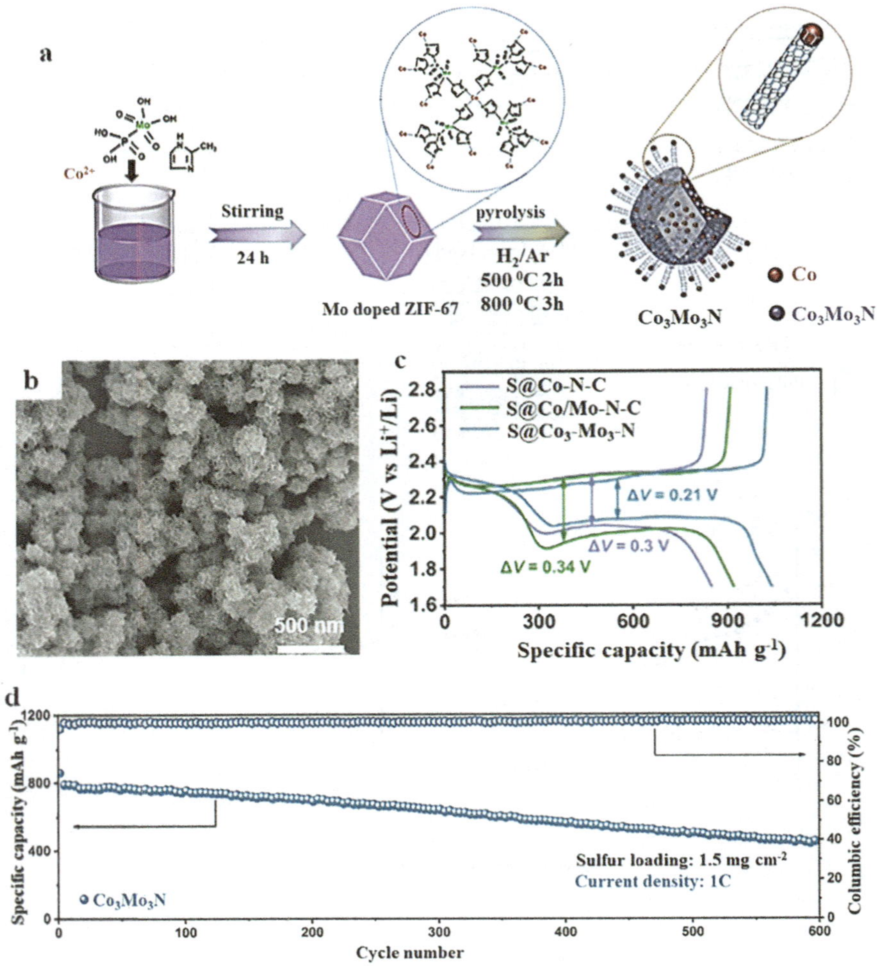

Fig. 8.14 (**a**) Co_3Mo_3N. (**b**) SEM image. (**c**) Charge-discharge profiles. (**d**) Long-term cycling performance of Li-S batteries with S@Co_3Mo_3N at 1 C (Yan et al. 2021)

desired electrochemical performance, sulfur derivatives have been used in the manufacture of different sulfur composites as cathode materials. So, we summarized it into different steps of sulfur that fulfill the upcoming researches.

- The particular existing literature on elemental sulfur, nanosulfur, and polymeric sulfur has been limited to solve fundamental problems regarding LSBs, while their indorsing effects on other constituents were extensively concerned. Due to various sizes, sulfur can be fulfilling the future energy demands in various electrical devices and vehicles by modifying various nanomaterials sulfur-based cathodes to get efficient electrochemical performance. In this area of research, more studies have been required to explore more potent materials.

- Sulfur nanocomposites knowledge is the main focused point for cathode materials and host materials, although their changes and final electronic performance were mainly concerned. Further understanding requires to develop more prevailing cathode for lithium-sulfur batteries.

Conclusion

In these days, Li-S batteries are the most attracted and discussing topic for the research community to develop potential cathode material which can be applicable to use at commercial level. The fuel and other energy crises increased day by day, researchers are able to develop numerous types of the highly efficient and low-cost sulfur cathode material but unable to reduce some limitations which cause capacity decline and low utilization of sulfur due to polysulfides (shuttle effect) during the discharge/charge process. It is concluded that in-depth research toward the morphological design, numerous nanomaterial architectures, and the exposure of other nanomaterial-based sulfur cathode composites should be taken practical engineering conditions into account for commercial perspective of lithium sulfur batteries. Here, we summarized various types of nanomaterial-based sulfur cathode catalysts and elaborated in the easy way for better understandings of the readers. Furthermore, carbon-sulfur or other polymer-sulfur composite cathodes explored which reduced the shuttle effect and solution of this problem by these composites. Additionally, porous material, hollow/core-shell, nanosheets/nanorods, and hierarchical nanomaterials have been elaborated as well. Size effects on Li^+/electron transfer and their morphological impact on electrochemical performance.

References

Bai L, Chao D, Xing P, Tou LJ, Chen Z, Jana A, Shen ZX, Zhao Y (2016) Refined sulfur nanoparticles immobilized in metal–organic polyhedron as stable cathodes for Li–S battery. ACS Appl Mater Interfaces 8(23):14328–14333

Barchasz C, Mesguich F, Dijon J, Leprêtre J-C, Patoux S, Alloin F (2012a) Novel positive electrode architecture for rechargeable lithium/sulfur batteries. J Power Sources 211:19–26

Barchasz C, Leprêtre J-C, Alloin F, Patoux S (2012b) New insights into the limiting parameters of the Li/S rechargeable cell. J Power Sources 199:322–330

Bhargav A, Chang C-H, Fu Y, Manthiram A (2019) Rationally designed high-sulfur-content polymeric cathode material for lithium–sulfur batteries. ACS Appl Mater Interfaces 11(6): 6136–6142

Cai L, Zhang Q, Mwizerwa JP, Wan H, Yang X, Xu X, Yao X (2018) Highly crystalline layered VS2 nanosheets for all-solid-state lithium batteries with enhanced electrochemical performances. ACS Appl Mater Interfaces 10(12):10053–10063

Cheng X-B, Huang J-Q, Zhang Q, Peng H-J, Zhao M-Q, Wei F (2014) Aligned carbon nanotube/sulfur composite cathodes with high sulfur content for lithium–sulfur batteries. Nano Energy 4:65–72

Cheon S, Ko K, Cho J, Kim S, Chin E, Kim H-T (2003) Rechargeable lithium sulfur battery. Electrochem Soc 150:A796–A799

Díez N, Sevilla M, Fuertes AB (2020) N/S-Co-doped porous carbon nanoparticles serving the dual function of sulfur host and separator coating in lithium–sulfur batteries. ACS Appl Energy Mater 3(4):3397–3407

Dirlam PT, Glass RS, Char K, Pyun J (2017) The use of polymers in Li-S batteries: a review. J Polym Sci A Polym Chem 55(10):1635–1668

Elazari R, Salitra G, Talyosef Y, Grinblat J, Scordilis-Kelley C, Xiao A, Affinito J, Aurbach D (2010) Morphological and structural studies of composite sulfur electrodes upon cycling by HRTEM, AFM and Raman spectroscopy. J Electrochem Soc 157(10):A1131

Fan H, Tang Q, Chen X, Fan B, Chen S, Hu A (2016) Dual-confined sulfur nanoparticles encapsulated in hollow TiO2 spheres wrapped with graphene for lithium–sulfur batteries. Chem Asian J 11(20):2911–2917

Feng J, Sun X, Wu C, Peng L, Lin C, Hu S, Yang J, Xie Y (2011) Metallic few-layered VS2 ultrathin nanosheets: high two-dimensional conductivity for in-plane supercapacitors. J Am Chem Soc 133(44):17832–17838

He X, Ren J, Wang L, Pu W, Jiang C, Wan C (2009) Expansion and shrinkage of the sulfur composite electrode in rechargeable lithium batteries. J Power Sources 190(1):154–156

He J, Chen Y, Manthiram A (2018) MOF-derived cobalt sulfide grown on 3D graphene foam as an efficient sulfur host for long-life lithium-sulfur batteries. iScience 4:36–43

Ji Q, Li C, Wang J, Niu J, Gong Y, Zhang Z, Fang Q, Zhang Y, Shi J, Liao L (2017) Metallic vanadium disulfide nanosheets as a platform material for multifunctional electrode applications. Nano Lett 17(8):4908–4916

Lee JS, Jun J, Jang J, Manthiram A (2017) Sulfur-immobilized, activated porous carbon nanotube composite based cathodes for lithium–sulfur batteries. Small 13(12):1602984

Li H, Sun M, Zhang T, Fang Y, Wang G (2014) Improving the performance of PEDOT-PSS coated sulfur@ activated porous graphene composite cathodes for lithium–sulfur batteries. J Mater Chem A 2(43):18345–18352

Li Z, Huang Y, Yuan L, Hao Z, Huang Y (2015) Status and prospects in sulfur–carbon composites as cathode materials for rechargeable lithium–sulfur batteries. Carbon 92:41–63

Li M, Carter R, Douglas A, Oakes L, Pint CL (2017) Sulfur vapor-infiltrated 3D carbon nanotube foam for binder-free high areal capacity lithium–sulfur battery composite cathodes. ACS Nano 11(5):4877–4884

Li X, Zhang Y, Wang S, Liu Y, Ding Y, He G, Zhang N, Yu G (2019a) Hierarchically porous C/Fe3C membranes with fast ion-transporting channels and polysulfide-trapping networks for high-areal-capacity Li–S batteries. Nano Lett 20(1):701–708

Li J, Chen C, Chen Y, Li Z, Xie W, Zhang X, Shao M, Wei M (2019b) Polysulfide confinement and highly efficient conversion on hierarchical mesoporous carbon nanosheets for Li–S batteries. Adv Energy Mater 9(42):1901935

Li Z, Xiao Z, Li P, Meng X, Wang R (2020) Enhanced chemisorption and catalytic effects toward polysulfides by modulating hollow nanoarchitectures for long-life lithium–sulfur batteries. Small 16(4):1906114

Li D, Li H, Zheng S, Gao N, Li S, Liu J, Hou L, Liu J, Miao B, Bai J (2022) CoS2-TiO2@ C Core-Shell fibers as cathode host material for high-performance lithium-sulfur batteries. J Colloid Interface Sci 607:655–661

Liang X, Wen Z, Liu Y, Wang X, Zhang H, Wu M, Huang L (2011) Preparation and characterization of sulfur–polypyrrole composites with controlled morphology as high capacity cathode for lithium batteries. Solid State Ionics 192(1):347–350

Liang X, Zhang M, Kaiser MR, Gao X, Konstantinov K, Tandiono R, Wang Z, Liu H-K, Dou S-X, Wang J (2015) Split-half-tubular polypyrrole@ sulfur@ polypyrrole composite with a novel three-layer-3D structure as cathode for lithium/sulfur batteries. Nano Energy 11:587–599

Liang J, Sun Z-H, Li F, Cheng H-M (2016) Carbon materials for Li-S batteries: functional evolution and performance improvement. Energy Storage Mater 2:76–106

Liu Y, Jiang L, Wang H, Wang H, Jiao W, Chen G, Zhang P, Hui D, Jian X (2019) A brief review for fluorinated carbon: synthesis, properties and applications. Nanotechnol Rev 8(1):573–586

Liu X, He Q, Yuan H, Yan C, Zhao Y, Xu X, Huang J-Q, Chueh Y-L, Zhang Q, Mai L (2020) Interface enhanced well-dispersed Co9S8 nanocrystals as an efficient polysulfide host in lithium–sulfur batteries. J Energy Chem 48:109–115

Liu P, Zhong W, Du W, Guo B, Qi Y, Bao S-J, Xu M (2021) Suppressed shuttling effect of polysulfides using three-dimensional nickel hydroxide polyhedrons for advanced lithium-sulfur batteries. J Colloid Interface Sci 593:89–95

Manthiram A, Fu Y, Chung S-H, Zu C, Su Y-S (2014) Rechargeable lithium–sulfur batteries. Chem Rev 114(23):11751–11787

Manthiram A, Chung SH, Zu C (2015) Lithium–sulfur batteries: progress and prospects. Adv Mater 27(12):1980–2006

Meng L, Yao Y, Liu J, Wang Z, Qian D, Zheng L, Su B-L, Wang H-E (2020) MoSe2 nanosheets as a functional host for lithium-sulfur batteries. J Energy Chem. 47:241–247

Mikhaylik YV, Akridge JR (2004) Polysulfide shuttle study in the Li/S battery system. J Electrochem Soc 151(11):A1969

Oakes L, Carter R, Pint CL (2016) Nanoscale defect engineering of lithium–sulfur battery composite cathodes for improved performance. Nanoscale 8(46):19368–19375

Pang J, Mendes RG, Bachmatiuk A, Zhao L, Ta HQ, Gemming T, Liu H, Liu Z, Rummeli MH (2019) Applications of 2D MXenes in energy conversion and storage systems. Chem Soc Rev 48(1):72–133

Qin J-L, Li B-Q, Huang J-Q, Kong L, Chen X, Peng H-J, Xie J, Liu R, Zhang Q (2019) Graphene-based Fe-coordinated framework porphyrin as an interlayer for lithium–sulfur batteries. Mater Chem Front 3(4):615–619

Su YS, Fu Y, Guo B, Dai S, Manthiram A (2013) Fast, reversible lithium storage with a sulfur/long-chain-polysulfide redox couple. Chemistry Eur J 19(26):8621–8626

Tang H, Yao S, Mi J, Wu X, Hou J, Shen X (2017) Ketjen Black/Mg0. 6Ni0. 4O composite coated separator for lithium-sulfur batteries with enhanced electrochemical performance. Mater Lett 186:127–130

Wang J, Yang J, Xie J, Xu N, Li Y (2002) Sulfur–carbon nano-composite as cathode for rechargeable lithium battery based on gel electrolyte. Electrochem Commun 4(6):499–502

Wang H, Yang Y, Liang Y, Robinson JT, Li Y, Jackson A, Cui Y, Dai H (2011) Graphene-wrapped sulfur particles as a rechargeable lithium–sulfur battery cathode material with high capacity and cycling stability. Nano Lett 11(7):2644–2647

Wang W, Wang X, Tian L, Wang Y, Ye S (2014) In situ sulfur deposition route to obtain sulfur–carbon composite cathodes for lithium–sulfur batteries. J Mater Chem A 2(12):4316–4323

Wang Y, Chen F, Liu Z, Tang Z, Yang Q, Zhao Y, Du S, Chen Q, Zhi C (2019) A highly elastic and reversibly stretchable all-polymer supercapacitor. Angew Chem 131(44):15854–15858

Wang C, Yi Y, Li H, Wu P, Li M, Jiang W, Chen Z, Li H, Zhu W, Dai S (2020) Rapid gas-assisted exfoliation promises V2O5 nanosheets for high performance lithium-sulfur batteries. Nano Energy 67:104253

Wei W, Li J, Wang Q, Liu D, Niu J, Liu P (2020) Hierarchically porous SnO2 nanoparticle-anchored polypyrrole nanotubes as a high-efficient sulfur/polysulfide trap for high-performance lithium–sulfur batteries. ACS Appl Mater Interfaces 12(5):6362–6370

Wild M, O'Neill L, Zhang T, Purkayastha R, Minton G, Marinescu M, Offer GJ (2015) Lithium sulfur batteries, a mechanistic review. Energy Environ Sci 8(12):3477–3494

Wu H, Huan Y, Wang D, Li M, Cheng X, Bai Z, Wu P, Peng W, Zhang R, Ji Z (2019) Hierarchical VS2 nano-flowers as sulfur host for lithium sulfur battery cathodes. J Electrochem Soc 166(2): A188

Wu Z, Yuan L, Han Q, Lan Y, Zhou Y, Jiang X, Ouyang X, Zhu J, Wang X, Fu Y (2020) Phosphorous/oxygen co-doped mesoporous carbon bowls as sulfur host for high performance lithium-sulfur batteries. J Power Sources 450:227658

Wu L, Dai Y, Zeng W, Huang J, Liao B, Pang H (2021) Effective ion pathways and 3D conductive carbon networks in bentonite host enable stable and high-rate lithium–sulfur batteries. Nanotechnol Rev 10(1):20–33

Yan Y, Li H, Cheng C, Yan T, Gao W, Mao J, Dai K, Zhang L (2021) Boosting polysulfide redox conversion of Li-S batteries by one-step-synthesized Co-Mo bimetallic nitride. J Energy Chem 61:336–346

Yang Y, Zheng G, Cui Y (2013) Nanostructured sulfur cathodes. Chem Soc Rev 42(7):3018–3032

Ye Y, Wu F, Liu Y, Zhao T, Qian J, Xing Y, Li W, Huang J, Li L, Huang Q (2017) Toward practical high-energy batteries: a modular-assembled oval-like carbon microstructure for thick sulfur electrodes. Adv Mater 29(48):1700598

Yu M, Ma J, Xie M, Song H, Tian F, Xu S, Zhou Y, Li B, Wu D, Qiu H (2017) Freestanding and sandwich-structured electrode material with high areal mass loading for long-life lithium–sulfur batteries. Adv Energy Mater 7(11):1602347

Yuan J, Wu J, Hardy WJ, Loya P, Lou M, Yang Y, Najmaei S, Jiang M, Qin F, Keyshar K (2015) Facile synthesis of single crystal vanadium disulfide nanosheets by chemical vapor deposition for efficient hydrogen evolution reaction. Adv Mater 27(37):5605–5609

Zeng L-C, Li W-H, Jiang Y, Yu Y (2017) Recent progress in Li–S and Li–Se batteries. Rare Metals 36(5):339–364

Zhang B, Qin X, Li G, Gao X (2010) Enhancement of long stability of sulfur cathode by encapsulating sulfur into micropores of carbon spheres. Energy Environ Sci 3(10):1531–1537

Zhang Z, Kong LL, Liu S, Li GR, Gao XP (2017) A high-efficiency sulfur/carbon composite based on 3D graphene nanosheet@ carbon nanotube matrix as cathode for lithium–sulfur battery. Adv Energy Mater 7(11):1602543

Zhang Y, Tang W, Zhan R, Liu H, Chen H, Yang J, Xu M (2019a) An N-doped porous carbon/MXene composite as a sulfur host for lithium–sulfur batteries. Inorg Chem Front 6(10): 2894–2899

Zhang H, Zou M, Zhao W, Wang Y, Chen Y, Wu Y, Dai L, Cao A (2019b) Highly dispersed catalytic Co3S4 among a hierarchical carbon nanostructure for high-rate and long-life lithium–sulfur batteries. ACS Nano 13(4):3982–3991

Zheng W, Liu Y, Hu X, Zhang C (2006) Novel nanosized adsorbing sulfur composite cathode materials for the advanced secondary lithium batteries. Electrochim Acta 51(7):1330–1335

Zheng G, Yang Y, Cha JJ, Hong SS, Cui Y (2011) Hollow carbon nanofiber-encapsulated sulfur cathodes for high specific capacity rechargeable lithium batteries. Nano Lett 11(10):4462–4467

Zhou J, Wang L, Yang M, Wu J, Chen F, Huang W, Han N, Ye H, Zhao F, Li Y (2017) Hierarchical VS2 nanosheet assemblies: a universal host material for the reversible storage of alkali metal ions. Adv Mater 29(35):1702061

Zhu X, Zhao W, Song Y, Li Q, Ding F, Sun J, Zhang L, Liu Z (2018) In situ assembly of 2D conductive vanadium disulfide with graphene as a high-sulfur-loading host for lithium–sulfur batteries. Adv Energy Mater 8(20):1800201

Chapter 9
Design and Optimization of Nanostructures for Sulphur Cathodes

Nimra Arshad, Adnan Khalil, Saima Nazir, M. B. Tahir, and M. Sagir

Abstract Recently, rechargeable lithium-sulphur has gained remarkable attention because of its low cost and specific energy. For different applications like grid-level energy storage, portable electronics, and electric vehicles, they are auspicious candidates. But, the main technical issues are low power capability and poor cycle life. To solve these problems, various nanostructured sulphur cathodes have designed, as they give greater trapping of soluble polysulfide, better resistance to pulverization, and an accelerated reaction kinetics. We will study the designs of nanostructured sulphur cathodes and their mechanism behind their operation.

Keywords Rechargeable lithium-sulphur · Nanostructures · Sulphur cathodes

---

N. Arshad · A. Khalil (✉)
Institute of Physics, Khwaja Fareed University of Engineering and Information Technology, Rahim Yar Khan, Pakistan
e-mail: adnan.khalil@kfueit.edu.pk

S. Nazir
Nawaz Sharif medical College, University of Gujrat, Gujrat, Pakistan

University of Health Sciences, Lahore, Pakistan
e-mail: saimanazir3344@kfueit.edu.pk

M. B. Tahir
Institute of Chemical and Environmental Engineering, Khwaja Fareed University of Engineering and Information Technology, Rahim Yar Khan, Punjab, Pakistan
e-mail: m.bilaltahir@kfueit.edu.pk

M. Sagir
Nawaz Sharif medical College, University of Gujrat, Gujrat, Pakistan
e-mail: dr.msagir@kfueit.edu.pk

M. S. Tahir et al. (eds.), *Lithium-Sulfur Batteries: Key Parameters, Recent Advances, Challenges and Applications*, Springer Tracts in Electrical and Electronics Engineering, https://doi.org/10.1007/978-981-99-2796-8_9

Introduction

High-performance rechargeable batteries are expected to fulfil current environmental and energy demands. Previously, portable electronic market was dominated by lithium ion (Li-ion) batteries, but, due to low energy density and high cost, these batteries were unable to fulfil their demand. Moreover, currently, the energy density limit of 300–350 W.h/Kg put a stopper on the usage of Li-ion batteries with intercalation mechanism. Cell voltage and electrode capacity are helpful in determining the energy density limit (Manthiram et al. 2014). The demand for high energy density paves the way for urgent formation of alternative batteries which are characterized with wide electrochemical stability window and high capacity or both (Pan et al. 2018). Among the emerging battery technology, lithium-sulphur (Li-S) batteries have promising and attractive characterization. These batteries have 1672 mAh/g specific capacity and up to 2600 W h/Kg of theoretical energy density, which is five times greater than Li-ion battery's theoretical energy density (Wang et al. 2015a).

In 1962, Herbert and Ulam first time used a cathode material made of sulphur in storage batteries and electric dry cell. Later, in 1967, a high-temperature Li-S system was developed by Argonne national laboratory. Until E. Peled's group reported the electrochemical characterization of sulphur in organic electrolytes at room temperature, the safety problems of high-temperature system remained challenging and of limited applications (Cao et al. 2020). After that, a large number of researches were focused on the development of safe Li-S system with long life span and high energy density (cathode material). However, the isolation nature and soluble compounds do not allow sulphur to be used as cathode individually, which demands a second element to act as a sulphur host and a conducting matrix. Hence, designing of microstructure with numerous materials, such as inorganic compounds, carbon material and their hybrids, and polymers, imparts diversification to sulphur cathodes (Chen et al. 2015a). The designing material should fulfill some significant requirements, like high electrical conductivity of electrode materials, highly stable incorporation with sulphur, and loose frame to accommodate volume expansion of cathode during charge-discharge mechanism (Gupta et al. 2019).

The usual Li-S battery design is composed of a metallic lithium anode separated by an organic electrolyte, a positively charged sulphur electrode, binder along with carbon additives (Wang et al. 2018). Although the Li-S batteries have significant advantages in terms of cost and energy density, there are several commercialization related. The primary challenge is concerned with change in volume of sulphur throughout charging and discharging process. The density of sulphur is 2.03 g/ cm^3, while density of Li_2S is 1.66 g/cm^3; thus Li_2S is lighter than sulphur (Yoo et al. 2016). Consequently, there is a volume expansion of 80%, when sulphur is completely converted to Li_2S. This change in volume results in crushing of active moieties; thus, a rapid capacity decay occurs. Ionic and electronic insulating property of Li_2S is the secondary challenging attribute of Li-sulphur batteries. The electronic resistivity of Li_2S is greater than 1014 ohm.cm, and the Li^+ diffusivity in Li_2S is a low value of 10^{-15} cm^2/s (Wang et al. 2019a). When the whole electrode is covered by the thin layer of Li_2S, there is a hindrance in lithiation and rapid decrease in voltage. As a result, full transformation of sulphur to Li_2S is challenging, and a discharge capacity of less than 80% of the theoretical limit is reported. The tertiary challenge of making Li-S batteries is concerned with solubility of polysulfide in the electrolyte (Li et al. 2018). At the lithium anode surface, these soluble species can be reduced, making anode inert by coating it and causes both an increase in hindrance and material loss. In addition, the morphology of cathode is altered in each cycle by precipitation and dissolution process, which results in internal straining of electrode and degradation of life cycle (Wang et al. 2015b). Shuttle effect is also the result of dissolution of polysulfide, in which diffusion of LCPs (long-chained polysulfides) to lithium anode surface results in their reduction to SCPs (short-chained polysulfides). The SCPs can be converted to LCPs on their return to cathode by oxidation. It is a continuous process which creates an inside "shuttle" episodes. Consequently, it lowers the utilization of active moiety during the discharging and significantly decreases the coulomb efficiency. Apart from the challenges concerned with sulphur cathode, the anode of lithium metal causes energy density penalties and safety concerns due to excess usage of lithium. These challenges demand for vigilant future examination (Fotouhi et al. 2017).

Rational design of the electrode structure is important to overcome these challenges. Some studies believe that ideal sulphur electrode structure must have some attributes, such as enough area to facilitate volume expansion of sulphur, to prevent crushing a limited range of active moiety; small transport pathways for both Li-ion and electrons to attain high power capability and high capacity; a huge conductive space for deposition of insulating Li_2S and Li_2S_2, for the purpose of protecting morphology of electrode; sufficient trapping of polysulfides through chemical and physical ways; and suitable additives in electrolyte to coat the surface of lithium to curtail the shuttle effect. Various attributes demand for a self-contradictory structural design. For instance, a huge space for Li_2S and Li_2S_2 deposition results in an open structure and causes ineffective polysulfides trapping (Ma et al. 2015).

The first step to show the capability of Li-S batteries, the rational design of composite cathode with long life span and high utilization of sulphur, is required. With the growth in developments of nanomaterials, the addition of sulphur to these nanomaterials imparts them the 3D interlinked electron pathways and interconnected ion-diffusion channels. Conductive nanomaterials have large surface area, adjustable porous structure, small size, and a distinctive chemical bonding with absorbents, which provide the whole complex adequate interfaces for multi-electron conversion reaction by meeting of Li-ions and electron with active sulphur (Lyu et al. 2015). On this account, the idea of optimized design of conductive nano-structured sulphur cathodes is widely accepted. Additionally, the formation of complex of sulphur cathode is divided into three families. The first family is comprised of a conductive host material for sulphur, such as nano-structured carbon in the form of micro-porous carbon, meso-porous carbon, micro-meso-macro- porous carbon, carbon spheres, carbon nano-tubes, carbon fibres, graphene, or their hybrids. The carbon materials are very potent in providing a physical barrier and hosting sulphur in C-S complex cathode, due to their 3D interconnected nature and remarkable electron pathways (Zhao et al. 2020). It notably improves the utilization of sulphur and renders C-S cathodes for practical Li-S batteries. However, there are limited sites for conjugate nonpolar carbon planes to strongly bind with polar molecules (lithium disulfides and lithium polysulfides). The incorporation of polar sites on the carbon planes was proposed to give attachment sites to polysulfide intermediates for binding chemically. Doped carbon with effective electron pathways and adjustable polar sites represents the second family of sulphur cathode complex. For example, graphene oxide, nano-structured carbon material (graphene, graphene oxide, carbon nano-tubes, and so on) doped with heteroatoms, and, additionally, conductive polymers were broadly investigated for Li-S batteries (Liu et al. 2017a). The conductive framework was contributed by carbon chains, and the chemical attachment sites were rendered by the functional groups or doped sites on the conductive polymers. Furthermore, the nano-structured polar inorganic compounds like sulphides, transitional metal oxides, and carbides emerged as polar host elements towards lithium (poly) sulfides and increased the tap density of electrodes along with regulating the binding energy with (poly) sulfides (Ma et al. 2016). The third family is represented by the framework which is made of nano-structured inorganic compounds. These compounds (TiC, Ti_2C, CoS_2, and Ti_4O_7) have a much greater adsorption ability than (poly) sulfides and impart the Li-S battery with long life span and high utilization of sulphur (Maihom et al. 2018).

It is proved from various researches that high performance is provided by nano-sized sulphur particles. Hence, an optimized design approach will give solution for Li-S cathode with long life cycle, high energy density, and remarkable practical performance (Zhang et al. 2015a).

Reaction Mechanism

For current energy storage application like EV (electronic vehicle) and SES (stationary energy storage), on conversion mechanism, the lithium-sulphur batteries are likely to be a good alternative because of their energy density which potentially is higher compared with presently available lithium-sulphur batteries. Usually, lithium-sulphur batteries are made up of lithium anode (metal), a sulphur cathode (composite), an electrolyte, and a porous polymer-based (PPB) separator (Chen et al. 2015a). While carbonate-based electrolytes (CBE) are generally used by lithium-ion batteries. Although, carbonate based electrolytes are mostly used by lithium-ion battries. In media of polar aprotic, high solubility demonstrated by polysulfides like generally used combination 1,2- DME (dimethoxyethane) and 1,3-DOL(dioxolane) supports the solid S_8 having electrochemical reaction and solid Li_2S with particularly minimum electronic conductivity. With two Li-ion by reduction process (two electron), the octa-cyclo-sulphur reacts to form an intermediate of polysulfide, and at the end (discharge), it is slowly converted into Li_2S (dilithium sulfide). In the reverse reaction, at a specific voltage with Li_2S oxidation, the electrons are extracted into S_8 and Li (lithium). Voltage profile of charge and discharge profiles contains two voltage level happening for charging 2.3 V/2.4 V, and for discharge at about 2.3 V/2.1 V, from polarization, the potential difference for charge/discharge emerges.

The solid sulphur (S_8) for dissolution region (upper voltage plateau) slowly decreases to species of polysulfide (soluble), and in the last, it converted into Li_2S_4, wherever it has specific total capacity (quarter) (416 mAh g^{-1}). For precipitation region (lower-voltage plateau), polysulfide (soluble) intermediates in the end decrease to Li_2S (solid), wherever it has specific total capacity (three quarters) (1255 mAh g^{-1}). The whole reaction is written as follows:

$$S_8 + 16Li \leftrightarrow 8Li_2S$$

This reaction explains that, with lithium ions and electrons, the polysulfide (soluble) reacts chemically and also electrochemically (by a reaction disproportionation) due to characteristics of diffusion and with polysulfide (soluble) (Cheng et al. 2017).

Sulphur Cathode

When sulphur cathode has observed, some challenges have been recognized. The main and most clear challenge is nature of sulphur which is electrically insulating (~10^{-30} S cm^{-1}) and intermediary products which are insoluble like Li_2S (~10^{-13} S cm^{-1}) (Zhang et al. 2015a). During reaction process, their menial results of electronic conductivity are due to meagre use of active materials and confine the electrons to transfer, and it causes slow kinetics moderately. One more challenge which is due to the diffusion of sulphur species (soluble) is the well-known shuttle effects.

Few years ago, to deal with abovementioned problems, the confinement of sulphur in well-made design of NPP (non-polar porous) host was introduced which has been verified to be useful strategy to alleviate, somewhat, the diffusion of intermediates of polysulfide (soluble). To further improve the interaction between substrates and sulphur species, the idea of chemical adsorption has been frequently presented to better put off the shuttle effect (Liang et al. 2015). With PP (polar polysulfides), either on binders or hosts, the concept of polar sites can go ahead to strong bond (chemical); hence, the well subdue their dissolution (irreversible) into electrolyte (Wang et al. 2019b).

Separator

In liquid battery systems, a necessary element is separator which acts as path to let ion transport and as a barrier to restrict electron (internal) transfer. In lithium-sulphur

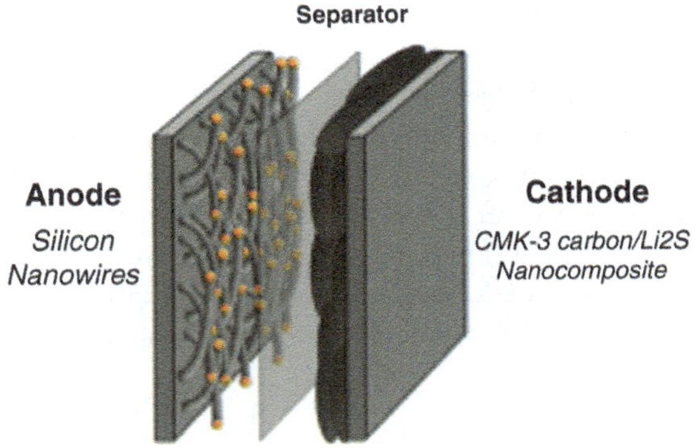

Fig. 9.1 Carbon hosts with outer micro-pores and inner macro/mesopores (Liu et al. 2016)

batteries, the more frequently separators which utilized are PP (polypropylene)/PE (polypropylene) multilayered composite, PP (polypropylene), and PE (porous polyethylene) (Huang et al. 2020). The advantages of separators based on non-toxic TPP (thermoplastic polyolefin) have excellent chemical stability, excellent performance for electrical insulation, good resistance of low temperature (the minimum service temp. can be from 100 °C to approx. −70 °C) (Maihom et al. 2018). Moreover, during the manufacturing and designing of sulphur cathodes and to get better performance (electrochemical) of lithium-sulphur batteries, the effectual and feasible strategy is functionalized separators. To manufacture new separators, mostly the two procedures have been used, firstly, coating of surface and secondly fictionalization (Lee et al. 2017). For procedure of surface coating, between separator and sulphur cathode, the interlayer coating can increase separator's ion transport and, by the use of conversion of electro-catalytic and chemical and physical confinement, adjust the PS (polysulfide) ions (Ren et al. 2019) (Fig. 9.1).

Electrolyte

For long-term procedure of lithium-sulphur batteries, it is demonstrated that performance of lithium anode (metal) and sulphur cathode in traditional electrolytes (liquid organic) is still adequate. To improve lithium anode (reversibility) and to enhance utilization of sulphur, the current analysis on formulation of lithium salts and solvents (electrolyte components) has gained huge progress (Zhang et al. 2015b). To regulate the stability of lithium anode (metal) and solubility of polysulfide, the choice of solvents is very important. Still, now, in lithium-sulphur batteries, the different solvents like carbonates, ILs, sulfones, and ethers have been explored. In lithium-ion batteries, in spite of fact, electrolytes which are carbonate-based have effectively applied, and majority are mismatched with cathodes (S_8) and lithium anodes (metal). For advance development in lithium sulphur batteries, between carbonates and lithium metals/polysulfides (soluble), the side (severe) reactions greatly impeded (Shi et al. 2017).

Nano-structured Sulphur Cathodes

On the bases of structure and composition of electrodes, the nano-structured sulphur cathode's fabrication can be classified into various categories: (1) porous oxide additives, (2) conductive polymer–sulphur composites, (3) nanoporous carbon–sulphur composites, (4) nano-structured Li_2S cathodes, (5) graphene–sulphur composites, and (6) one-dimensional (1D) carbon–sulphur composites. Some of composites used carbon as a matrix to trap PS (polysulfides) and as a conductive pathway. Still, various structures of carbon have possessed their limitations and advantages (Chen et al. 2015b). Rather than carbon, soft polymers used by composite of polymer–sulphur hold more strain and more flexible. Recently, in sulphur cathodes, to enhance capacity retention, the porous oxides have been established, though the mechanisms understanding has been involved (Liu et al. 2017b). To explore these, the lithium-sulphur cathode is raising area of research. To avoid efficiency (low coulombic) and saved concerns of lithium anode (metal), it can be combined with anodes of free lithium (Li). Hence, for improving overall performance of lithium sulphur batteries and fabrication of electrolytes, the binders and electrolyte additives play an essential role (Yang et al. 2013).

Carbon–Sulphur Nano-composites

In broad range of applications, the materials of nano-structured carbon have gained great attention due to dimensional structures (flexible) (0D to 3D), high chemical inertness, low cost, intriguing textures (e.g. large pore volume, pore size distribution, high specific surface area), light weight, easy procedure, non-toxicity and natural abundance. To solve the sulphur cathodes problems, these characteristics are favourable (Jin et al. 2018). On composite of carbon-sulphur cathodes, the increasing research is due to the early work using carbon (meso-porous) which has showed important improvement in utilization of sulphur performance (cycling) of lithium-sulphur cells. The carbon matrices in a perfect composite arrangement act as (1) accessibility of ions (electrolyte) allowed by porous architecture to the inner materials (active); (2) for loading (high mass) of sulphur with exclusive disciplining the energy density (whole) of a fuel cell; (3) to stabilize sulphur, a structure of solid without outlets (large), additionally, adjust volume change and trap PS (polysulfides); and (4) from sulphur (insulating), a conductor (straightforward) made for electrons transporting. The characteristics of nanoscale carbon structure and all of above-mentioned functions have play important role for construction of high-performance lithium-sulphur batteries (Wang et al. 2015c) (Fig. 9.2).

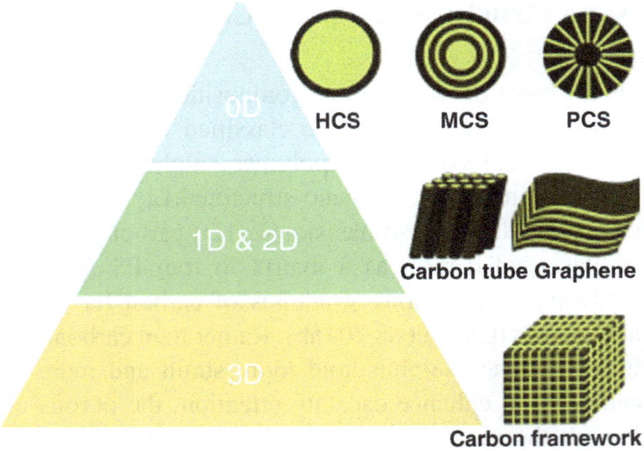

Fig. 9.2 Carbon hosts for sulphur cathodes (Wang et al. 2019c)

Nano-porous Carbon–Sulphur Composite

A combination of porous carbon and sulphur is a nano-porous carbon–sulphur composite where inside the pores of carbon the sulphur exists mainly. It (porous carbon) helps to trap polysulfides (dissolved) and enhance the electrode composite's electronic conductivity (Jung et al. 2014). Inside the structure (strain), volume expansion is accommodated by nano-pores, if in the design an empty pore space is involved. Wang et al. performed research in this area, where, with the pore size of about 2.5 nm the carbon (activated) played the role of conductive matrix. A 400 $mAh.g^{-1}$ capacity (reversible) was attained (Xu et al. 2015).

In this method to trap polysulfides, the uses of ordered mesoporous (CMK-3) are due to the quantum leap. The carbon CMK-3 contains a group of 6.5 nm-thick rods of hollow carbon detached by unfilled 3–4 nm-wide (void) channel (Zhang et al. 2018). The little pore size enables fast electronic and ionic transport along with a reliable trap PS (polysulfides) by heating, at 155 °C, to both the CMK-3 carbon and sulphur, as the sulphur has low viscosity; thus, it melts. The amount of sulphur in the composite can be optimized to leave enough space for volume expansion. In the work of Ji et al., the weight ratio of sulphur to C K-3 carbon is 7:3. The mass of sulphur in the final electrode is B1.1 mg cm^2, which accounts for 59% of the total weight of the electrode. This heating method has been widely adopted to incorporate sulphur into various carbon or polymer structures (Werner et al. 2015).

The sulphur cathode life cycle improved frequently by the matrix of porous carbon. Carbon spheres of bimodal porous (close-packed structure) synthesized to increase the capability (power) at contents of high sulphur. In this combination, a closed packed structure (interconnected), carbon (meso-porous) nano-sphere of diameter 300 ± 40 nm, formed (Bu et al. 2016). In the nano-sphere, the pores of two types are of diameter 3 and 6 nm. The infiltration of electrolyte facilitates large space between nano-spheres, for sulphur storage, as reservoirs the small pores act, and leads at high power capability (Choudhury et al. 2017). To enhance the

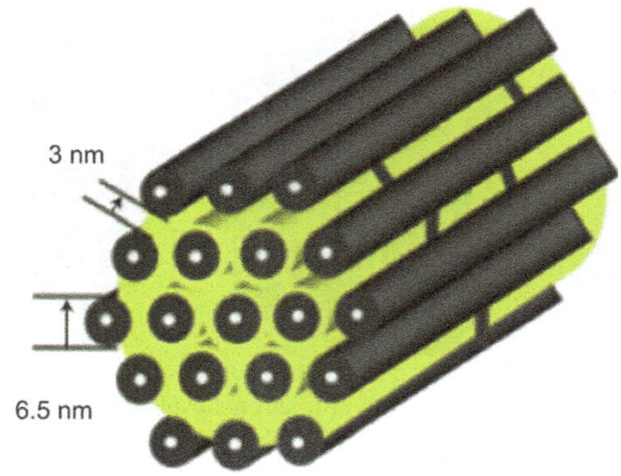

Fig. 9.3 Associated pore structure of mesoporous carbon (CMK-3) confined sulphur (yellow) within it (Zeng et al. 2022)

performance of electrode, from surface to remove the sulphur of carbon nanospheres. The amount of sulphur decreases by removal from surface and into electrolyte it steadily dissolves. Inside pores of carbon, these conclusions determine the importance of sulphur trapping (Choudhury et al. 2015). To improve the production of sulphur cathode, the carbon-sulphur (nanoporous) composites show a method which is mentioned above. To understand improved performance, the important features should be considered carefully: (1) trap PS (polysulfides) by closed structure, pore size optimized and pulverization can be minimized; (2) inside pores of carbon, the sulphur amount maximized; and (3) for practical applications, the pore volume (large) to enhance sulphur content (Gong et al. 2016) (Fig. 9.3).

Hollow Porous Carbon as Sulphur Host

Rational design of C-S cathode has been fabricated with specific hollow porous carbon structure to attain high energy density Li-S batteries. The C-S host not only shifts maximum quantity of sulphur but also effectively hinders the shuttle effect of polysulfides. Sulphur is soaked in nano-cubes of individually associated multi-chambered micro-meso-porous carbon, which also behaves as physical restraint of manifolds for soluble polysulfides (Zhao et al. 2020). The successful performance is attributed to adequate space that shifts maximum quantity of sulphur in mesopores, dual confinement strategy by multifold micro-mesoporous barriers, and cushion effect for large volume. The dissolution of polysulfides is expected to be halted by porous shell of hollow carbon nano-structures which have large internal space. However, various studies reported that maximum amount of sulphur is retained in the porous shell instead of interior space because of the absence of potential adsorption sites. Recently, Sun et al. proposed solution to this problem by designing an integrated hollow core-shell carbon sphere (Xiao et al. 2015). The distinctive carbon nano-framework was made of three interconnected segments such as an internal cavity covered by a mesoporous shell that was further preserved by microporous shell, which acts as storage for encapsulation of sulphur. As a result, a

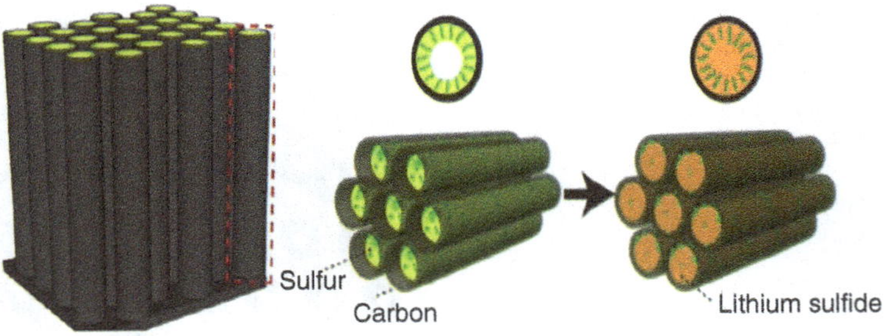

Fig. 9.4 Hollow carbon fiber encapsulated sulphur (Wei et al. 2017)

significant improvement was observed in the performance and showed an efficient utilization of sulphur. It exhibited a long life span of more than 200 cycles and a high capacity of 960 mAh/g. The superb performance was credited to peculiar hollow in hollow nano-structure (Tan et al. 2018).

1-D hollow carbon structure was another technique that involves treatment of sulphur-hollow carbon tubes complex at high temperature and caused an insertion of amorphous carbon in the walls and sulphur into graphitic clusters (Yang et al. 2013). A stable specific capacity up to 700 mAh/g for about 80 cycles was attained by the sulphur electrode through this technique. A weight content of 40% of sulphur is involved in this process (Liu et al. 2016) (Fig. 9.4).

Porous Carbon Nanofibers

Porous carbon nano-fibers are used in sulphur electrodes to decrease the exposure of sulphur. In this procedure, carbon nano-fibers of activated carbon electrospun or polymeric precursor and cloth (Lin et al. 2016) are soaked in sulphur. Narrower pore size distribution makes commercial activated carbon fibres more attractive, and most of them have less than 2 nm of micro-porous structure. A thorough contact between the conductive matrix and sulphur is facilitated by the small pore size of carbon fibre. Meanwhile the polysulfide dissolution in electrolyte matrix is reduced along with decrease in the sulphur exposure area (Li et al. 2019). The initial discharge capacity of activated carbon nano-fiber cloth-sulphur complexes is about 1000 mAh/g, and decay capacity over 80 cycles was 20%. Moreover, the discharge capacity of activated carbon nanofiber-sulphur cathode increased because of progressive activation of sulphur, during the initial cycling in the micropores (Zhou et al. 2019).

Graphene–Sulphur Composite

To date, multiple graphene-sulphur complexes have been formed as cathode in Li-S batteries. This kind of cathodes illustrates enhanced electrochemical properties by chemical absorption of dissolved polysulfides and physical trapping (Yan et al. 2019).

Many researchers used graphene oxide (GO) in the initial development of sulphur cathode. Reported by Ji et al. during functionalization of oxygen on the basal plane of graphene sheets, the soluble polysulfides showed a polar interaction with GO. Consequently, to immobilize soluble polysulfides, this functionalization of GO is approached (Shi et al. 2017).

Another method of preparing graphene-wrapped sulphur particles was proposed by Wang et al. by coating polyethylene glycol-sulphur (PEG-S) with moderately oxidized graphene oxide sheets. Trapping polysulfides and increasing electrical conductivity are the attribute of graphene coating, but coating with PEG gives another remarkable advantage of cushion property which accommodates change in volume. Thus, complexes without graphene wrapping or PEG coating are less advantageous than those complexes with graphene wrapping as they provide greater cyclic stability (Han et al. 2018).

Xu et al. developed a core shell design of graphene-encapsulated sulphur (GES) compounds through a solution-chemical reaction and deposition method. The GES compounds contain 83.3 wt% of sulphur that gives 915 mAh/g at the rate of 0.75 C and during 160 cycles, 86% of cycle retention (Chen et al. 2015b) (Fig. 9.5).

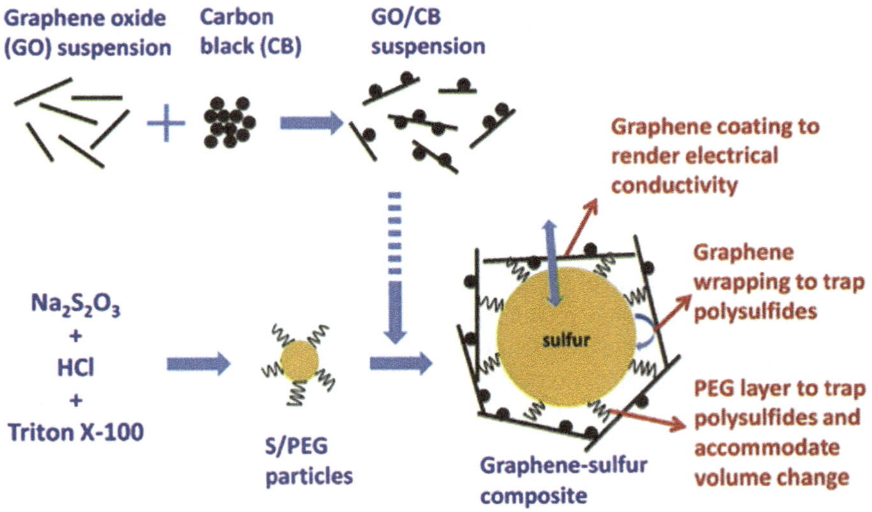

Fig. 9.5 Method for graphene-sulphur complex, along with schematic diagram (Zyga 2011)

One-Dimensional (1-D) Carbon Nano-structure/Sulphur Electrodes

The traditional way of synthesizing sulphur electrode comprised of some conductive matrix like conductive polymer or carbon black is mixed with elemental sulphur, and then this mixture is heated to obtain a definite design. In such kind of complexes, some micro-sized sulphur moieties are precipitated and, without any coating in pores, retained in the electrode. During charge and discharge process, the crucial structural changes of sulphur occur which results in loss of electrical contact between the current collector and the active material, and it becomes the initial step in capacity decay (Ma et al. 2015). There is a solid-liquid-solid interaction between lithium and sulphur which also involves the reaction of lithium ions with dissolved polysulfides in the electrolyte. The product of this reaction is lithium sulfide, which is then settled on the conductive matrix. As the cycle continues, there is a decrease in active ingredient utilization because of genesis of porous electrode and nonuniform settling of lithium sulfide on it. For dealing with all these challenges, one-dimensional (1-D) carbon nano-structure is a suitable approach (Yang et al. 2020). A progressive electrical connection with the active materials is created in the conductive matrix of 1-D carbon nano-structure. Hence, in the formation of sulphur cathodes, the novel frameworks of 1-D carbon nano-structures can play a significant role as they tackle the above specified problems (Jiang et al. 2020).

Sulphur-carbon nanotube (CNT) complex is another application of 1-D carbon nano-structure and improves in retention capacity and utilization of sulphur. Its benefits are obtained when a core-shell structure is formed by coating the sulphur on to CNT. Although there are various opinions about the electrochemical functioning, sulphur-CNT electrodes can achieve about 1000 mAh/g of specific capacity, and there is a betterment in retention capacity in comparison to simple mixing of carbon and sulphur particles. Alterations in the surface of CNT give a beneficial impact of increased retention capacity (Xu et al. 2018).

Polymer–Sulphur Nanocomposites

In the protective matrix, to summarize sulphur (confined), the most fabricate choice is nano-composite system of sulphur polymer. At appropriate low temperature below 100 °C and low sulphur's melting point 118 °C, different polymer methods are useful to evade any damage (thermal) to sulphur (Kong et al. 2016). Generally, during the process of lithiation/delithiation in the electrode, to accommodate sulphur's volume change, the polymer matrix is due to its structure (soft) is favourable. Though because of the sulphur and polymer bad ionic and electronic conductivity, the operation of sulphur electrode containing composite of sulphur-polymer is very low. Hence, to reduce sulphur's bad electronic conductivity, the conducting polymers (electrically) like PPY (polypyrrole), PANI (polyaniline), PEDOT (polyethylene dioxythiophene), PTH (polythiophene), and PDm (polydopamine) have been used for composite of sulphur composite (Wang et al. 2016).

Nano-structured Sulfides and Other Compounds

For lithium-sulphur cells on earth, many metal oxides amply existing with more than one S^{2-} ions attached to different cations have been researched. As an active material, the most demonstrated sulfides are Li_2S for progress of electrode of Li_2S; important efforts have been made; and well-organized utilization of Li_2S has recently reported. After early process of activation, as sulphur electrode (conventional), to the electrode of Li_2S behaves the same and to that electrode, the material changes approach relatively same. Now, as sulphur-catalytic additive, for use in sulphur electrode, different transition metal sulfides have been studied. Firstly, in 1970 to 1980s as storage material of active lithium, the metal chalcogenides were studied due to its graphite-like structure (attractive layered) which takes in Li-ion through intercalation reactions with no structure change (Yang et al. 2017).

For sulphur species, some metal (transition) oxides have good chemical empathy, which helps to increase sulphur electrode's electro-chemical performance. Furthermore, to requite the bad conductivity (electronic) of sulphur, some metal sulfides having good electronic conductivity are extensively used. The Co-S, Fe-S, Sn-S, Mo-S, WS, and Ni-S are metal sulphides that have catalytic properties in lithium-sulphur cells. For the sulphur electrode, the CoS_2, Co_9S_8, Co_3S_4, and CoS have been considered as well-designed additive (Zhu et al. 2015). In summary, to gain specific (high) energy lithium-sulphur cells, different sulfides (metal) have been explored as Li-PS absorbers or lithium (Li) active materials. For lithium-sulphur cells, the most likely active material is Li_2S; meanwhile as the sulphur electrode reacts the same if during early stage the Li_2S adequately activated. In the sulphur electrode, as well-designed additive, some more metal sulfides are considered due to good electronic conductivity and chemical affinity (Yang et al. 2018).

Conclusion

From many decades till date, to increase the performance of sulphur cathodes, different nanostructures have been recognized, like various nano complexes of carbon-sulphur. For persistent stability and high loading of sulphur cathodes, the most efficient design approach is carbon hosts with inner macro/mesopores and outer micro-pores.

The rational and creative designs of Li-S batteries are helpful in solving the challenges of volumetric expansion, poor electronic and ionic movement, and dissolution of polysulfide, hence significantly increasing the life cycle, power capability, and capacity of sulphur cathodes. This evolution illustrates a successful gate towards innovations in lithium-sulphur batteries. It is needed to explore further about the alteration and improvement of microstructure, such as macro-, meso-, and micro-pores. For improving the tap density and reducing the amount of inert material, conductive ingredients other than carbon are needed to develop to facilitate active sulphur. Hence, it is the need of time to put efforts in the development of lithium-sulphur batteries with greater sulphur content and high mass loading in the presence of weak electrolyte. Moreover, it will boost technology of lithium sulphur batteries.

References

Bu Y et al (2016) Sandwich-type porous carbon/sulfur/polyaniline composite as cathode material for high-performance lithium–sulfur batteries. RSC Adv 6(106):104591–104596

Cao Y et al (2020) Lithiation of covalent organic framework nanosheets facilitating lithium-ion transport in lithium-sulfur batteries. Energy Storage Mater 29:207–215

Chen H et al (2015a) Monodispersed sulfur nanoparticles for lithium–sulfur batteries with theoretical performance. Nano Lett 15(1):798–802

Chen S et al (2015b) Multi-chambered micro/mesoporous carbon nanocubes as new polysulfides reserviors for lithium–sulfur batteries with long cycle life. Nano Energy 16:268–280

Cheng X-B, Huang J-Q, Zhang Q (2017) Review—Li metal anode in working lithium-sulfur batteries. J Electrochem Soc 165(1):A6058–A6072

Choudhury S et al (2015) Nanoporous cathodes for high-energy Li–S batteries from gyroid block copolymer templates. ACS Nano 9(6):6147–6157

Choudhury S et al (2017) Electroless formation of hybrid lithium anodes for fast interfacial ion transport. Angew Chem Int Ed 56(42):13070–13077

Fotouhi A et al (2017) Lithium-sulfur battery technology readiness and applications—a review. Energies 10(12):1937

Gong Z et al (2016) A hierarchical micro/mesoporous carbon fiber/sulfur composite for high-performance lithium–sulfur batteries. RSC Adv 6(44):37443–37451

Gupta A, Bhargav A, Manthiram A (2019) Highly solvating electrolytes for lithium–sulfur batteries. Adv Energy Mater 9(6):1803096

Han D-D et al (2018) Lithiophilic gel polymer electrolyte to stabilize the lithium anode for a quasi-solid-state lithium–sulfur battery. J Mater Chem A 6(38):18627–18634

Huang L et al (2020) Electrode design for lithium–sulfur batteries: problems and solutions. Adv Funct Mater 30(22):1910375

Jiang J-H et al (2020) P(VDF-HFP)-poly(sulfur-1,3-diisopropenylbenzene) functional polymer electrolyte for lithium–sulfur batteries. J Energy Chem 46:114–122

Jin C et al (2018) Metal oxide nanoparticles induced step-edge nucleation of stable Li metal anode working under an ultrahigh current density of 15mAcm−2. Nano Energy 45:203–209

Jung DS et al (2014) Hierarchical porous carbon by ultrasonic spray pyrolysis yields stable cycling in lithium–sulfur battery. Nano Lett 14(8):4418–4425

Kong W et al (2016) Binder-free polymer encapsulated sulfur–carbon nanotube composite cathodes for high performance lithium batteries. Carbon 96:1053–1059

Lee Y-G et al (2017) Dendrite-free lithium deposition for lithium metal anodes with interconnected microsphere protection. Chem Mater 29(14):5906–5914

Li B, Xiao Q, Luo Y (2018) A modified synthesis process of three-dimensional sulfur/graphene aerogel as binder-free cathode for lithium-sulfur batteries. Mater Des 153:9–14

Li X et al (2019) High-performance Li–SeSx all-solid-state lithium batteries. Adv Mater 31(17): 1808100

Liang X et al (2015) A highly efficient polysulfide mediator for lithium–sulfur batteries. Nat Commun 6(1):5682
Lin D et al (2016) Layered reduced graphene oxide with nanoscale interlayer gaps as a stable host for lithium metal anodes. Nat Nanotechnol 11(7):626–632
Liu Y et al (2016) Influence of micropore and mesoporous in activated carbon air-cathode catalysts on oxygen reduction reaction in microbial fuel cells. Electrochim Acta 214:110–118
Liu X et al (2017a) Nanostructured metal oxides and sulfides for lithium–sulfur batteries. Adv Mater 29(20):1601759
Liu Y et al (2017b) An artificial solid electrolyte interphase with high Li-ion conductivity, mechanical strength, and flexibility for stable lithium metal anodes. Adv Mater 29(10):1605531
Lyu Z et al (2015) Hierarchical carbon nanocages confining high-loading sulfur for high-rate lithium–sulfur batteries. Nano Energy 12:657–665
Ma L et al (2015) Nanomaterials: science and applications in the lithium–sulfur battery. Nano Today 10(3):315–338
Ma L et al (2016) Enhanced Li–S batteries using amine-functionalized carbon nanotubes in the cathode. ACS Nano 10(1):1050–1059
Maihom T et al (2018) Lithium bond impact on lithium polysulfide adsorption with functionalized carbon fiber paper interlayers for lithium–sulfur batteries. J Phys Chem C 122(13):7033–7040
Manthiram A et al (2014) Rechargeable lithium–sulfur batteries. Chem Rev 114(23):11751–11787
Pan H et al (2018) Addressing passivation in lithium–sulfur battery under lean electrolyte condition. Adv Funct Mater 28(38):1707234
Ren W et al (2019) Recent advances in shuttle effect inhibition for lithium sulfur batteries. Energy Storage Mater 23:707–732
Shi L et al (2017) High-safety lithium-ion sulfur battery with sulfurized polyacrylonitrile cathode, prelithiated SiOx/C anode and carbonate-based electrolyte. J Alloys Compd 723:974–982
Tan KW et al (2018) Synthesis and formation mechanism of all-organic block copolymer-directed templating of laser-induced crystalline silicon nanostructures. ACS Appl Mater Interfaces 10(49):42777–42785
Wang Q et al (2015a) Layer-by-layer assembled C/S cathode with trace binder for Li–S battery application. ACS Appl Mater Interfaces 7(45):25002–25006
Wang H et al (2015b) Large-scale synthesis of ordered mesoporous carbon fiber and its application as cathode material for lithium–sulfur batteries. Carbon 81:782–787
Wang J-G, Xie K, Wei B (2015c) Advanced engineering of nanostructured carbons for lithium–sulfur batteries. Nano Energy 15:413–444
Wang X et al (2016) Structural and chemical synergistic encapsulation of polysulfides enables ultralong-life lithium–sulfur batteries. Energy Environ Sci 9(8):2533–2538
Wang L et al (2018) Development and challenges of functional electrolytes for high-performance lithium–sulfur batteries. Adv Funct Mater 28(38):1800919
Wang D-Y et al (2019a) Tuning the electrochemical behavior of organodisulfides in rechargeable lithium batteries using N-containing heterocycles. J Mater Chem A 7(13):7423–7429
Wang J et al (2019b) An electronegative modified separator with semifused pores as a selective barrier for highly stable lithium–sulfur batteries. Ind Eng Chem Res 58(31):14538–14547
Wang M et al (2019c) Porous carbon hosts for lithium–sulfur batteries. Chemistry 25(15): 3710–3725
Wei Q et al (2017) Porous one-dimensional nanomaterials: design, fabrication and applications in electrochemical energy storage. Adv Mater 29(20):1602300
Werner JG et al (2015) Carbon–sulfur composites from cylindrical and gyroidal mesoporous carbons with tunable properties in lithium–sulfur batteries. Chem Mater 27(9):3349–3357
Xiao S et al (2015) Polyurethane-derived N-doped porous carbon with interconnected sheet-like structure as polysulfide reservoir for lithium–sulfur batteries. J Power Sources 293:119–126
Xu Y et al (2015) Confined sulfur in microporous carbon renders superior cycling stability in Li/S batteries. Adv Funct Mater 25(27):4312–4320

Xu S et al (2018) Three-dimensional, solid-state mixed electron–ion conductive framework for lithium metal anode. Nano Lett 18(6):3926–3933
Yan M et al (2019) Interfacial design for lithium–sulfur batteries: from liquid to solid. EnergyChem 1(1):100002
Yang Y, Zheng G, Cui Y (2013) Nanostructured sulfur cathodes. Chem Soc Rev 42(7):3018–3032
Yang W et al (2017) Pyrrole as a promising electrolyte additive to trap polysulfides for lithium-sulfur batteries. J Power Sources 348:175–182
Yang W et al (2018) A polypyrrole-coated acetylene black/sulfur composite cathode material for lithium–sulfur batteries. J Energy Chem 27(3):813–819
Yang X, Luo J, Sun X (2020) Towards high-performance solid-state Li–S batteries: from fundamental understanding to engineering design. Chem Soc Rev 49(7):2140–2195
Yoo S et al (2016) Well-dispersed sulfur wrapped in reduced graphene oxide nanoscroll as cathode material for lithium–sulfur battery. J Electroanal Chem 780:19–25
Zeng L et al (2022) Catalytic effects of electrodes and electrolytes in metal–sulfur batteries: progress and prospective. Adv Mater 34(49):2204636
Zhang Q et al (2015a) Understanding the anchoring effect of two-dimensional layered materials for lithium–sulfur batteries. Nano Lett 15(6):3780–3786
Zhang S et al (2015b) Recent advances in electrolytes for lithium–sulfur batteries. Adv Energy Mater 5(16):1500117
Zhang Q et al (2018) Pathways to mesoporous resin/carbon thin films with alternating gyroid morphology. ACS Nano 12(1):347–358
Zhao M et al (2020) Lithium–sulfur batteries under lean electrolyte conditions: challenges and opportunities. Angew Chem Int Ed 59(31):12636–12652
Zhou J et al (2019) 8-Acetonyldihydronitidine inhibits the proliferation of human colorectal cancer cells via activation of p53. Eur J Pharmacol 854:256–264
Zhu L et al (2015) Interconnected carbon nanotube/graphene nanosphere scaffolds as free-standing paper electrode for high-rate and ultra-stable lithium–sulfur batteries. Nano Energy 11:746–755
Zyga L (2011) Rechargeable lithium-sulfur batteries get a boost from graphene 13. PhysOrg.com 1–5

Chapter 10
Metallic Li Anode: An Introduction

Afaq Hassan, Saima Nazir, M. Sagir, Tausif Ahmad, and M. B. Tahir

Abstract The generation and storage of clean energy are important to access the global development, due to the depletion of the fossil fuels and high prices of oil and the need for sustainable social development. This chapter focuses on the one of the solutions for the storage of the energy in the form of lithium-based batteries. Li-ion batteries are prone to the dendritic growth thus effecting the efficiency of the batteries; thus, the Li-ion insertion method is discussed in detail that has revolutionized lithium-ion battery systems. The exceptional characteristics of lithium metal, including its lightweight nature and high power-to-weight ratio, position Li-ion batteries as a robust contender. It discusses the widespread use of the Li-ion batteries in mobile phones, electric cars, computers, and drones, thanks to research efforts over the last two decades. This chapter serves as a comprehensive resource for understanding the significance of Li anode and its implications for the future of energy storage.

A. Hassan · M. Sagir
Institute of Chemical and Environmental Engineering, Khwaja Fareed University of Engineering and Information Technology, Rahim Yar Khan, Punjab, Pakistan
e-mail: afaqhassan898@kfueit.edu.pk; dr.msagir@kfueit.edu.pk

S. Nazir
Nawaz Sharif medical College, University of Gujrat, Gujrat, Pakistan

University of Health Sciences, Lahore, Pakistan

T. Ahmad (✉)
Institute of Chemical and Environmental Engineering, Khwaja Fareed University of Engineering and Information Technology, Rahim Yar Khan, Punjab, Pakistan

Institute of Energy and Environmental Engineering, University of the Punjab, Lahore, Pakistan
e-mail: tausif.ieee@pu.edu.pk

M. B. Tahir
Institute of Physics, Khwaja Fareed University of Engineering and Information Technology, Rahim Yar Khan, Punjab, Pakistan
e-mail: m.bilaltahir@kfueit.edu.pk

M. S. Tahir et al. (eds.), *Lithium-Sulfur Batteries: Key Parameters, Recent Advances, Challenges and Applications*, Springer Tracts in Electrical and Electronics Engineering, https://doi.org/10.1007/978-981-99-2796-8_10

Keywords Lithium ions batteries · Dendritic growth · Green energy generation · Energy storage

Introduction to Li Anode

Green energy generation and storage is an important metric for gauging global progress. This is owing to the depletion of fossil fuels, rising oil prices, and the strong need for social development that is sustainable (Kumar et al. 2021a, b). Due to their large energy storage capacity and high density power, second batteries based on lithium (Li metal) are now being researched.

When Li is placed in a tree-like dendritic morphology, a big issue arises. Lithium dendritic growth has been extensively investigated thus far, as Li dendrites can pass through electrolyte, causing a separator break to reach the electrode of the opposite area, resulting in a short battery circuit due to the excess current flowing through the installed lithium. As a result, these Li dendrites are the primary source of the problem, putting cells with short battery circuits at risk. In 1990, Sony Corporation created the Li-ion insertion method employing a graphite anode as a Li-ion handle, which revolutionized lithium-ion (LIB) battery systems. For the first time, this success brought attention to the problem of Li dendritic development (Yasin et al. 2021).

Li is the lightest metal on the Earth's crust, with a mass of 0.534 g cm^3 and an atomic number of 6.941. This was the motivation for considering Li metal-based secondary technology a few decades ago. It appears as a soft alkaline metal with a silver-white tint. Because lithium has a low atomic weight, it charges faster and has a higher power-to-weight ratio. Under normal circumstances, Li is classified as a low-density material. Li is the most potent element of electricity generation (3.04 V) compared to traditional hydrogen electrodes, offering a very high energy storage system and high power density. Furthermore, the Li has a specific capacity of 3860 (Ibraheem et al. 2019a).

The mAh g^1 of this battery is higher than most rechargeable anode batteries. One way to boost the system's energy capacity is to add a thin layer of Li to the anode material, which may absorb huge numbers of ions and hence result in high cell density. An efficient and continuous electrode object with a high voltage window and high power should be researched in order to develop a robust battery cell. At the graphite anode, Li is currently presenting around ten times the ultrahigh theory, whereas the use of the lithium anode in modern battery systems is hampered by safety concerns, volume extensions, low bike performance, and high performance (Ibraheem et al. 2019b, 2020).

Although there has been a lot of study done in recent years to address these concerns, there are still no practical and viable answers. As a result, basic understanding and protective mechanisms must be thoroughly researched in order to make Li anode work in advanced LMB. We outlined the fundamental understanding of some of the primary obstacles facing the Li anode and provided some effective defense tactics when security issues arise in the Li anode as a helpful hand. Li-based batteries recharged between steel batteries are regarded the most promising battery technology for achieving high gravity/volumetric density and density due to their low density and low ionic radii (Li). Lithium batteries are greatly used in our lives, thanks to research efforts over the last two decades, which include a cell phone, electrical car, computers, and even the drones. Lately, a rapidly expanding market has demanded Li and re-charged Li-ion batteries that are lighter, thinner, and less expensive. Commercially used lithium ion batteries have higher limits of 250 Wh kg^{-1} and 680 Wh L^{-1}. Theoretical power (372 mAh g^{-1}) and volume power (735 mAh cm^{-3}) of graphite-based anodes are limited, limiting the continual rise in the energy received by Li-ion batteries. As a result, the development of battery components that include both the cathode and the anode has mostly focused on increasing the power density of the new coming Li-ion batteries, which will include low weight and size batteries for a range of uses (Ibraheem et al. 2021). Lithium is fully anode suitable for developing and recharging batteries among high-volume anodes, because it has a high-power capacity (3860 mAh g^{-1} and 2061 mAh cm^{-3}) and a low electrochemical negative charge (−3.04 V vs. standard hydrogen electrode). According to earlier predictions concerning future Li and Li-ion battery potential dissipation, the battery system's density could reach a point where the direct electrode resembles a Li anode (beating Si and graphite anodes) (Nadeem et al. 2018, 2020, 2021).

Lithium-based batteries, i.e., lithium cathode battery with high voltage, lithium-fluoride Li-Fi battery, lithium-oxygen battery, and lithium-sulfur battery, are the most promising candidates in the following to produce high-capacity rechargeable batteries suitable for the long distance covering up to 40 miles in a single recharge, thus proving to be the best solution for the nowadays problem (Kumar et al. 2021c; Liu et al. 2019). Metallic Li anode is now being explored extensively due to the high-density properties of lithium metal batteries, though lithium metal anode technology has a number of issues that make it too tiny to be used in lithium metal batteries. The remaining issues with separate Li metal anode approaches are discussed and researched in this article, from basic to practical. Finally, the roadmap and major trends in Li metal anode future development are explored.

How Lithium Battery Were Manufactured

Depending on whether or not they are charged, LMBs are classified into basic and secondary batteries. In the 1970s, the most powerful LMBs, such as $Li\text{-}I_2$, $Li\text{-}MnO_2$, Li-CuO, and Li-(CF) batteries, were developed and are now widely employed in timepieces, clocks, and moveable medicinal devices. The density of the secondary LMB is generally lower than that of the basic LMB. Secondary lithium-based battery, on the other hand, has a wide range of uses due to their ability to be charged again. Secondary LMBs are referred to in the following definitions. The first LMBs, developed at Exxon in the 1970s by Stanley Whittingham, use lithium metal as an anode and titanium sulfide as a cathode (Lin et al. 2017). Though due to the considerable security risks posed by the formation of Li dendrites, research on LMBs has been halted since the 1980s. Armand developed the LIB concept, also known as moveable seat cells, in the late 1970s, which combined the cathode and anode.

Li ions are moved from one side to the other using the discharge of LIB charges concept, which eliminates the requirement to decrease Li ions to Li atoms during charging and eliminates the formation of Li dendrites. LIBs were reintroduced to the market in 1991. However, LIB capacity has essentially reached a limit that can only be overcome by significant market expansion. Lithium batteries have recently gained a lot of research attention due to the high need for power (Wu et al. 2018).

Li Metal Batteries Electrolytes

Li-based anode has highest recorded theoretical specific capacity of all metallic elements due to its lowest specific (0.59 g cm^{-3}) (3860 mAh g^{-1}). The electrochemical potential of the Li is lowest among other used metals, i.e., 3.04 V. As a result, it is assumed to be the potential element for the electrode material. The development of the dendrite while the charging and discharging cycles on the other hand not only creates major safety issues such as interior short circuits; it also results in "dead lithium" and a damaged solid electrolyte interfacial (SEI) coating, resulting in poor CE. Electrolyte, cathode material, anode material, separator, and current collector are all components of a lithium metal battery in general. Improving the performance of the anode via cathode design may not be simple. Current research is mostly focused on the separator, electrolyte, lithium metal anode, and current collector improvements. Based on aforementioned four characteristics, we outline recent investigations of highly efficient Li-based anode material in the study. The design of these devices, as well as their influence on electrochemical performance, is primarily examined. Finally, various novel avenues in this subject, as well as potential development problems, are discussed.

The SEI film is formed by Li metal, which has a strong chemical activity and is unsustainable in organic solvents (Aurbach et al. 2002). Due to the development of the SEI films, lithium dendrites dispose the SEI film and bring the fresh lithium on the surface of the electrode. This repeated process wastes Li metal and electrolyte, while it is affecting the performance of battery (Lu et al. 2014). Liquid electrolytes, on the other hand, offer Li^+ conductivity is high, and it is already extensively used in the commercial LIBs. As a result, electrolytes liquids remain a viable option for Li metal batteries. Solid-state-based electrolytes are an excellent choice for lithium batteries because of the high shear modulus, which inhibits dendrite development. However, they have limited Li^+ conductivity and weak electrode contact. Electrolytes with a composite structure have gotten a lot of interest in recent years. They have the ability to combine the benefits of several kinds of materials and also have real applications.

Liquid-Based Electrolytes

An electrolyte that is liquid is typically composed of organic solvent, lithium salt, and additives. Almost every component of the liquid electrolyte may affect the SEI film in some way (Kim et al. 2011). As a result, by regulating the electrolyte's components, a stable SEI layer may be created. Furthermore, according to the models presented and the experimentation, the amount of the potential that is reduced, concentration of the lithium salt, size of anions that are present in the salt, and the viscosity of the electrolyte all have a role in the nucleation and formation of Li dendrites (Qian et al. 2015). These findings lead us to develop a liquid electrolyte that may inhibit dendrite formation while also stabilizing the SEI layer.

Lithium Salt and Various Solvent in the Liquid Electrolytes

Organic compounds based on carbonate or ether serve as general solvents. Carbonate-based electrolytes, such as diethyl carbonate (DEC), ethylene, and propylene carbonate (EC and PC), are often employed in $LiPF_6$ with LIBs and lithium perchlorate ($LiClO_4$) as a salt, although they are considered good solvents for the lithium battery because of the dendrites which result in poor performance (Stone et al. 2012). Ether-based electrolytes are often used in the lithium-sulfur battery because of the reduced potential (Min et al. 2014). Thus, due to the less interaction of the lithium metal and electrolyte, lithium batteries show higher columbic efficiency (CE) as compared to the ether-based electrolyte. Ideally, a good solvent must have few properties such as lower potential, larger anodic window, and high solubility for the Li salt. Thus to have such properties, combination of two or three solvents may perform better. Miao et al. (Miao et al. 2016) developed a novel ether-based electrolyte using solvents DME and 1, 4-dioxane (DX).

High Concentration of Lithium Salt in Liquid Electrolytes

Brissot and Rosso's model (Brissot et al. 1999) illustrates that the nucleation of dendrite causes the exhaustion of the lithium anode during the plating process. Due to the reduced concentration variation surface during the lithium exhaustion at the anode, higher concentrations of Li salt may limit the formation of dendrite. For excellent acceptable viscosity and ionic conductivity of electrolytes, the concentration of the lithium salt should be about 1 mol L^{-1}. Higher ionic conductivity is not, however, the sole desirable attribute. Great performance in LIBs has shown because of high electrolyte concentration (Yamada et al. 2014); they are predicted to be suitable for Li metal batteries. Despite the fact that high concentration liquid electrolytes increase cyclic stability and CE for Li metal anodes, Li salts' exorbitant cost will severely limit their practical usage. The electrolytes' high viscosity might be another stumbling block in any future usage.

Additives in Liquid Electrolytes

To develop the desired SEI layer on the electrode, additives play an important role. To obtain stable, flexible, compact, thin, high ionic conductivity, the desired SEI film should be developed. Due to the less reaction among the anode and the electrolyte, the performance of the lithium anode increases. Some additives have previously been used to stabilize the SEI coating of the Si anode in LIBs, such as fluoroethylene carbonate (FEC) and vinylene carbonate (VC) (Jm and Balbuena 2014; Xu et al. 2015). Because both Li and Si suffer large volume changes during charge-discharge cycles that cause the exhaustion of the SEI film, the additives help the film to stabilize and reduce the chance of the depletion (Ota et al. 2004; Wandt et al. 2015).

Electrolytes containing LiF additions have been shown to reduce interfacial resistance by increasing Li^+ interfacial transfer. The SEI film in the main species is LiF, which explains why the SEI layer is stable and compact with higher ionic conductivity. Similarly, they also predicted that LiBr might be used as an addition for Li metal batteries, which are subsequently proved. Other additions, like as CO_2, 2-methylfurane, triacetoxy vinylsilane, and trace amounts of H_2O, may aid in the creation of the stable SEI film and hence increase Li metal battery performance. Cations, such as Cs^+ and Ru^+, are another form of additive with a lower reduction potential than Li^+'s. Through a unique method, these additional cations may prevent dendritic development (Wandt et al. 2015).

Ionic Liquid

Because of their desired qualities such as low vapor pressure, a large electrochemical stability window, adequate electrical conductivity, and non-flammability, ionic liquids are a promising choice for the electrolytes in both Li metal batteries and LIBs. Furthermore, the ionic liquid electrolyte has been shown to reduce dendrite formation. As a result, ionic liquids, particularly at room temperature, are a safer option for carbonate- and ether-based liquid electrolytes. Yet, when compared to ordinary liquid electrolytes, the greater cost is a barrier to commercial applications (Aurbach et al. 1994). The nonvolatile electrolyte N-methyl-N-butylpyrrolidinium dicyanamide is an ionic liquid. Batteries using lithium iron phosphate ($LiFePO_4$) as cathode and lithium acting as anode demonstrate great cycle stability when using this electrolyte. However, such an ionic liquid's operating temperature is 50 °C, limiting its practical applicability.

Solid Electrolytes

Because of their higher mechanical strength, exceptional thermal stability, and large electrochemical stability window, solid electrolytes are a hot topic. Solid electrolytes have a high shear modulus, which may prevent dendrite development. In addition, most solid electrolyte batteries do not need separators.

Inorganic Electrolytes

Solid electrolytes are most effective to solve the difficulties with Li metal batteries and LIBs. However, there are two major issues that severely restrict their applicability. The first is low ion conductivity at the ambient temperature, and second is the electrolyte-electrode contact issue. The compatibility of Li metal anodes with Li metal should be neglected for inorganic solid electrolytes, and the interface issue has to be investigated further. Even if high ionic conductivity at ambient temperature has been attained, inorganic solid electrolytes still have a long way to go in terms of scalable applications.

Organic Electrolytes

Solid electrolytes, which consist of polymer electrolytes (GPE), solid electrolytes gel, such as block copolymer electrolytes, are examples of organic solid electrolytes. Because of its immobility of electrolyte molecules, strong ionic conductivity GPE bridges the gap between solid electrolytes and liquid (Isken et al. 2013). It has also been proven that dendritic development in GPE may be slowed (Tatsuma et al. 2001). Typically, GPE is made up of host polymers and liquid electrolytes (Huang et al. 2013). However, owing to the leaking of liquid electrolytes, traditional host polymer materials lose their stability over time (Gerbaldi et al. 2012). Another form of host polymer material is polymeric ionic liquids (PILs). Covalent bonding exists between the polymer backbone and ions. PILs have higher thermal stability, excellent electrode compatibility, and effective ionic conductivity (Balducci et al. 2011).

Composite Electrolytes

Combination of both organic and inorganic electrolytes results in the formation of the composite having the advantages and properties of both electrolytes. Zhou et al. developed a polymer/ceramic/polymer sandwich electrolyte (PCPSE) that solved both ceramic electrolyte and polymer issues (Stone et al. 2012). The ceramic layer in the sandwich construction prevents anion transit, which solves the Li anode-polymer electrolyte interface issue of a double-layer electric field. This alignment of the electrolyte helps in reducing the chemical decomposition and dendrite formation, thus improving the columbic efficiency (CE) (Bouchet et al. 2013; Zhou et al. 2016; Sheng 2007).

Current Collectors for Li Batteries

Copper is the mostly used current collector in the LIBs but is not practicably used with the lithium, because it causes inhomogeneous deposition of the lithium metal on the electrode surface. Modified current collectors are used in primarily two ways to enhance the activity of Li metal anodes.

Modified Lithium Anodes

The most basic way to enhance the performance of Li metal batteries is to use an advanced Li anode design. Without any extra electrochemical processing, the improved Li anode may be utilized directly in lithium-sulfur and Li-oxygen batteries. Three major Li anode modifications have been presented: (1) in situ creation of a stable SEI layer on the Li metal surface; (2) covering Li metal with a functional coating; and (3) using a composite nanostructured Li metal anode have all been reported as Li anode modifications.

Li Metal Anode with an In Situ Formed SEI Layer on Surface

The most important aspect of a higher-performance Li metal anode is a stable SEI layer. However, because of its extremely reactive nature, Li metal may react with Li salts and organic solvents in typical solvent-based electrolytes, forming an unbalanced SEI layer that breaks down during cycling. A straightforward and industrially relevant technique to tackle the issue is to develop a stable SEI layer on the surface of the Li anode before battery construction. As previously mentioned, ionic liquid electrolytes are a possible contender for use in Li metal batteries and have already shown promising results.

Li Metal Anode with a Functional Coating

A desired functional coating may inhibit Li dendrite development and help build a limited side reaction and a stable SEI layer between the electrolytes and Li anode. The use of a conductive polymer layer as a coating on the sulfur cathode has been shown to increase Li-S battery performance (Zhang et al. 2016). The conductive polymer covering has also been shown to improve the functioning of Li anodes (Li et al. 2013; Ma et al. 2014).

The diffusion of lithium polysulfide may be inhibited by the conductive polymer layer in Li-S batteries, whereas the Li^+ can be transmitted by the ion-conductive phase, such as a polymer (ethylene glycol) (Liang et al. 2016; Liu et al. 2016a; Lin et al. 2016). Moreover, simple approaches, such as immediately dipping the Li foil in the polymer solution, may be used to create such modified Li anodes. The protective layer not only efficiently inhibits dendrite development but also supports the production of a stable and less resistant SEI coating on the surface of the Li anode.

Manufacturing of the Lithium Metal Anode

The synthesis process for the negative electrodes containing graphite and silicon composite can be carried out in mesmerizing area, synthesized electrodes are utilized in modern lithium batteries. The cathode material as well as electrolyte work in the lithium batteries are stable when exposed to dry climate. Empowering manufacturing process like stacking, enclosing, and filling of lithium batteries proceeds in dry laboratory environment (Flynn and Marsh 1998; Hawley and Li 2019; Guo et al. 2016). It is proved that lithium metal is exponentially reactive toward all the components present in atmospheric air, the lithium metal anode after reacting with air try to make more complex anode materials. Due to the formation of complex materials in anodic form, these lithium anodes are usually preferred to manufacture in isolated inert atmosphere. The infrastructure built-up for the manufacturing of lithium batteries is unstable, especially for the lithium negative electrodes that unstable process leads to complexity and caused more cost (Li et al. 2018). As a result, the developing lithium anode in controlled isolated inert environment leads to economical procedure with more practical anode material.

Air Stable Lithium Anode

Further two main techniques brought into practical utilization for manufacturing air-stable lithium anodes. The surface energy of fluid proved as a determining factor for attraction toward any material. The fluids, having surface completely wet, more surface energy, exhibit considerable attraction to draw the fluid down. Synthesizing interfacial layer, of lithium anode, which repel air molecules, results in prominent improvement of lithium air environment. In recent couple of years, the interfacial layers with said characteristics developed and brought advancement in the lithium metal anode. Favorable, economical method for the synthesizing of lithium metal anode was described by the Yan's team, and prepared anode was stable in dry atmosphere (Shen et al. 2019). A layer of carbon monofluoride and lithium fluoride was created on the surface of lithium anode. The low surface energy of fluoride caused the long-term stability of anode in dry atmosphere and avoids electrode from contacting directly with the electrolyte used in the cycling process. After covering surface with carbon and lithium fluoride, the lithium anode was capable enough to perform well even lying in wet atmosphere more than 24-h. Different scientists selected this covering layer as protective layer for the isolation of the different materials; additionally, the fluorine layer is more aqua phobic than other carbon compounds. The batteries with surface covering layer of fluorine performed distinguishing task with long-term stability even the prepared remodeled lithium anode exposed to humid atmosphere for more than 12 h (Xu et al. 2020). Another technique to stable the anode in wet atmosphere was to wrap the lithium anode in thick insulation. The insulated covered layer on the lithium anode was providing insulation at such extent that electrolyte could not permeate through it. The one most promising factor of insulated layer was that Li^+ in it enhanced the conductivity of the battery.

Water-Stable Lithium Anode

Currently existing lithium batteries have limited requirement of the moisture requirement, e.g., should be less than 10 ppm. When the concentration of the moisture contents exceeds certain limit in modern batteries, it disturbed the chemical equilibrium maintained in the battery; SEI vanished in the electrodes and results in unsatisfying aging characteristics. The effect of water contents in lithium batteries is distinguishing, and its performance is also disturbed. The reaction of water contents with battery is possible due to which LiOH, H_2, and heat generated; destruction of battery and generating more safety issues can be the results. Due to the consequences of more attraction of water by electrolyte, the corrosion of electrode made impossible the development of open system for lithium metal batteries. To prevent the protective layer from breaking, it must also prevent the piercing of lithium dendrites during charge/discharge. Many powerful protective coatings have lately been suggested in this area to keep metallic lithium from being harmed by liquid water during the battery construction and cycling process. It has been discovered that fabricating a durable lithium metal anode that may be utilized in an open environment is as simple as establishing an artificial protective layer. However, we cannot overlook the fact that adding more layers to the battery system has implications for total energy density and prices. As a result, some proposed approaches for protecting lithium anodes by erecting a thick covering would lessen the considerable benefit of LMBs' high energy density, putting the wagon before the horse.

To completely exclude omitted side effects, strategies to construct this functional layer by in situ conversion reaction or mutually stable materials seem to be more favorable (Janek and Zaire 2016; Schnell et al. 2019). The most promising technique to shield the lithium anode from air corrosion and dendritic puncture during cycling is to use a solid electrolyte protective layer that does not reduce the battery's energy density (Schnell et al. 2019). Of course, the solid electrolyte layer thickness should be kept to a maximum of 25 μm (in comparison to a conventional Celgard separator), which places a greater demand on the manufacturing process. As a result, advances in solid electrolyte processing technologies will aid in the construction of a reliable lithium anode. In the future, ASSLMBs will be the dominant lithium battery shape.

Thermal-Stable Lithium Anode

The protected activity of lithium batteries at raised temperatures has forever been an objective sought after by researchers. Later on, ASSLMBs will take the battery's security and steadiness to the powerful in the field of the auxiliary battery. By then, at that point, in any case, the low dissolving point (MT) of metallic lithium (180.5 °C) will undoubtedly be a short board that keeps ASSLMBs from working at higher temperatures (>200 °C). The framed SEI on the lithium surface will be obliterated when the lithium liquefied because of the base's liquidity. The strong surface pressure of liquid metals makes usually overlaid cells incapable to oblige it at raised temperatures. The liquid lithium will in general expel from the edge, causing battery disappointment. Furthermore, the liquid lithium is profoundly responsive and may respond fiercely with strong electrolytes straightforwardly, in this manner overheating and terminating (Chung and Kang 2017; Wu et al. 2021). These days, lithium metal's thermal resistance is additionally direly required for specific applications in extraordinary essential batteries. For instance, the essential batteries give capacity to the direction and correspondence arrangement of rockets, and rockets require super high activity steadiness where warm resistance is a huge precondition (Xie et al. 2016). Subsequently, planning a strong lithium metal anode with fantastic warm resistance is imperative to improve battery security and soundness. As of recently, alloying and space imperatives are two primary procedures that can work on the warm resistance of lithium metal anode. Alloying is a compelling system to change the liquefying of metals (Zhang 2011). Escalated consideration has been given to lithium amalgams as an option in contrast to metallic lithium for a considerably longer time, since they could bring down electrolyte decrease while permitting Li^+ to move into the terminal (Heligman and Manthiram 2021; Huggins 1999; Besenhard et al. 1997). The lithophilic elements may generally create alloys with additional lithium, according to a general concept. This quality is also important for the LMBs' steady functioning. The lithophilic phase in alloy anodes may minimize the Li^+ nucleation barrier by uniforming the local current density of the electrode's surface (Pathak et al. 2021). It was also shown that a lithophilic alloy anode may work as an excellent current collector and host for future stable lithium plating by serving as a highly electronic and ionic conduit to enable Li^+ to move within the electrode materials. Furthermore, adding alloys into a lithium metal anode helps control the SEI production process, lowering side effect reactions with electrolytes and improving anode stability.

Li Metal Anode Failure Mechanisms

Throughout the course of recent many years, broad examination has been led to acquire a more profound and more fundamental comprehension of (a) Li-particle battery and (b) Li-metal battery. The disappointment of the Li metal anode is perceived to be because of three factors: (a) high Li metal execution, (b) persistent volume variety among cycling, and (c) dendrite development brought about by the initial two elements.

Li Metal High Reactivity

Li is a sort of soluble metal component, with a little particle range between a wide range of metal. This component makes Li an extremely high limit (3860 mAh g^{-1}) and, right now, exceptionally quick transmission energy. Li metal, then again, has a much lower intensity among all ongoing terminal materials, making it exceptionally productive. Due to these variables, Li metal can undoubtedly respond with different electrolytes (the two salts containing Li, e.g., carbonate and ethylene carbonate) to shape a purported strong electrolyte (SEI). In the event that this SEI isn't areas of strength for artificially sufficiently able to safeguard the Li metal surface against additional response, the electrolyte will be utilized ceaselessly with Li. the metal will rust in the future, which unavoidably prompts battery exhaustion, loss of usefulness, and, over the long haul, quick battery demise.

Infinite Volume Variation During Cycling

As a matter of fact, practically a wide range of anode objects experience volume changes when charged and released. Contrasted with these materials, the immediate stripping/depriving of the lower Li Liar, nonetheless, will bring about a consistent volume change in the ideal express that all Li metal in the anode will be striped and joined to the cathode, which will start breaks in SEI. This outcomes in extremely high anode structure, SEI, what not battery suspension. For instance, in genuine applications, a little region the business anode of uneven is 3 mAh cm^{-2}, comparing to the thickness of ca. 14.6 μm of Li metal anode. If this Li volume is totally debased (e.g., consistent volume drop to 0), at least this solidness will happen between the ongoing anode authority and the separator, which can cause terminal underlying flimsiness and essentially change the real battery setup.

Development of Dendrite on Li Metal

Li dendrites are delivered by neighborhood polarization during Li plating on the outer layer of the Li metal or current gatherers, which presents two central issues: (1) potential partition of the separator and prompts a short inside pattern of the phone, making the battery risky (e.g., "Moli"), and (2) the development of "dead Li," which brings about the division of Li dendrites from the substrate and will prompt a diminishing in battery limit and productivity.

By and large, it has been seen that the arrangement of Li dendrites is firmly connected with the lopsided surface of the surface during Li plating. In particular, in this cycle, a cation fixation slope is shaped in the electrolyte between two terminals. When basic the flow thickness (J^*, which is the greatest flow that can be conveyed by a given electrolyte) is reached, the quantity of cations in the electrolyte diminishes, prompting a nonpartisan electrical aggravation close to the terminal region. This outcomes in disparity current appropriation in the terminal region, prompting the arrangement of Li metal dendrites here. The focus angle ($\partial C/\partial x$) and J^* in the battery can be resolved utilizing the accompanying estimations (Liu et al. 2016b; Wang et al. 2017):

$$\frac{\partial C}{\partial x}(x) = \frac{J\mu_a}{eD(\mu_a + \mu_{Li^+})}$$

$$J^* = \frac{2eC_0D}{t_aL} \quad \text{with} \quad t_a = \frac{\mu_a}{\mu_a + \mu_{Li^+}}$$

where C_0 is the principal component, L is the distance between the terminal, t_a is the quantity of anionic vehicle, μ_a and μ_{Li+} anionic and Li^+ nobilities, J is the dynamic anode, D diff dispersion steady, and e-electronic charging. It has been seen that the development of Li dendrites is self-restricting, implying that bigger dendrites will generally be bigger than more modest ones because of the solid electric field on top of huge dendrites with enormous arch. In this manner, Li metal stores specifically in regions with enormous layers (e.g., at the foundation of existing dendrite or breaks in the metal surface), bringing about consistent development of Li dendrites.

Engineering of the Electrolyte

The SEI movie on Li surface, shaped by the programmed response between Li metal and electrolytes when in direct contact or during charging/release processes, decides the electrochemical presentation of the Li metal anode for an enormous scope. Electrolyte designing can straightforwardly influence the SEI climate and consequently change the Li anode/electrolyte connector, which can essentially affect the exhibition of the Li metal anode. In electrolyte, natural solvents and Li-containing salts are distinguished as a wellspring of natural and inorganic species in the SEI film, separately. Tuning these parts to the electrolyte can fundamentally control the genuine arrangement of SEI. Subsequently, throughout recent many years, an assortment of natural solvents and Li salts has been examined, prompting critical enhancements in the presentation of Li metal anode.

Organic Solvent Design

Natural solvents, which break down Li salt to frame a fluid electrolyte, fundamentally affect Li salt disintegration and ionic electrolyte conductivity, which can additionally add to cycling battery productivity, sustainable power, and well-being. What's more, regular solvents additionally decide the fundamental substance and synthetic/actual properties of the SEI film (Xu 2014), which is frequently expected for high Li particle stream and great adaptability to adjust to volume changes during cycling to keep away from dendrite development. Electrolyte solvents concentrated so far for the most part incorporate carbonic ester, carboxylic ester, and alkyl carbonates, which contain LiPF6, LiBF4, LiClO4, LiN $(SO_2CF_3)_2$, and so on as Li salt (Kanimozhi et al. 2022).

Li Salts Design for Electrolytes

Notwithstanding the natural solvents of electrolytes, Li salt additionally essentially affects the presentation of Li metal anode. Li salts can commonly be partitioned into inorganic and natural species. Inorganic Li salt contains $LiClO_4$, $LiAsO_6$, and $LiPF_6$, and natural mixtures incorporate LiN $(CF_3SO_2)_2$ (LiTFSI), LiN (SO_2CF_3), LiC $(SO_2CF_3)_3$, and others (Kanimozhi et al. 2022).

Li Salts as Electrolyte Solutes

Basically, Li salts have been utilized as electrolyte solutes to give high ionic adaptability to keep up with execution without disturbing the redox response of the battery (Zhang et al. 2021; Wang et al. 2022; Li et al. 2022). In LIB, $LiClO_4$, $LiAsF_6$, $LiBF_4$, and $LiPF_6$ are utilized as electrolyte solutes. Among them, $LiPF_6$ is the most generally utilized electrolyte dissolvable to date with its restricted properties and high particle stream (Wang et al. 2022). LiPF6 issues with temperature precariousness. Strong intensity boundary above (ca. 50 °C) ought to be set during battery activity to stay away from any conceivable response among LiPF6 and solvents (Kanimozhi et al. 2022). In light of $LiPF_6$, $LiPF_6$-n (CF_3) and $LiPF_3$ $(C_2F_5)_3$ are related with better warm soundness, as an option in contrast to $LiPF_6$.

Notwithstanding LIBs, these Li salts, like LiPF6 and $LiBF_4$, have additionally been broadly utilized in Li-S or Li-O_2 battery research. Particularly for Li-S batteries, LiTFSI is in many cases utilized as an electrolyte solute, which can forestall the arrangement of polysulfides in dissolvable high-request and smother the unfortunate impact of transporting (Li et al. 2022).

Li Salts as Electrolyte Additives

Li liquid utilized as an enhancement for the most part incorporates $LiNO_3$, LixSy, and its mixtures. During cycling, $LiNO_3$ can be diminished straightforwardly to LiNxOy, which surpasses Li, while LixSy can expand on Li_2S and Li_2S_2, which can create thick film to smother Li dendrite development. On account of concurrent utilization of these two kinds, a synergistic impact can be accomplished, bringing about a more viable Li anode insurance. To meet the particular requirements of various battery frameworks, cautious determination and plan of the most suitable additional items should be made.

Notwithstanding the inorganic Li salt added substances, a few natural Li salts are likewise added to electrolytes to upgrade the presentation of Li metal anode. For instance, lithium 2-bromo ethane sulfonate can respond with its Li metal surface SEI made out of LiBr and other Li-based organometallic, slender, and uniform and can be utilized to balance out Li-O_2 batteries from two significant variables (Li et al. 2017). To begin with, it can give the appropriate security of Li anode, and second, it can likewise go about as a compelling redox converter to diminish the potential for oxygen development response (OER) in the Li-O_2 cathode.

Non-Li Additives for Electrolyte Design

To additionally change the outer layer of Li metal, other dynamic added substances, other than different Li salts, have been embraced into electrolytes. These added substances can spoil, polymerize, or stick to the outer layer of metallic Li to emulate SEI film structures (Liu et al. 2016b; Wang et al. 2017; Li et al. 2017), for example, solidness and consistency, to work on the presentation of all LMBs. These added substances include inorganic parallel oxides, fluorinated compounds, soluble base metal cations, and different added substances to be indicated.

References

Aurbach D, Zinigrad E, Cohen Y et al (2002) A short review of failure mechanisms of lithium metal and lithiated graphite anodes in liquid electrolyte solutions. Solid:148:7355–7367

Aurbach D, Weissman I, Zaban A et al (1994) Correlation between surface chemistry, morphology, cycling efficiency and interfacial properties of Li electrodes in solutions containing different Li salts. Electrochim Acta 39(1):51–71

Balducci A, Jeong SS, Kim GT et al (2011) Development of safe, green and high performance ionic liquids-based batteries (ILLIBATT project). J Power Sources 196(22):9719–9730

Besenhard JO, Yang J, Winter M (1997) Will advanced lithium-alloy anodes have a chance in lithium-ion batteries? J Power Sources 68:87–90

Bouchet R, Maria S, Meziane R, Aboulaich A, Lienafa L, Bonnet JP, Phan TN, Bertin D, Gigmes D, Devaux D, Denoyel R, Armand M, B.A.B. (2013) Single-ion, triblock copolymers as highly efficient electrolytes for lithium metal batteries. Nat Mater 12(5):452–457

Brissot C, Rosso M, Chazalviel JN et al (1999) Dendritic growth mechanisms in lithium/polymer cells. J Power Sources 81–82(9):925–929

Chung H, Kang B (2017) Mechanical and thermal failure induced by contact between a Li 1.5 Al 0.5 Ge 1.5 $(PO_4)_3$ solid electrolyte and Li metal in an all solid-state Li cell. Chem Mater 29: 8611–8619

Flynn JC, Marsh C (1998) Development and experimental results of continuous coating technology for lithium-ion electrodes. In: Thirteenth annual battery conference on applications and advances. Proceedings of the conference, pp 81–84

Gerbaldi C, Nair JR, Ferrari S et al (2012) New electrolyte membranes for Li-based cells: methacrylic polymers encompassing pyrrolidinium-based ionic liquid by single step photo-polymerization. J Member Sci 423–424(12):459–467

Guo XQ, Wang M, Meng F, Tang YF, Tian S, Yang HL, Jiang GQ, Zhu JL (2016) Rational design and synthesis of an amino-functionalized hydrogen-bonded net- work with an ACO zeolite-like topology for gas storage. CrystEngComm 18:5616–5619

Hawley WB, Li J (2019) Electrode manufacturing for lithium-ion batteries—analysis of current and next generation processing. J Energy Storage 25:100862

Heligman BT, Manthiram A (2021) Elemental foil anodes for lithium-ion batteries. ACS Energy Lett 6:2666–2672

Huang W, Zhu Z, Wang L et al (2013) Quasi-solid-state rechargeable lithium-ion batteries with a calix quinone cathode and gel polymer electrolyte. Angew Chem Int Ed 52(35):9162–9166

Huggins RA (1999) Lithium alloy negative electrodes. J Power Sources 81–82:13–19

Ibraheem S, Chen S, Li J, Li W, Gao X, Wang Q, Wei Z (2019a) Three-dimensional Fe, N decorated carbon-supported NiFeP nanoparticles as an efficient bifunctional catalyst for rechargeable zinceO2 batteries. ACS Appl Mater Interfaces 11(1):699–705

Ibraheem S, Chen S, Li J, Wang Q, Wei Z (2019b) In situ growth of vertically aligned FeCoOOH-nanosheets/nanoflowers on Fe, N co-doped 3D-porous carbon as efficient bifunctional electrocatalysts for rechargeable zinceO2 batteries. J Mater Chem 7(16):9497–9502

Ibraheem S, Chen S, Peng L, Li J, Li L, Liao Q, Shao M, Wei Z (2020) Strongly coupled iron selenides-nitrogen-bond as an electronic transport bridge for enhanced synergistic oxygen electrocatalysis in rechargeable zinc-O2 batteries. Appl Catal B Environ 265:118–569

Ibraheem S, Li X, Shah SSA, Najam T, Yasin G, Iqbal R, Hussain S, Ding W, Shahzad F (2021) Tellurium triggered formation of Te/Fe-NiOOH nanocubes as an efficient bifunctional electrocatalyst for overall water splitting. ACS Appl Mater Interfaces 13(9):10972–10978

Isken P, Winter M, Passerini S et al (2013) Methacrylate based gel polymer electrolyte for lithium-ion batteries. J Power Sources 225(225):157–162

Janek J, Zaire WG (2016) A solid future for battery development. Nat Energy 1:16141

Jm MDLH, Balbuena PB (2014) Reduction mechanisms of additives on Si anodes of Li-ion batteries. Phys Chem Chem Phys 16(32):17091–17098

Kanimozhi G, Naresh N, Kumar H, Satyanarayana N (2022) Review on the recent progress in the nanocomposite polymer electrolytes on the performance of lithium – ion batteries. Int J Energy Res 46:7137–7174

Kim SP, Duin ACTV, Shenoy VB (2011) Effect of electrolytes on the structure and evolution of the solid electrolyte interphase (SEI) in Li-ion batteries: a molecular dynamics study. J Power Sources 196(20):8590–8597

Kumar A, Ibraheem S, Anh Nguyen T, Gupta RK, Maiyalagan T, Yasin G (2021a) Molecular-MN4 vs atomically dispersed M_N4_C electrocatalysts for oxygen reduction reaction. Coord Chem Rev 446:214–122

Kumar A, Vashistha VK, Das DK, Ibraheem S, Yasin G, Iqbal R, Nguyen TA, Gupta RK, Rasidul Islam M (2021b) M-N-C-based single-atom catalysts for H2, O2 & CO2 electrocatalysis: activity descriptors, active sites identification, challenges and prospects. Fuel 304:121–420

Kumar A, Das DK, Vashistha VK, Ibraheem S, Yasin G, Gautam S, Sharma V (2021c) A novel CoN4-driven self-assembled molecular engineering for oxygen reduction reaction. Int J Hydrog Energy 46(52):26499–26506

Li W, Zhang Q, Zheng G et al (2013) Understanding the role of different conductive polymers in improving the nanostructured sulfur cathode performance. Nano Lett 13(11):5534–5540

Li Y et al (2017) Atomic structure of sensitive battery materials and interfaces revealed by cryo-electron microscopy. Science 358:506–510

Li T, Liu H, Shi P, Zhang Q (2018) Recent progress in carbon/lithium metal composite anode for safe lithium metal batteries. Rare Metals 37:449–458

Li C, Li L, He B, Ling Y, Pu J, Wei L, Sun L, Zhang Q, Yao Y (2022) Roadmap for flexible solid-state aqueous batteries: from materials engineering and architectures design to mechanical characterizations. Mater Sci Eng R Rep 148:100671

Liang Z, Lin D, Zhao J et al (2016) Composite lithium metal anode by melt infusion of lithium into a 3D conducting scaffold with lithophilic coating [J]. Proc Natl Acad Sci U S A 113:2862–2867

Lin D, Liu Y, Liang Z et al (2016) Layered reduced graphene oxide with nanoscale interlayer gaps as a stable host for lithium metal anodes. Nat Nanotechnol 11:626–632

Lin D, Liu Y, Cui Y (2017) Reviving the lithium metal anode for high-energy batteries. Nat Nanotechnol 12:194

Liu Y, Lin D, Liang Z et al (2016a) Lithium-coated polymeric matrix as a minimum volume-change and dendrite-free lithium metal anode. Nat Commun 7:10992

Liu YY, Lin DC, Liang Z, Zhao J, Yan K, Cui Y (2016b) Lithium-coated polymeric matrix as a minimum volume-change and dendrite-free lithium metal anode. Nat Commun 7:10992

Liu J, Bao Z, Cui Y, Dufek EJ, Goodenough JB, Khalifah P, Li Q, Liaw BY, Liu P, Manthiram A, Meng YS, Subramanian VR, Toney MF, Viswanathan VV, Whittingham MS, Xiao J, Xu W, Yang J, Yang X-Q, Zhang J-G (2019) Pathways for practical high-energy long-cycling lithium metal batteries. Nat Energy 4(3):180–186

Lu D, Shao Y, Lozano T et al (2014) Failure mechanism for fast-charged lithium metal batteries with liquid electrolytes. Adv Energy Mater 5(3):1400993

Ma G, Wen Z, Wang Q et al (2014) Enhanced cycle performance of a Li-S battery based on a protected lithium anode. J Mater Chem A 2(45):19355–19359

Miao R, Yang J, Xu Z et al (2016) A new ether-based electrolyte for dendrite-free lithium-metal based rechargeable batteries. Sci Rep 6:21771

Min SP, Sang BM, Dong JL et al (2014) A highly reversible lithium metal anode. Sci Rep 4(1): 3815–3815

Nadeem M, Yasin G, Bhatti MH, Mehmood M, Arif M, Dai L (2018) Pt-M bimetallic nanoparticles (M¼Ni, Cu, Er) supported on metal organic framework-derived N-doped nanostructured carbon for hydrogen evolution and oxygen evolution reaction. J Power Sources 402:34–42

Nadeem M, Yasin G, Arif M, Bhatti MH, Sayin K, Mehmood M, Yunus U, Mehboob S, Ahmed I, Flörke U (2020) Pt Ni@PC900 hybrid derived from layered-structure Cd-MOF for fuel cell ORR activity. ACS Omega 5(5):2123–2132

Nadeem M, Yasin G, Arif M, Tabassum H, Bhatti MH, Mehmood M, Yunus U, Iqbal R, Nguyen TA, Slimani Y, Song H, Zhao W (2021) Highly active sites of Pt/Er dispersed N-doped hierarchical porous carbon for trifunctional electrocatalyst. Chem Eng J 409:128205

Ota H, Sakata Y, Otake Y et al (2004) Structural and functional analysis of surface film on Li anode in vinylene carbonate-containing electrolyte. J Electrochem Soc 151(11):A1778

Pathak R, Chen K, Wu F, Mane AU, Bugga RV, Elam JW, Qiao Q, Zhou Y (2021) Advanced strategies for the development of porous carbon as a Li host/current collector for lithium metal batteries. Energy Storage Mater 41:448–465

Qian J, Henderson WA, Xu W et al (2015) High rate and stable cycling of lithium metal anode. Nat Commun 6:6362

Schnell J, Tietz F, Singer C, Hofer A, Billot N, Reinhart G (2019) Prospects of production technologies and manufacturing costs of oxide-based all-solid-state lithium batteries. Energy Environ Sci 12:1818–1833

Shen X, Li Y, Qian T, Liu J, Zhou J, Yan C, Goodenough JB (2019) Lithium anode stable in air for low-cost fabrication of a dendrite-free lithium battery. Nat Commun 10:900

Sheng SZ (2007) A review on the separators of liquid electrolyte Li-ion batteries. J Power Sources 164(1):351–364

Stone GM, Mullin SA, Teran AA et al (2012) Resolution of the modulus versus adhesion dilemma in solid polymer electrolytes for rechargeable lithium metal batteries. J Electrochem Soc 159(3): A222

Tatsuma T, Taguchi M, Oyama N (2001) Inhibition effect of covalently cross-linked gel electrolytes on lithium dendrite formation. Electrochim Acta 46(8):1201–1205

Wandt J, Marino C, Gasteiger HA et al (2015) Operando electron paramagnetic resonance spectroscopy – formation of mossy lithium on lithium anodes during charge–discharge cycling. Energy Environ Sci 8(4):1358–1367

Wang C, Gong Y, Liu B, Fu K, Yao Y, Hitz E, Li Y, Dai J, Xu S, Luo W, Wachsman ED, Hu L (2017) Conformal, nanoscale ZnO surface modification of garnet-based solid-state electrolyte for lithium metal anodes. Nano Lett 17:565

Wang D, Li Q, Zhao Y, Hong H, Li H, Huang Z, Liang G, Yang Q, Zhi C (2022) Insight on organic molecules in aqueous Zn-ion batteries with an emphasis on the Zn anode regulation. Adv Energy Mater 12:2102707

Wu F, Yuan Y-X, Cheng X-B, Bai Y, Li Y, Wu C, Zhang Q (2018) Perspectives for restraining harsh lithium dendrite growth: towards robust lithium metal anodes. Energy Storage Mater 15: 148–170

Wu Y, Wang S, Li H, Chen L, Wu F (2021) Progress in thermal stability of all-solid-state-Li-ion-batteries. InfoMat 3(8):827–853

Xie S, Deng Y, Mei J, Yang Z, Lau WM, Liu H (2016) Facile synthesis of CoS_2/CNTs composite and its exploitation in thermal battery fabrication. Compos Part B Eng 93:203–209

Xu K (2014) Electrolytes and interphases in Li-ion batteries and beyond. Chem Rev 114:11503–11618

Xu C, Lindgren F, Philippe B et al (2015) Improved performance of the silicon anode for Li-ion batteries: understanding the surface modification mechanism of fluoroethylene carbonate as an effective electrolyte additive. Chem Mater 27:7

Xu Q, Lin J, Ye C, Jin X, Ye D, Lu Y, Zhou G, Qiu Y, Li W (2020) Air-stable and dendrite-free lithium metal anodes enabled by a hybrid interphase of C 60 and Mg. Adv Energy Mater 10:1–9

Yamada Y, Furukawa K, Sodeyama K et al (2014) Unusual stability of acetonitrile based super concentrated electrolytes for fast-charging lithium-ion batteries. J Am Chem Soc 136(13): 5039–5046

Yasin G, Ibrahim S, Ibraheem S, Ali S, Iqbal R, Kumar A, Tabish M, Slimani Y, Nguyen TA, Xu H, Zhao W (2021) Defective/graphitic synergy in a heteroatom-interlinked triggered metal-free electrocatalyst for high-performance rechargeable zinc-air batteries. J Mater Chem 9(34): 18222–18230

Zhang WJ (2011) A review of the electrochemical performance of alloy anodes for lithium-ion batteries. J Power Sources 196:13–24

Zhang Y, Xia X, Wang X et al (2016) Graphene oxide modified metallic lithium electrode and its electrochemical performances in lithium-sulfur full batteries and symmetric lithium-metal coin cells. RSC Adv 6:66161-66168

Zhang L, Wang H, Zhang X, Tang Y (2021) A review of emerging dual – ion batteries: fundamentals and recent advances. Adv Funct Mater 31:2010958

Zhou W, Wang S, Li Y et al (2016) Plating a dendrite-free lithium anode with a polymer/ceramic/polymer sandwich electrolyte. J Am Chem Soc 138(30):9385–9388

Chapter 11
Advanced Carbon-Based Nanostructured Framework for Li Anodes

Asadullah Dawood, Junaid Ahmad, Sami Ullah, Zeenat Jabeen, and Zeeshan Asghar

Abstract In light of rising global resource consumption and the looming environmental crises, renewable energy sources are increasingly being explored in industries, transportations, and households. Novel energy materials and related high-energy-density storage technologies have received a lot of attention. Lithium (Li) metal is the most promising anode for implementing a high-performance energy storage system due to its elevated theoretical capacity. Uncontrollable dendrite formation, high volume change, and unstable solid electrolyte interface layer raise a slew of problems (low Coulombic efficiency, serious safety hazard, short lifetime, and so on) that make practical use of Li-metal anodes a pipe dream. To address these issues, researchers have focused their efforts on accommodating and guiding Li deposition as well as stabilizing the interface using a variety of carbon materials, which have proven to be highly effective due to their wide variety and excellent tunability of the structure–property relationship. This chapter primarily provides an overview of the crucial function of carbon-based hosts in increasing the overall performance of Li-metal anodes. The basic failure process of lithium metal anodes is described using linked mainstream models. The benefits and techniques for dealing with the accompanying issues of carbon-based hosts are discussed in general. The function, limitations, and current research advances of essential carbon-based host materials for Li-metal anodes are discussed. Finally, a conclusion is offered, as well as a future perspective for carbon-based host research is given. This chapter aims to summarize current achievements in carbon-based materials hosts and to serve as a guide for future development of carbon-based hosts for improved Li-metal anodes.

A. Dawood (✉)
Department of Physics, National Excellence Institute (University), Islamabad, Pakistan

J. Ahmad · S. Ullah · Z. Asghar
Department of Physics, Division of Science and Technology, University of Education Lahore, Lahore, Pakistan

Z. Jabeen
Department of Physics, COMSATS University Lahore Campus, Lahore, Pakistan

M. S. Tahir et al. (eds.), *Lithium-Sulfur Batteries: Key Parameters, Recent Advances, Challenges and Applications*, Springer Tracts in Electrical and Electronics Engineering, https://doi.org/10.1007/978-981-99-2796-8_11

Keywords Environmental crises · Li-metal anodes · High-energy density

Introduction

In recent years, the search of renewable and clean energy, as well as the electrification of lifestyles, has heightened the need for innovative energy materials and highly efficient energy conversion and storage technologies. Lithium-ion batteries (LIBs) are one of the most popular types of batteries, with applications in electric cars, portable gadgets, and other power sources. However, because of the limiting capabilities of both lithium-metal-oxide (LMO) cathode and graphite anode, traditional LIBs have been locked in an inefficient energy density, unable to meet the societal energy needs (Bruce et al. 2012; Ji and Nazar 2010; Aurbach et al. 2009; Xie et al. 2021; Park et al. 2010). As a result, new anode materials having boosted capacities are needed for next-generation energy batteries.

Lithium has the smallest atomic radius (0.534 g/cm^3), the highest theoretical capacity (3860 mAh/g), the lowest density (0.534 g/cm^3), as well as redox potential (−3.04 V vs. the standard hydrogen electrode), of all metals, endowing it with boosted energy density and rapid charge transfer properties, correspondingly. The LMO/Li batteries may achieve a specific energy density of 400 Wh/kg by substituting graphite with Li metal. The lighter atomic mass, on the other hand, denotes a lesser amount of extranuclear electrons, implying lower binding energy as well as reduced surface diffuse energy between Li-atoms. As a result, when used as an anode, Li metal likes to generate Li nuclei in "hot spots" and grow vertically into Li dendrites rather than spreading evenly over the region (Ling et al. 2012).

The resulting Li dendrites will pierce separators and come into touch with cathodes, causing thermal runaway or short-circuiting. Furthermore, high-activity Li metal can always cause low Coulombic efficiency (CE) and an unstable solid electrolyte interface (SEI), both of which are caused by spontaneous side reactions among electrolytes and Li metal. Furthermore, unlike graphite's intercalation/deintercalation process, Li metal's stripping/plating conversion reaction is hostless. As a result, during long-term cycling, unregulated Li deposition and dendritic Li will cause significant volumetric change (Cheng et al. 2017).

Massive efforts have been made to take advantage of the above benefits while avoiding the associated disadvantages, with a particular focus on electrolytes (super-concentrated electrolytes (Suo et al. 2013), additives (Lin et al. 2017; Xiong et al. 2014), solid-state electrolytes (Bates et al. 1993), etc.), electrolyte-electrode interfaces (modified separators (Hao et al. 2016), artificial SEI (Lin et al. 2018)), and structured anodes (Jin et al. 2018). Those solutions that depend on interface or electrolytes coatings are primarily concerned with establishing a stable interface. Even if an ideal interface were created on the surface of plane Li foil or conventional current collectors like stainless steel or Cu foil, the poor affinity with Li and limited effective surface area (ESA) would still result in a high local current density and broken interface, causing the corresponding electrodes to fail into dendritic Li. With Li deposition, a single current collector channel might slow ion/electron transfer.

In contrast to 2D current collectors, 3D hosts, such as carbon-based hosts (C-Hosts), can assist in stabilizing the SEI layer and enhance electron/ion transmission while suppressing volume expansion and reducing local current density. As a result, fabricating anodes having 3D framework is thought to be another useful technique for limiting Li dendrites and improving electrochemical performance (Liu et al. 2018). Both the features of collect collectors (high mechanical strength with sturdy lightweight skeleton, wide space, quick electrical and thermal transport for electron exchange and heat dissipation, electrochemical inertness) and electrode materials (including large ESAs, elevated ion conductivity, morphological plasticity, cheap and abundant resources) must be included into 3D hosts.

Conductive hosts such as metallic hosts (Yang et al. 2015; Chen et al. 2019), or C-Hosts (Niu et al. 2019; Zhang et al. 2017), and nonconductive hosts such as polymers (Zheng et al. 2020), or glass fibers are commonly used 3D hosts (Yang et al. 2018). By lowering the local current density, insulating frameworks have been shown to be unfavorable in managing the distribution of electric field over the surface and the concentration gradient of Li-ion near the surface (Monroe and Newman 2003). Conductive hosts, on the other hand, have a greater chance of being used because of their capacity to be modified by external components as well as their natural capability to modify electron and ion distribution at the substrate-electrolyte interface (Cheng et al. 2018). Metal-based hosts have excellent shear module, electric conductivity, and elastic strength as compared to other conductive hosts. Nonetheless, obstacles like as increased density and higher costs make large-scale manufacture of Li-metal anodes (Li-MAs) difficult. Furthermore, the metal skeletons are susceptible to hydrofluoric acids (HF) produced by the dissolution of the commonly employed F-based electrolytes in Li-MAs, resulting in structural deterioration (Myung et al. 2011).

Carbon-based hosts, on the other hand, have a greater economic potential because of the intriguing benefits of flexible and lightweight architectures. These architectures combine the strength provided by cross-linking covalent connections with the robustness provided by the potential of slippage under external stress, as a result of weak interactions between hydrogen bonds and the van der Waals force (Ye et al. 2017). The market for emerging energy supplies, such as next-generation wearable smart gadgets, may benefit from such unique benefits as a solid ultralight structure, elevated ESA, abundant space, and a quick transit method (Mo et al. 2020).

Several C-Hosts with exceptional features have been anticipated in recent years to help with the production of Li-MAs. We've compiled a list of the major roadblocks to Li-MAs use, as well as the failure mechanisms that go along with them. Also identified and clarified are the corresponding carbon-based host solving strategies. Furthermore, current advancement on C-Hosts (carbon nanotubes (CNTs), graphene, carbon-fibers (CFs), carbon derived from biomass (and so on), and the associated characteristics and modifications in stabilization of Li-MAs is reviewed and deliberated in depth. Future research on Li-MAs will benefit from the knowledge gained from these carbon-based hosts.

Electrochemical Reactions and Challenges of Li-S Batteries

A lithium–sulfur battery (Li-S), made up of a lithium anode, polymer, or liquid electrolyte, and a sulfur composite cathode, is an enthralling nominee for highly efficient energy devices. Figure 11.1 depicts a schematic illustration of a Li-S battery. It's based on the Li/S redox process, which may be calculated as follows:

$$16\,Li + S_8 \rightarrow 8Li_2S \tag{11.1}$$

It shows a high specific capacity theoretical 1675 mA h/g theoretically) and a high specific energy (2600 Wh/kg theoretically) assuming 100% conversion (Marmorstein et al. 2000; Sun et al. 2008).

Basics of the Li-S Battery

Because of its elevated theoretical specific capacity of roughly 1675 mA h/g, sulfur is a viable positive electrode for LIBs. The Li/S reaction has a Gibbs energy of 2600 Wh/kg, which is more than five times the theoretical energy of a Li-ion system. The idea of using sulfur as the positive electrode in an metal anode battery for electrochemical energy conversion and storage dates back to the 1960s

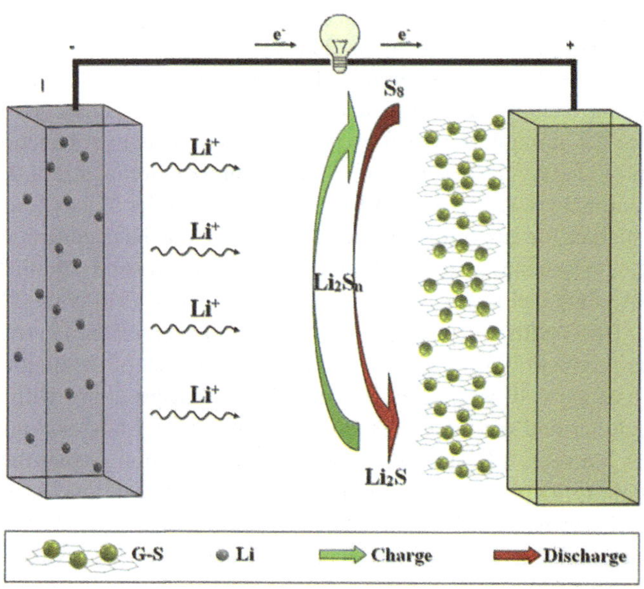

Fig. 11.1 A schematic illustration of Li-S battery with a graphene-sulfur composite (G-S) cathode (Barghamadi et al. 2013)

(Demir-Cakan 2015). There has been a considerable push in recent decades to produce a rechargeable Li/S battery (Mikhaylik and Akridge 2004; Barchasz et al. 2012). There have been several publications published on the electrochemical parameters of the Li–S cell, such as discharge capacity (Ryu et al. 2005; Lee et al. 2003) cycling and self-discharge (Cheon et al. 2003).

Because of its large theoretical capacity, wide temperature range of operation, and elevated theoretical power density, this system is a particularly appealing contender among the numerous types of rechargeable batteries. Furthermore, low cost, natural abundance, exceptional safety owing to an inherent overload prevention system, and nontoxicity are also advantages of elemental sulfur (Wang et al. 2008). The most stable form of sulfur is ring-structural cyclooctasulfur (S_8), which may be found in nature in over 30 different allotropes (Zhang et al. 2010). During the discharge of a new Li-S battery, a S–S covalent bond of S_8 is broken first to generate a chain-structural polysulfide (PS) anion ($S_x{}^{2-}$, $x = 8$), which is then further reduced to Li_2S via multistage reactions (Kolosnitsyn and Karaseva 2008; Choi et al. 2008).

Li-S batteries have a different reaction mechanism than secondary lithium batteries, which use an intercalation-deintercalation process in Li-metal oxide and graphite. In a Li-S battery, sulfur reduction is a multistep electrochemical process involving many intermediate species. Lithium polysulfides with the formula Li_2S_n are formed when metal combines with sulfur (S_8). Long-chain polysulfides, such as Li_2S_6 and Li_2S_8, are formed first and then shortened when the sulfur is reduced further. Lithium sulfide (Li_2S) is the discharge's end product, and Eq. 11.1 (Barchasz et al. 2012; Jung and Kim 2007) describes the complete process. Sulfur takes electrons from an open-circuit voltage (OCV) to 2.1 V in this process, generating lithium polysulfide, which is then reduced.

I:. Reaction between S and Li is given by:

$$S_8 + 2Li^+ + 2e^- \rightarrow Li_2\,S_8 \tag{11.2}$$

II:. A reaction between Li and dissolved $Li_2\,S_8$ is written as:

$$Li_2\,S_8 + 2Li^+ + 2e^- \rightarrow 2Li_2\,S_4 \tag{11.3}$$

III:. A transition from the dissolved $Li_2\,S_4$ to insoluble $Li_2\,S_2$ or $Li_2\,S$ by the coexistence of Eqs. 11.4 and 11.5:

$$Li_2\,S_4 + 2Li^+ + 2e^- \rightarrow 2Li_2\,S_2 \tag{11.4}$$

$$Li_2\,S_4 + 6Li^+ + 6e^- \rightarrow 4Li_2\,S \tag{11.5}$$

IV:. An equilibrium reaction of insoluble $Li_2\,S_2$ and $Li_2\,S$ is described as:

$$Li_2\,S_2 + 2Li^+ + 2e^- \rightarrow 2Li_2\,S \quad (11.6)$$

Because it is influenced by both the solubility of the polysulfides in the electrolyte and the chemical equilibrium between each kind of polysulfide in the solution, Eq. 11.3 is the most difficult of the four phases. As a result, the kind of electrolyte solvents has a significant impact on this process. The battle between Eqs. 11.4 and 11.5 determines the result of stage III. The final discharge products are mostly Li_2S_2 and Li_2S mixtures. The Li-S cell has a large capacity with somewhat lower discharge voltages and a shorter stage IV, because Eq. 11.5 is the primary reaction. Because of the nonconductive nature of Li_2S_2 and Li_2S, Stage IV is kinetically sluggish and suffers from excessive polarization. Stage IV is kinetically sluggish and suffers from excessive polarization due to the nonconductive nature of Li_2S and Li_2S_2.

Challenges of Li-S Batteries

Despite its numerous advantages, the Li-S battery is plagued by a number of issues that have limited its practical implementation. During the second discharge process, the discharge products precipitate, enclosing the positive electrode surface and creating poor capacity and electrode rechargeability. This is mostly due to the positive electrode's passivation. In reality, the discharge comes to a halt when these insulating species have completely coated the surface (Barchasz et al. 2012; Cheon et al. 2003). Li-polysulfides dissolve in the electrolyte and migrate to the anode, where they react with lithium to generate lower-order polysulfides, which then diffuse back to the cathode to form a longer chain.

This results in the so-called internal "shuttle" effect, which causes lithium anode corrosion, as a result, low efficiency and limited cycle life in rechargeable Li-S batteries. Furthermore, ongoing reactivity of the soluble polysulfide with the Li anode causes considerable self-discharge and deposition of solid Li_2S_2 and Li_2S on the cathode, resulting in capacity fading and active mass loss. On the Li anode, some of the insoluble Li_2S and Li_2S_2 collect. Another issue is the volume change that happens in the electrode when the active sulfur dissolves and the end products precipitate, which is followed by morphological changes. These issues lead to the rapid aging of electrodes and the loss of the battery's practical specific charge (Jayaprakash et al. 2011; Schneider et al. 2012; Bruce et al. 2011).

As a result, the Li-S battery is inappropriate for a primary battery with a high energy density and a long service life. Another issue is sulfur's extremely low electrical conductivity, which results in poor electrochemical contact and inadequate usage of active elements in the cathode. As a result, while producing the sulfur cathode, a high number of conductive agents is required. The electrical conductivity of a sulfur cathode can be greatly improved by combining elemental sulfur with carbon (Choi et al. 2008; Wang et al. 2003) and conducting polymers (Yang et al. 2011). Extensive study on the electrochemical behavior under various operating circumstances is required for the successful creation of a Li-S battery, which is considered fascinating for the next-generation batteries (Marmorstein et al. 2000; Strauss et al. 2000).

A great deal of study has gone toward reducing the harmful effects of the polysulfide shuttle. The protection of the lithium anode (Lee et al. 2003) and the limiting of the ionic mobility of the polysulfide anions have been the focus of much of this research (Wang et al. 2002; He et al. 2005). However, because the lithium anode's protection creates a delayed response rate at the anode during the discharge cycle owing to passivation, the battery's power density suffers. The use of gel electrolytes and solid electrolytes to reduce the ionic mobility of the electrolytes has been described as a way to slow down the polysulfide shuttle (Huggins 2006). In addition, conductive additives with a wide ESA and active adsorbent agents must be added to the cathode. The shuttle effects have been reported to be softened by the fabrication of sulfur-carbon or sulfur-polymer composites (Lai et al. 2009).

Carbon-Based Materials for Lithium Metal Anodes

The commercialization of lithium–sulfur batteries has been boosted by recent advances in sulfur cathodes. Furthermore, research into the chemistry and performance of metallic lithium anodes has grown in recent years in an effort to understand the chemistry of lithium metal and develop more efficient metallic aluminum anodes. The formation of the dendrites of lithium and the disintegration of SEI layers on the material of anode pose severe safety problems and lower Coulombic efficiency and cycle life spans for lithium metal anodes, which limit their practical applicability. The lithium-electrolyte interface must be preserved, and the lithium plating/stripping mechanism improved in order to reduce the development of dendrites of lithium and promote uniform lithium distribution during cycling in order to enhance lithium metal anode performances (Bruce et al. 2012; Ji and Nazar 2010).

Anode stabilization via a variety of prospective protective approaches, such as stabilizing interfaces, building lithium hosts, and creating nucleation sites, is currently being investigated. By using the material's mechanical qualities, SEI layers that are stable need to have sufficient mechanical strength to restrict lithium dendrites from growing. Lithium anodes with uniform lithium deposition aid to reduce dendrite development during cycling and increase performance. This is a critical step in creating high-performance lithium metal anodes. Lithium hosts with high specific surface areas have been created as a means of reducing local current densities and increasing lithium's affinity for substrates (Aurbach et al. 2009; Xie et al. 2021; Park et al. 2010). Additionally, carbon-based materials have shown to be multifunctional in all of these solutions. This study will cover their usage in the shelter of lithium metal anodes.

Lithium Hosts Based on Carbon

Carbon elements have greater specific surface areas and electrical conductivities, which may minimize localized current densities and lessen the likelihood of dendrite development (Aurbach et al. 2009; Ling et al. 2012). Mukherjee et al. (Cheng et al. 2017) developed a porous graphene framework utilizing a thermal shock technique using the paper of graphene oxide and observed that defect places in the graphene may function as seed points and that Li prefers to drop near divacancies creating high Li content interpolated states. The porous graphene network's lithium capacity improved significantly as a result of the increasing density of flaws. Lithium was discovered to be enclosed inside the graphene pores, and the formation of Li dendrites was successfully prevented. LiTFSI-LiFSI dual-salt electrolytes were also employed by Zhang et al. (Xie et al. 2021) to generate an SEI layer in a separate study. The resultant non-layered graphene has a greater specific surface area (up to 1666 m^2 g^{-1}) and a conductivity of 435 S/cm, hence lowering surface current densities and inhibiting dendrite development. LiTFSI–LiFSI dual-salt electrolyte has also been described to assist the creation of a sustainable and resilient SEI layer, which resulted in an enhanced Coulombic efficiency of 93% at 2 mA/cm^2 for the anode.

Additionally, Lin et al. (Suo et al. 2013) produced anodes that had high electrochemical performance by synthesizing spark-gap rGO films and injecting molten Li into the films (Fig. 11.2b). A spark interaction between GO films and molten Li converted the GO films to rGO films, resulting in homogeneous nanogaps that may support lithium deposition concurrently in this work. The synergetic effects of rGO's lithiophilicity and the capillary forces of the nanogaps aided in the uniform dispersion and deposition of Li in this case. Additionally, the films of rGO have been shown to be capable of accommodating volume variations during lithium stripping/plating and may contribute significantly to the creation of a steady artificial interface that stabilizes the SEI layer. Consequently, the resultant anode demonstrated excellent electrochemical characteristics (Fig. 11.2c).

Stabilization of SEI Layers

Lithium anodes may be protected against undesirable side reactions by the creation of SEI layers on their surface during electrochemical processes. SEI layers, on the other hand, are often brittle and susceptible to fracture during cycling. This shattering of the SEI layer may result in the formation of dendrites of lithium and continual electrolyte usage. As a consequence of these phenomena, Coulombic efficiencies are reduced, and capacity losses occur rapidly. As a result, the SEI layer must be balanced to inhibit lithium dendrite formation and limit electrolyte consumption, hence assuring high-performance batteries (Bruce et al. 2012; Ji and Nazar 2010; Lin et al. 2017). On lithium metal anodes, the ideal SEI layer should

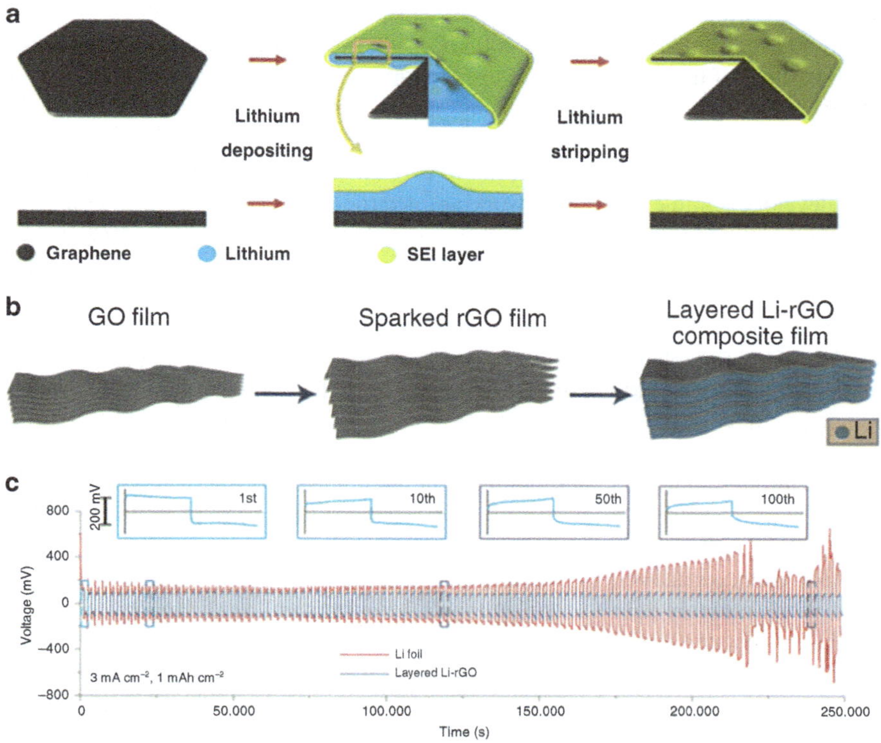

Fig. 11.2 (**a**) The depositing/stripping of lithium on graphene flakes is shown schematically. (Permission obtained for reproduction from Xie et al. 2021) (**b**) Fabrication of the layered Li–rGO film; (**c**) cycle life of the symmetric Li–rGO electrode (blue) and bare Li foil (red) in the first 250,000 cycles at 3 mA cm^{-2}. (Permission was obtained from Suo et al. 2013)

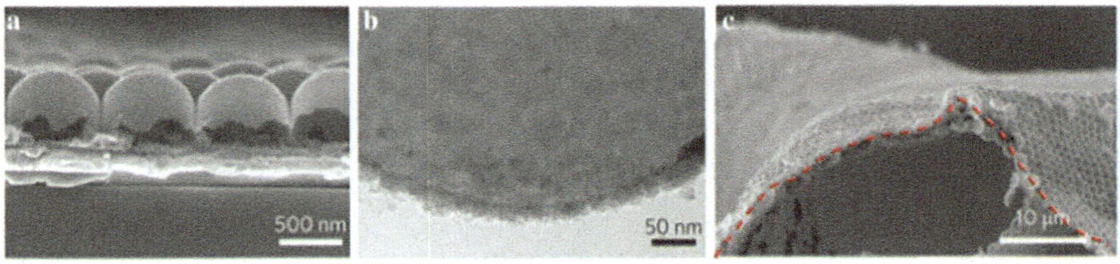

Fig. 11.3 (**a**) SEM picture of hollow carbon nanospheres in cross-section, (**b**) TEM image of hollow carbon nanospheres, and (**c**) SEM image of a hollow carbon nanospheres thin film peeling away from the Cu substrate. (Permission acquired to reproduce from Park et al. 2010)

exhibit mechanical flexibility, chemical stability, and ionic conductivity (Bruce et al. 2012). Then Zheng et al. (Park et al. 2010) came up with the idea of making an anode that was made of metal and covered in carbon that would help make the SEI layer more stable and make it easier for lithium metal to be spread out evenly beneath (Fig. 11.3 illustrates the shape of the hollow carbon spheres). As a consequence, a steady SEI layer developed on the upper layer of the spheres of carbon, with the amorphous carbon layer's having 200 GPa value of Young's modulus, preventing the formation of lithium dendrites. Furthermore, the amorphous carbon layer did not block charge movement during cycling, and the poor contact among the layer of carbon and the foil of copper enabled Li to deposit in the vacuum regions among the layer of carbon and the current collector of copper. Furthermore, as a consequence of this developed framework for a working electrode, uncontrolled development of dendrites of lithium was stopped after 50 cycles having 1 mA cm^{-2} current density, and the electrode displayed an enhanced Coulombic efficiency of 99% for more than 150 cycles at 0.25 mA cm^{-2}.

Apart from amorphous layers of carbon, Yan et al. (Xiong et al. 2014) show in their work that graphene may also be used as amended layers on copper substrates and, like amorphous carbon, graphene is chemically firm and has a high mechanical toughness having a Young's modulus approaching 1.0 TPa. Additionally, the graphene hexagonal ring's pore width is too tiny to enable electrolytes or Li metals to flow through, but point and line flaws in the graphene plane may permit for the entrance of Li ions, which can be fastened by SEI layers to stop electrolyte infiltration.

As a result, a steady SEI layer may be assembled on top of the graphene, with Li depositing smoothly between the latter and the current collector. According to this, anodes of copper/graphene may achieve a Coulombic efficiency of 93% after 50 cycles having 1.0 mA/cm^2 current density. Kim et al. (Bates et al. 1993) have found that the regulated development of Li may be accomplished by joining the benefits of multilayered graphene-modified Li foil with electrolyte additives. The scientists propose that multilayered graphene may be used to prevent head-on contact among SEI layers and Li metal, hence increasing the battery's Coulombic efficiency, and that the addition of Cs ions could increase lithium ion transport and reduce the creation of dendrites of lithium. As a consequence, bilayer graphene enhanced with Cs ion additions exhibits collective properties and increased cycle stability.

Lithiophilicity of Carbon-Based Composites

The affinity of Li for substrates may be used to minimize nucleation barriers and create homogeneous lithium dispersion and deposition (Suo et al. 2013). To improve the lithiophilicity of carbon fiber frameworks, Liang et al. (Aurbach et al. 2009) synthesized a Li composite with good gravimetric ability and cycle stability using CVD techniques on carbon fibers with silicon shells. The researchers claimed that, as compared to bare carbon frameworks, their silicon-decorated carbon networks demonstrated high wettability to molten Li, enabling Li to be evenly transferred onto the layer of carbon fiber frameworks, with the 3D conductive trails ensuring quick electron movement. As a consequence, the Li composite produced had a high gravimetric ability of 2000 mAh g^1 and an enhanced cycling stability, with a Li plating/stripping over potential of less than 90 mV at 3 mA cm^{-2} after 80 cycles. Furthermore, Yang et al. (Hao et al. 2016) revealed that carbon nanofibers coated with ultrafine silver nanoparticles that function as lithium metal nucleation sites may enable the homogeneous deposition of lithium onto the surface of 3D structures. Such modified 3D carbon networks, according to the findings of this study, may successfully inhibit the establishment of lithium dendrites and provide smooth lithium deposition layers on the anode while maintaining a low lithium plating/stripping over potential of less than 25 mV at 0.5 mA cm^{-2}. Furthermore, nitrogen-doped graphene is a lithiophilic material with distributed lithiophilic functional groups that promote lithium nucleation and inhibit dendrite formation (Lin et al. 2018). Additionally, simulations indicate that pyrrolic-N and pyridinic-N nitrogen have high binding energies with lithium, indicating that these nitrogen groups are not only lithiophilic but also electrochemically stable, meaning they will not react with li atoms. But since quaternary nitrogen is unable of providing additional lone-pair electrons necessary for lithium adsorption, interactions between it and the lithium atoms are weak. According to these results, nitrogen-doped graphene can inhibit dendrite formation while exhibiting high Coulombic efficiency and long-term cycling stability, attaining a Coulombic efficiency of 98% and staying stable for about 200 cycles at a current density of 1.0 mA cm^{-2}.

Conclusion and Outlook

Despite the fact that LMBs have been studied for decades, there is still a long way to go before they can be commercialized. The primary challenges with Li-MAs are the restricted cycling performance and the security issue. 3D C-Hosts might be a viable option for overcoming the drawbacks of Li-MAs discussed above. The key C-Hosts for Li-MAs, such as CNTs, graphene, biomass-derived carbon, CFs, and so on, have been reviewed and retrospect in this study. Because of the ultrahigh ESA and advantageous electron conductivity of carbon materials, a decreased and well-distributed local current density, as well as quick kinetics, may be tailored.

Meanwhile, C-Hosts are mostly lightweight and have ample void space, allowing Li anodes to have a fulfilled specific capacity and eased volume change. Furthermore, these carbon compounds may be composited with other materials to improve Li-ion transport and Li metal affinity. Defect engineering techniques like functional groups and intrinsic defects may also be used to activate carbon materials and lower the Li nucleation barrier. Based on theoretical dendritic development models, the aforesaid benefits of carbon-based materials can eventually result in good dendrite delay ability and improved electrochemical performance.

Some concerns and directions should be given greater consideration in order to construct 3D C-Hosts for Li-MAs in the future:

I. During initial cycles, when the SEI layer is largely produced, C-Hosts always consume extra liquid electrolyte and Li supply due to their ultrahigh ESA. The structure of C-Hosts must be properly optimized, including the use of appropriate ESA, pore size, and surface stabilization, among other things. For both extended cyclic life span and high energy density, a trade-off between elevated ESA and low electrolyte use may be necessary.

II. Short electron transfers on the top surface of C-Hosts can cause substantial Li metal deposition on the top rather than within, causing C-Hosts to be underutilized. To sustain a homogenous electron field and Li-ion flux across the hosts, novel designs and concepts, i.e., gradient lithiophilic and gradient conductive hosts, are required.

III. In the presence of excessive areal capacity or elevated current density, a stable and robust SEI layer is difficult to achieve. The porous C-Hosts, in particular, are always on limits to electrolytes, causing the SEI layer to become unstable. To stabilize the anode surface, a combination of 3D Li-MAs and artificial SEI engineering ought to be investigated more.

IV. The interaction between solid-state electrolyte and 3D Li-frameworks is still a significant issue. The use of 3D C-Hosts may cause anode surface flatness to degrade, resulting in weak interfacial contact and high resistance between solid-state electrolytes and anodes.

V. Ultra-flexible electronic gadgets have a large commercial potential, necessitating the development of next-generation rechargeable batteries. C-Hosts, including graphene, carbon cloth, and carbon nanotubes, are ideal for constructing flexible LMBs, because they can fold and aid to limit volume expansion while still having strong thermal diffusivity.

VI. A simple, low-cost, and environmentally friendly synthesis procedure is also essential for the use of C-Hosts.

To summarize, the C-Hosts are abundant and diverse, demonstrating multifunctional capabilities for Li-MAs. Defect engineering, structural design, hybridizing with other suitable materials, and in situ observation have all improved. More options for carbon materials in Li-metal anode applications should be investigated, and the aforementioned issues must be resolved entirely. We think that 3D C-Hosts can efficiently support Li-MAs for long-term cyclability and elevated energy density. The aforementioned discussions have the potential to speed up the commercialization of LMBs.

References

Aurbach D, Pollak E, Elazari R, Salitra G, Kelley CS, Affinito J (2009) On the surface chemical aspects of very high energy density, rechargeable Li–sulfur batteries. J Electrochem Soc 156: A694

Barchasz C, Leprêtre J-C, Alloin F, Patoux S (2012) New insights into the limiting parameters of the Li/S rechargeable cell. J Power Sources 199:322–330

Barghamadi M, Kapoor A, Wen C (2013) A review on Li-S batteries as a high efficiency rechargeable lithium battery. J Electrochem Soc 160:A1256

Bates J, Dudney N, Gruzalski G, Zuhr R, Choudhury A, Luck C, Robertson J (1993) Fabrication and characterization of amorphous lithium electrolyte thin films and rechargeable thin-film batteries. J Power Sources 43:103–110
Bruce PG, Hardwick LJ, Abraham KM (2011) Lithium-air and lithium-sulfur batteries. MRS Bull 36:506–512
Bruce PG, Freunberger SA, Hardwick LJ, Tarascon J-M (2012) Li–O2 and Li–S batteries with high energy storage. Nat Mater 11:19–29
Chen Q, Zheng H, Yang Y, Xie Q, Ma Y, Wang L, Peng DL (2019) Ion-and electron-conductive buffering layer-modified Si film for use as a high-rate long-term lithium-ion battery anode. ChemSusChem 12:252–260
Cheng X-B, Zhang R, Zhao C-Z, Zhang Q (2017) Toward safe lithium metal anode in rechargeable batteries: a review. Chem Rev 117:10403–10473
Cheng X-B, Yan C, Zhang X-Q, Liu H, Zhang Q (2018) Electronic and ionic channels in working interfaces of lithium metal anodes. ACS Energy Lett 3:1564–1570
Cheon S-E, Ko K-S, Cho J-H, Kim S-W, Chin E-Y, Kim H-T (2003) Rechargeable lithium sulfur battery: I Structural change of sulfur cathode during discharge and charge. J Electrochem Soc 150:A796
Choi Y-J, Chung Y-D, Baek C-Y, Kim K-W, Ahn H-J, Ahn J-H (2008) Effects of carbon coating on the electrochemical properties of sulfur cathode for lithium/sulfur cell. J Power Sources 184: 548–552
Demir-Cakan R (2015) Targeting the role of lithium sulphide formation for the rapid capacity fading in lithium-sulphur batteries. J Power Sources 282:437–443
Hao X, Zhu J, Jiang X, Wu H, Qiao J, Sun W, Wang Z, Sun K (2016) Ultrastrong polyoxyzole nanofiber membranes for dendrite-proof and heat-resistant battery separators. Nano Lett 16: 2981–2987
He X, Shi Q, Zhou X, Wan C, Jiang C (2005) In situ composite of nano SiO2–P (VDF-HFP) porous polymer electrolytes for Li-ion batteries. Electrochim Acta 51:1069–1075
Huggins RA (2006) Lithium electrode reactants containing hydrogen or water. J Power Sources 153:365–370
Jayaprakash N, Shen J, Moganty SS, Corona A, Archer LA (2011) Porous hollow carbon@ sulfur composites for high-power lithium–sulfur batteries. Angew Chem Int Ed 50:5904–5908
Ji X, Nazar LF (2010) Advances in Li–S batteries. J Mater Chem 20:9821–9826
Jin S, Jiang Y, Ji H, Yu Y (2018) Advanced 3D current collectors for lithium-based batteries. Adv Mater 30:1802014
Jung Y, Kim S (2007) New approaches to improve cycle life characteristics of lithium–sulfur cells. Electrochem Commun 9:249–254
Kolosnitsyn V, Karaseva E (2008) Lithium-sulfur batteries: problems and solutions. Russ J Electrochem 44:506–509
Lai C, Gao X, Zhang B, Yan T, Zhou Z (2009) Synthesis and electrochemical performance of sulfur/highly porous carbon composites. J Phys Chem C 113:4712–4716
Lee YM, Choi N-S, Park JH, Park J-K (2003) Electrochemical performance of lithium/sulfur batteries with protected Li anodes. J Power Sources 119:964–972
Lin D, Liu Y, Cui Y (2017) Reviving the lithium metal anode for high-energy batteries. Nat Nanotechnol 12:194–206
Lin C-F, Qi Y, Gregorczyk K, Lee SB, Rubloff GW (2018) Nanoscale protection layers to mitigate degradation in high-energy electrochemical energy storage systems. Acc Chem Res 51:97–106
Ling C, Banerjee D, Matsui M (2012) Study of the electrochemical deposition of Mg in the atomic level: why it prefers the non-dendritic morphology. Electrochim Acta 76:270–274
Liu B, Zhang J-G, Xu W (2018) Advancing lithium metal batteries. Joule 2:833–845
Marmorstein D, Yu T, Striebel K, McLarnon F, Hou J, Cairns EJ (2000) Electrochemical performance of lithium/sulfur cells with three different polymer electrolytes. J Power Sources 89:219–226

Mikhaylik YV, Akridge JR (2004) Polysulfide shuttle study in the Li/S battery system. J Electrochem Soc 151:A1969
Mo F, Liang G, Huang Z, Li H, Wang D, Zhi C (2020) An overview of fiber-shaped batteries with a focus on multifunctionality, scalability, and technical difficulties. Adv Mater 32:1902151
Monroe C, Newman J (2003) Dendrite growth in lithium/polymer systems: a propagation model for liquid electrolytes under galvanostatic conditions. J Electrochem Soc 150:A1377
Myung S-T, Hitoshi Y, Sun Y-K (2011) Electrochemical behavior and passivation of current collectors in lithium-ion batteries. J Mater Chem 21:9891–9911
Niu C, Pan H, Xu W, Xiao J, Zhang J-G, Luo L, Wang C, Mei D, Meng J, Wang X (2019) Self-smoothing anode for achieving high-energy lithium metal batteries under realistic conditions. Nat Nanotechnol 14:594–601
Park M, Zhang X, Chung M, Less GB, Sastry AM (2010) A review of conduction phenomena in Li-ion batteries. J Power Sources 195:7904–7929
Ryu H, Ahn H, Kim K, Ahn J, Lee J-Y, Cairns EJ (2005) Self-discharge of lithium–sulfur cells using stainless-steel current-collectors. J Power Sources 140:365–369
Schneider H, Garsuch A, Panchenko A, Gronwald O, Janssen N, Novák P (2012) Influence of different electrode compositions and binder materials on the performance of lithium–sulfur batteries. J Power Sources 205:420–425
Strauss E, Golodnitsky D, Peled E (2000) Study of phase changes during 500 full cycles of Li/composite polymer electrolyte/FeS2 battery. Electrochim Acta 45:1519–1525
Sun J, Huang Y, Wang W, Yu Z, Wang A, Yuan K (2008) Application of gelatin as a binder for the sulfur cathode in lithium–sulfur batteries. Electrochim Acta 53:7084–7088
Suo L, Hu Y-S, Li H, Armand M, Chen L (2013) A new class of solvent-in-salt electrolyte for high-energy rechargeable metallic lithium batteries. Nat Commun 4:1–9
Wang J, Yang J, Xie J, Xu N, Li Y (2002) Sulfur–carbon nano-composite as cathode for rechargeable lithium battery based on gel electrolyte. Electrochem Commun 4:499–502
Wang J, Yang J, Wan C, Du K, Xie J, Xu N (2003) Sulfur composite cathode materials for rechargeable lithium batteries. Adv Funct Mater 13:487–492
Wang J, Chew S, Zhao Z, Ashraf S, Wexler D, Chen J, Ng S, Chou S, Liu H (2008) Sulfur–mesoporous carbon composites in conjunction with a novel ionic liquid electrolyte for lithium rechargeable batteries. Carbon 46:229–235
Xie L, Tang C, Bi Z, Song M, Fan Y, Yan C, Li X, Su F, Zhang Q, Chen C (2021) Hard carbon anodes for next-generation Li-ion batteries: review and perspective. Adv Energy Mater 11: 2101650
Xiong S, Xie K, Diao Y, Hong X (2014) Characterization of the solid electrolyte interphase on lithium anode for preventing the shuttle mechanism in lithium–sulfur batteries. J Power Sources 246:840–845
Yang Y, Yu G, Cha JJ, Wu H, Vosgueritchian M, Yao Y, Bao Z, Cui Y (2011) Improving the performance of lithium–sulfur batteries by conductive polymer coating. ACS Nano 5:9187–9193
Yang C-P, Yin Y-X, Zhang S-F, Li N-W, Guo Y-G (2015) Accommodating lithium into 3D current collectors with a submicron skeleton towards long-life lithium metal anodes. Nat Commun 6:1–9
Yang Y, Xiong J, Zeng J, Huang J, Zhao J (2018) VGCF 3D conducting host coating on glass fiber filters for lithium metal anodes. Chem Commun 54:1178–1181
Ye H, Xin S, Yin YX, Guo YG (2017) Advanced porous carbon materials for high-efficient lithium metal anodes. Adv Energy Mater 7:1700530
Zhang SS, Foster D, Read J (2010) A high energy density lithium/sulfur–oxygen hybrid battery. J Power Sources 195:3684–3688

Zhang R, Li NW, Cheng XB, Yin YX, Zhang Q, Guo YG (2017) Advanced micro/nanostructures for lithium metal anodes. Adv Sci 4:1600445

Zheng H, Zhang Q, Chen Q, Xu W, Xie Q, Cai Y, Ma Y, Qiao Z, Luo Q, Lin J (2020) 3D lithiophilic–lithiophobic–lithiophilic dual-gradient porous skeleton for highly stable lithium metal anode. J Mater Chem A 8:313–322

Chapter 12
Recycling of Lithium Sulfur Batteries

Tausif Ahmad, Muhammad Sagir, Muhammad Arif, and M. B. Tahir

Abstract The high-energy battery systems are getting an indispensable significance in accordance to the exponentially growing needs of electronic appliances and electromotive vehicles, having excellent cycle stability. The lithium sulfur battery systems can be the best substitutes for future storage electromotive schemes than their rivals such as lithium-ion battery systems due to the high energy density and excellent life cycle. Hence keeping in view, the growing global demands of economically sustainability and motivations toward product recycling, there is strong need to develop and evaluate the new technologies for digging out the sustainable recycling potential. In this context, new recycling techniques should be focused on the recovery of precious metals along with valuable construction materials for achieving a decent vale of recycling efficiency (RE). The current chapter is planned with an aim to explain the thermal and hydrometallurgical processes in a safer

T. Ahmad (✉)
Institute of Chemical and Environmental Engineering, Khwaja Fareed University of Engineering and Information Technology, Rahim Yar Khan, Punjab, Pakistan

Institute of Energy & Environmental Engineering, University of the Punjab, Lahore, Pakistan
e-mail: tausif.ieee@pu.edu.pk

M. Sagir
Institute of Chemical and Environmental Engineering, Khwaja Fareed University of Engineering and Information Technology, Rahim Yar Khan, Punjab, Pakistan
e-mail: dr.msagir@kfueit.edu.pk

M. Arif
Institute of Chemical and Environmental Engineering, Khwaja Fareed University of Engineering and Information Technology, Rahim Yar Khan, Punjab, Pakistan

Centre for Thermal and Renewable Energy Research, Khwaja Fareed University of Engineering and Information Technology, Rahim Yar Khan, Punjab, Pakistan
e-mail: muhammad.arif@kfueit.edu.pk

M. B. Tahir
Institute of Physics, Khwaja Fareed University of Engineering and Information Technology, Rahim Yar Khan, Punjab, Pakistan
e-mail: m.bilaltahir@kfueit.edu.pk

M. S. Tahir et al. (eds.), *Lithium-Sulfur Batteries: Key Parameters, Recent Advances, Challenges and Applications*, Springer Tracts in Electrical and Electronics Engineering, https://doi.org/10.1007/978-981-99-2796-8_12

environment for achieving the maximum recycling efficiency of lithium sulfur battery systems. In addition to that, designing of recycling process for lithium sulfur battery systems has also been described and explained. The multiple steps-based process including sorting, discharging, vacuum distillation, mechanical separation, and hydro-treatment was selected for the recovery of elements in accordance to the chemical composition of lithium sulfur battery systems. Both thermal and mechanical steps were employed for the separation of black powder, having all the material resources, recovered from metal castings. Various basic and acidic leaching solutions were trialed for evaluation of their performances, followed by addition of additives for precipitation and adjusting the pH of aqueous solution-based black mass powder. Finally, chemical analysis was conducted for having the evaluation in terms of terms of yield and purity to recover the maximum amount of lithium sulfur battery fractions including lithium, sulfur, carbon, and aluminum.

Keywords High-energy battery systems · Recycling efficiency · Aqueous solution

Theoretical Background

Current society is in a bad need of sustainable technologies for the depletion of greenhouse gas, which is severely damaging our environment. Furthermore, it is ultimately affecting various industrial sectors such as automobile industry, which has caused the shifting of technology from combustion to electric engine (Al Rashdan et al. 2019; Kawamoto et al. 2019; Thackeray et al. 2012). One of the major key technologies, being widely employed in electric vehicle industry, is the lithium batteries that have also deeply rooted its branches in small electronic industry such as laptops and smartphones (Kawamoto et al. 2019; Helbig et al. 2018). The exponentially growing significance of electric automotive industry (56%) has also been reflected through production of 1 million electric vehicles in 2017 than 2016 (Kittner et al. 2020; Harper et al. 2019).

The lithium-ion battery (LIB) is a well-known form of lithium battery technology since the 1990s as shown in Fig. 12.1.

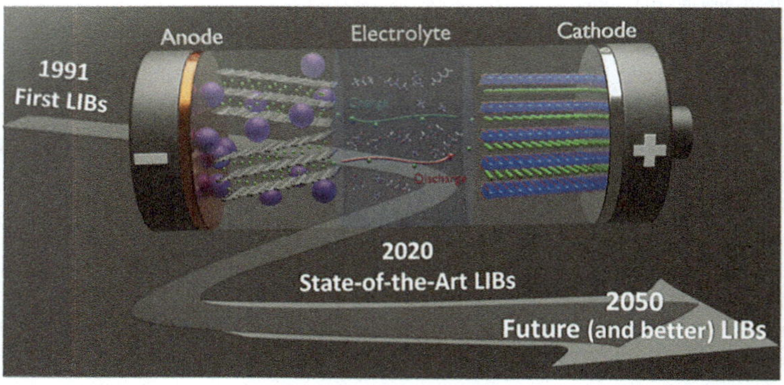

Fig. 12.1 Schematic illustration of the state-of-the-art lithium-ion battery (Armand et al. 2020)

However, the continuous development in LIB technology with different forms such as NMC, NCA, LMO, LPF, etc. has been facing a physical limit at a specific gravimetric energy density of 350–400 Wh/kg (Harper et al. 2019; Korthauer 2013; Kurzweil and Dietlmeier 2018). The production of batteries on a commercial scale is done via utilizing critical materials such as graphite (C) or cobalt (Co) and strategic raw materials including lithium (Li) or nickel (Ni) (Helbig et al. 2018; Harper et al. 2019; Bureau de Recherches Géologiques et Minières, Deloitte Sustainability 2017; Regett and Fischhaber 2017). The electromotive industry in European countries has been facing various issues such as the need of abovementioned materials, complex geographical positions of mining spots along with import channel barriers. Furthermore, the social critical aspects associated with other raw materials such as cobalt should also be kept in mind while working on development of electromotive industry (Helbig et al. 2018; Harper et al. 2019). The better substitute technologies to LIB are lithium-air (Li-A) and lithium-sulfur batteries (Li-SBs). The aforementioned approaches are yet in a birth stage, which will hopefully address all the associated issues and challenges associated with LIBs (Korthauer 2013; Bruce et al. 2011). Though both technologies have almost same structure containing electrodes and liquid electrolyte however are much different with respect to their fundamental cell chemistry. The Li-SBs contain an anode of pure lithium and a cathode, made of sulfur-carbon cathode, where lithium holds significant importance and carbon grips electrically nonconductive features (Korthauer 2013). The Li-SBs cell can contain different content ratio of sulfur in between 50% and 70%, while the rest of the space is filled by small quantities of binders along with carbon. Generally, lithium bis(rifluoromethanesulfonyl)imide ($LiN(SO_2\,CF_3)_2$, LiTFSI) is employed as conducting salts. The cell reaction occurred through the creation of various polysulfides such as (from S_8 over Li_2S_4 up to Li_2S) that makes its a perfectly polysulfides-based battery (Kurzweil and Dietlmeier 2018). The LiSBs can theoretically obtain 600 Wh/kg of specific energy density due to the materials employed in its construction. The specific energy density of 400 Wh/kg has already been achieved on the cell level of LiSBs (Dörfler et al. 2020). Another benefit other than high energy density associated with LiSBs is the employment of easily available sulfur as an active material, which contains toxicology nonhazardous and cost-effective features. However, the cycle stability of LiSBs has been considered as a major challenge, instead of having various advantages. The shuttle mechanism (an effect induced by the cycle in which cathodically dissolved polysulfides and diffuse undesirably to lithium anode. Where they get reduced into lower polysulfides $[S_{n\text{-}x}]^{2-}$ and get back to the cycle (Kurzweil and Dietlmeier 2018)) is another operational impediment associated with LiSBs, where cathode reaction is occurred at anode and causes the continuous discharging the cell, known as self-discharge of the cell. Currently, LiSBs can only achieve a few hundreds of charging cycles due to the aforementioned operational and construction materials-associated issues (Li et al. 2019; He et al. 2019; Liu et al. 2019). The charging LiSBs are composed of diffusion of Li ions from carbon-sulfur cathode to lithium anode, whereas discharging shows its reverse process. Different polysulfides with various state of charge (SOH) and charging sequence creations such as S_8, Li_2S_8, Li_2S_6, Li_2S_4, Li_2S_3, Li_2S_2, and lastly Li_2S. Therefore, relying on SOH, contribution of Li_2S_8 and Li_2S will be different having dissimilar features including stability and composition materials (Frey 2017).

Irrespective of which lithium approach will progress, used LiSBs are undoubtedly a great source important and value-added materials, which will not only reduce the burden of import dependencies but also enhance the growing production of electromotive industry. The major key objective for the recycling technique is to reduce the negative effects, generated by waste of batteries, as heavy metals with toxic nature should not be directly released into the environment. The circular economy can be achieved through recovery of the value-added resources from the batteries wastes. However, recycling of LiSBs is yet in the stage obscurity and underlined. In this context, development of recycling techniques with decent economical, environmental, and compatible features is great need of current time (Harper et al. 2019). There are various efficient and well-proved recycling methods including mechanical and hydrometallurgical techniques (Al Rashdan et al. 2019). However, no significant and well-established recycling approach has been developed for the LiSBs. Various European financed projects like HELIS (high-energy lithium sulfur cells and batteries) (European Commission 2015) or LISA (lithium sulfur for safe road electrification) (European Commission 2019) attempted battery evolution and circular economical techniques; however, no significant outcomes have been reported till today. Furthermore, several approaches with improved design strategy have shown their efficiency for the recycling of materials such as Co_3Mo_3C-separator (Zhang et al. 2020). Schwich et al. (2020) made the very first attempt with a desire to recover all the elements with a legal binding of 50 wt.% for the recycling of lithium sulfur battery waste. This legal requirement is implemented in Europe battery directives 2006/66/EG (Directive 2006) and 493/2012 (Commission Regulation 2012), where 50 wt.% is declared as the benchmark value for the recycling efficiency of battery, as mentioned in formula 12.1.

$$\mathrm{RE}[\%] = \frac{\sum m_{\mathrm{output}}}{m_{\mathrm{input}}} \times 100\ [\mathrm{mass.}\%] \qquad (12.1)$$

Therefore, current targets can only be achieved through developing an efficient and suitable recycling technique. The high reactivity of metallic lithium anode in LiSBs may lead toward exothermal oxidation, causing serious safety concerns (Placke et al. 2017; Yamaki et al. 1998), which can be handled through adopting a pretreatment approach before going to metallurgical process. Working in an inert atmosphere to get rid of oxidation derived environment is harder to be executed (Heimes et al. 2018). In addition to that, heating above 180 °C in air (Brode and Mark 1982) can cause the ignition of metallic lithium that can happen due to mechanical actions such as shredding (Yan 2015) or the presence of moisture (Xu et al. 2008). As LiSBs systems are very new and located in immature stage, no significant studies have been conducted on its recycling procedures. However, a limited literature exists on the recycling approaches, employed for getting the maximum recovery of LiSBs fractions, as described in below sections.

Recycling Techniques for LiSBs

Thermal Pretreatment

Pyrolysis is one of the forms of thermal pretreatments, where deactivation of cells occurred at high temperature of 600 °C in the nonappearance of oxygen (Martens and Goldmann 2016). An optimum value of temperature 550 °C is reported by Chen et al. (2018) via employing LCO cells for the removal of organic components such as binders. Pyrolysis and incineration in the presence of oxygen are also employed for the thermal pretreatment process to safely deactivate the cells in an efficient way for downstream recycling with no uncontrolled thermal runway (Kwade and Diekmann 2018; Schwich et al. 2020). The thermal pretreatment is also recommended for having an easy way of detaching substrate foils, active mass (Werner et al. 2020), and hydrometallurgical technique (Chen et al. 2019).

Metallurgical Recycling

The metallurgical process can be done through hydro- and pyrometallurgical techniques (Zheng et al. 2018).

Pyrometallurgical process is recommended for recovery of Co and Ni alloys (Pinegar and Smith 2019), whereas hydrometallurgical technique is mostly suggested for the separation graphite, aluminum, and lithium, as shown in Fig. 12.2 (Pinegar and Smith 2019; Lombardo et al. 2020; Becker et al. 2019). Therefore, hydrometallurgical is most preferred and suitable approach for achieving the circular economy. It gives high yield with low energy utilization, however having slower kinetics (Lombardo et al. 2020; Becker et al. 2019).

Wet Chemical Processing

Chemical processing is majorly composed of leaching, solvent extraction, and ion exchange other than physical techniques such as flotation, etc. (He et al. 2017; Chagnes and Swiatowska 2015). The recent studies have shown the employment hydrochloric acid, phosphoric acid, and nitric acid as inorganic solvents whereas citric acid, malic acid, and lactic acids as organic acids for lithium ions battery recycling process, as shown in Fig. 12.3.

All these abovementioned solvents need different operating temperatures, concentrations, and liquids to solid ratios (Or et al. 2020). Zou et al. have reported the 100% leaching performance via employing four-molar sulfuric acid along with 30% wt. hydrogen peroxide. Lithium was recovered 80% yield in the form lithium carbonate via employing sodium carbonate (Zou et al. 2013). Other than acidic

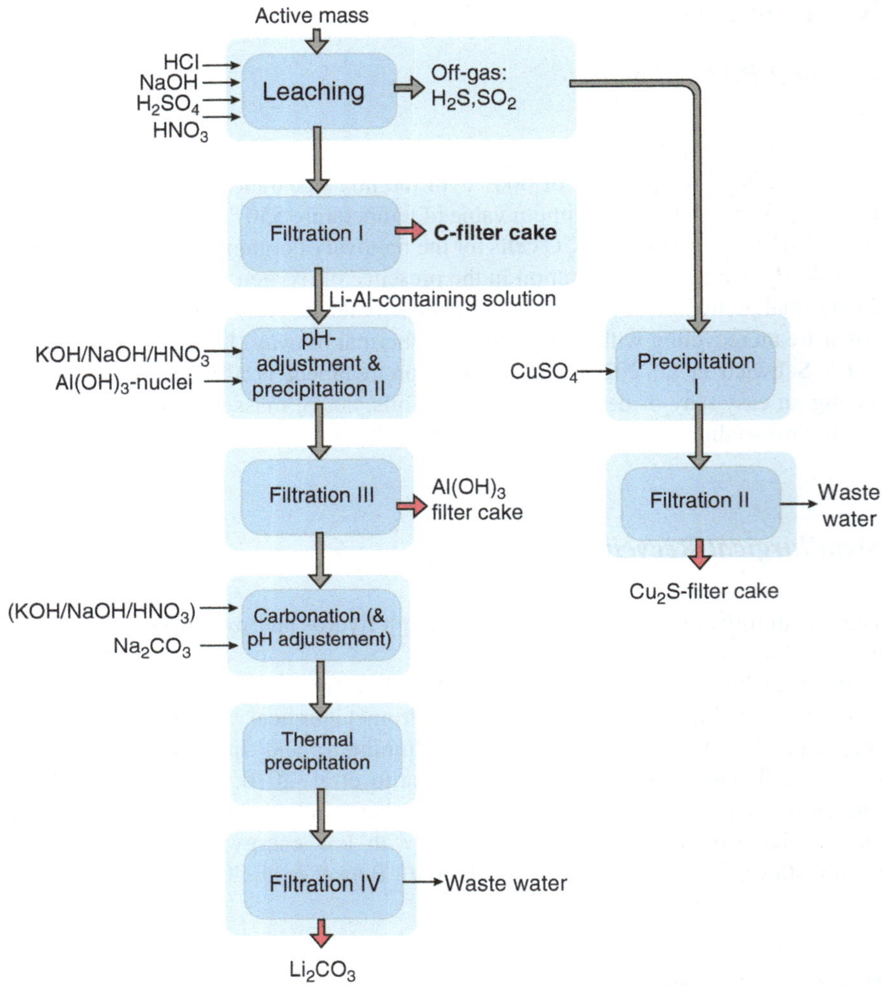

Fig. 12.2 Proposed pathway for hydrometallurgical process for lithium sulfur batteries (Schwich et al. 2020)

leaching, Ferreira et al. (2009) investigated the role of alkaline leaching, where aluminum depicted a decent recyclability performance (lithium recovery 100%) through maintaining the pH by adding sulfuric acid and hydrogen peroxide. This approach is quite new and innovative that was established with the help, received from hydrometallurgical technique done for recycling of lithium ions battery recycling. However, formation of various kinds of intermediates with lithium and sulfur during the recycling process is great challenge (Yan et al. 2016).

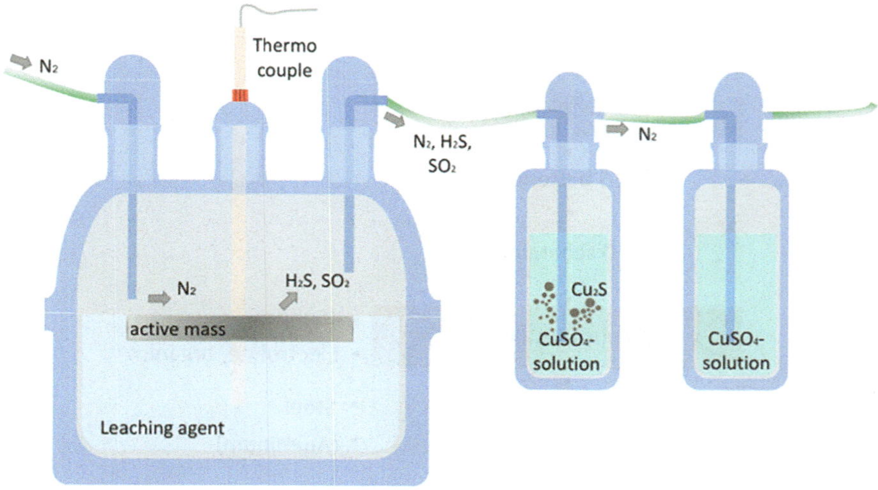

Fig. 12.3 A schematic diagram of leaching step for the lithium sulfur recycling

Design of Recycling Flowchart

The lithium sulfur battery recycling process, made by Accurec, has been given below (Fig. 12.4).

The recycling process is composed of various steps including sorting, hydrometallurgical metal separation, done at the final stage. It requires a continuous input material stream without any impurity that may have impact on the stability cycle of recycling. This kind of contamination may also be caused by other kinds of battery such as lead acid or lithium-ion battery. All the recycling steps have been explained in the below sections.

Disassembly The disassembling is performed for large battery systems to separate the fractions with reduced volume for having any easy and continuous mass flow toward next steps. Nowadays, sorting and disassembling of batteries are performed manually.

Discharging The lithium batteries, having high state of charge (SOC), are sent for thermal runway or a high explosion during the recycling process. The reactivities of batteries are reduced through discharging for the safety of the equipment. Accurec has developed an efficient discharging scheme for lithium-ion batteries after going through a series of experimental procedures for years that performs the discharging process through employing electric conductive liquid.

Vacuum Distillation The deactivation of lithium sulfur for its recycling is done at its final stage in vacuum distillation for having the smooth process, followed by the mechanical techniques to recover the organic fractions, which may have bad effect on downstream hydro-treatment. During the vacuum distillation, sulfur may be

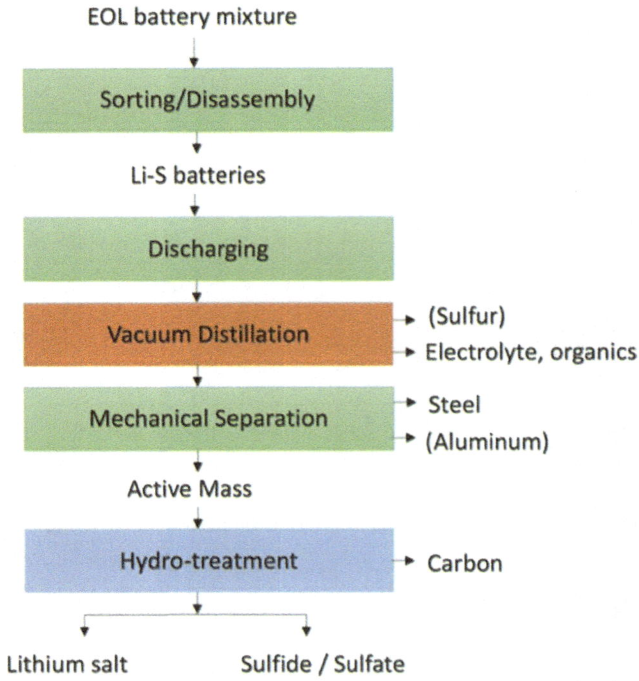

Fig. 12.4 LiSBs recycling flowchart planned by Accurec (Velázquez-Martínez et al. 2019)

formed in volatile state due to its excessive amount or its rapid reaction with metals. Its intensity be avoided or minimized through installing medium temperature condenser for the separation of sulfur.

The organic solvents and sulfur can be separated through operating vacuum distillation in two zones, determined via using vapor pressure curves, produced by materials, as shown in Fig. 12.5.

The zone 1 is operated at 1–10 mbar pressure, 400–500 °C temperature, and zone 2 is performed at 1–10 mbar pressure along with 100–150 °C temperature. A design scheme for the vacuum distillation furnace has also been shown in Fig. 12.6, where lithium sulfur battery is introduced in the upper section and processed under zone 1 conditions. Under zone 1 conditions, organic solvents are moderately degraded through evaporation, while sulfur is also converted into vapor dorm through evaporation. These gases will be shifted toward medium temperature condenser that was being operated at zone 2 conditions, where sulfur is made separate from organic solvents. The organic solvents are collected at the bottom section through condensation, occurring at low temperature. The uncondensed leftover gases have not been analyzed, which are sent for off-gas evaluation.

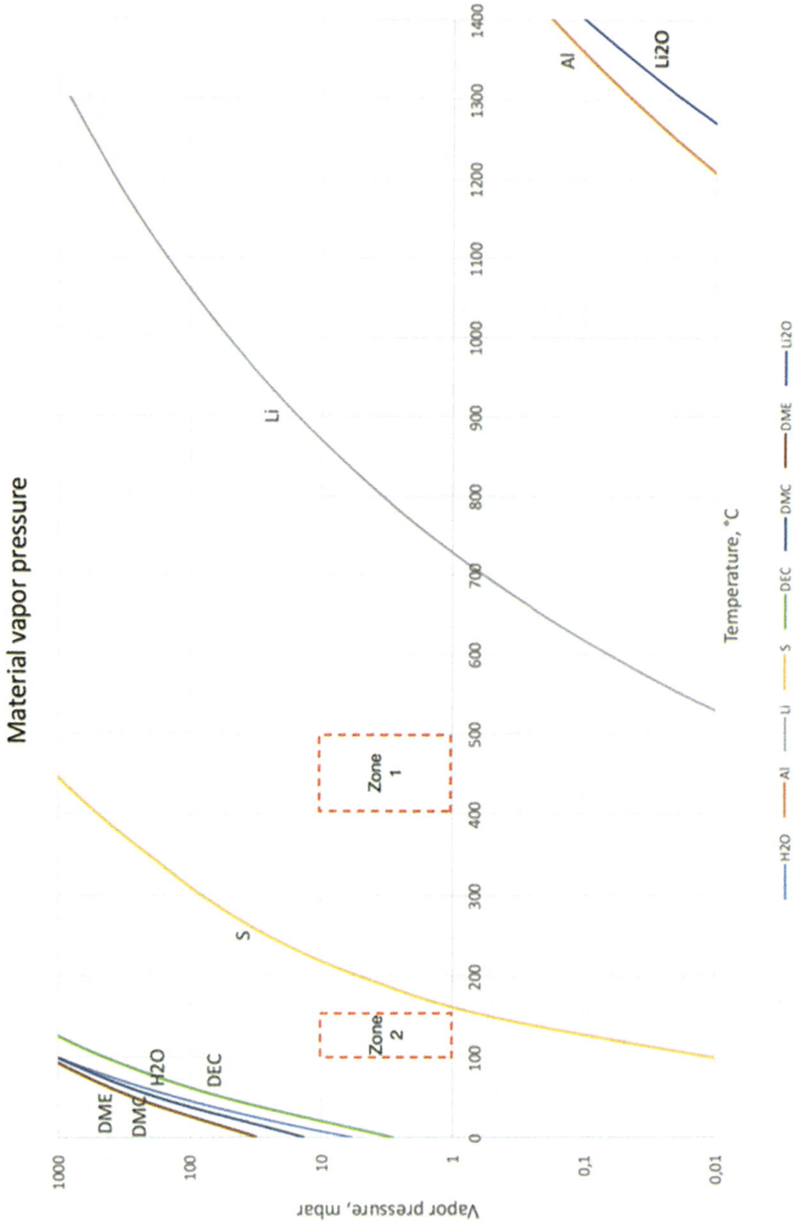

Fig. 12.5 Operating zones based on vapor pressure curves

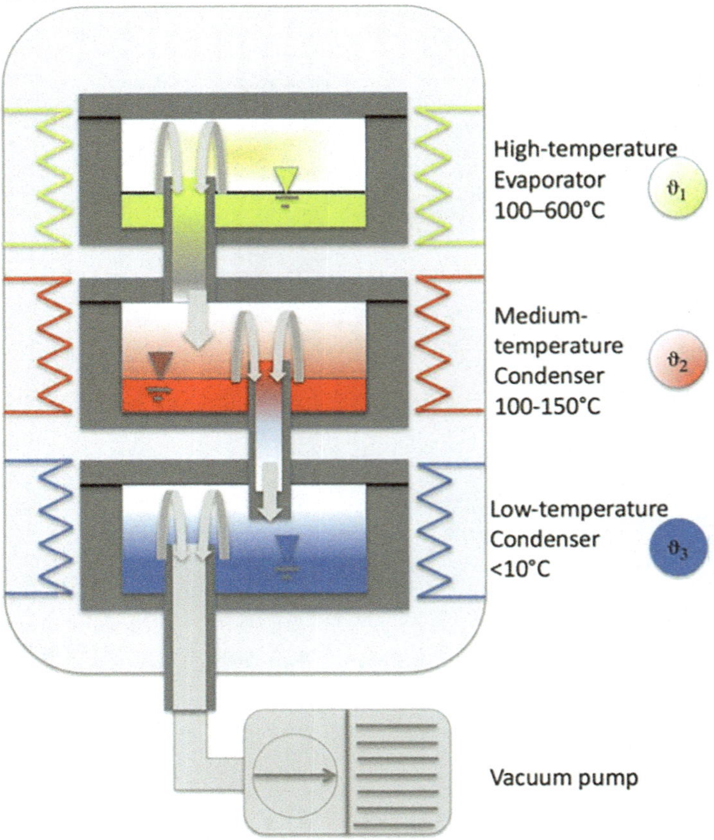

Fig. 12.6 Schematic diagram of vacuum distillation furnace

The formation of hazardous gases can be suppressed through introducing the suitable additives or catalysts. The off-gases, generated and released to the environment, can be sent for the adsorption on the surface of activated carbon.

Mechanical Separation The metal and electrode fractions are separated mechanically from deactivated batteries after its thermal treatment. The possible reactions of metal and electrode fractions with moisture or air are avoided through shredding the materials under vacuum conditions. Various mechanical separation approaches including magnetic separation and sieving are trialed for the recovery of steel and aluminum casings. The fine electrode powder, obtained from mechanical techniques, will be further sent for the hydro-treatment.

Hydro-Treatment The resulting fine electrode powder from the aforementioned techniques is consisted of lithium sulfide and carbon, which are sent further for the chemical analysis through analytical tools. The impurities in the form of iron or aluminum may also be present. The requirement of high energy and costly accessories by the pyro-metallurgy technique to recover lithium makes this process economical unviable due to high thermal constancy of lithium-based compounds. Hence, cost-effective hydro-treatment process will be applied for the efficient recovery of lithium and other value-added battery fractions in the form fine powder. A series of hydro-treatment-based subprocesses including filtration, precipitation, and solvent extraction are attempted for enhancing the recovery efficacy. The steel, aluminum, sulfur, lithium, and carbon-based fractions can be separated through employing the aforesaid approaches.

Hydrometallurgy It is a wet process, in which desired material fractions are dissolved in an aqueous solution, which are further separated and purified through employing various physical methods including filtration, solvent extraction, and precipitation. The resulting materials from vacuum distillation and mechanical separation are majorly composed of lithium oxide, sulfide, and carbon. The attachment of lithium sulfide with air or moisture might produce toxic gas using the below chemical equation.

$$Li_2S + 4H_2O \rightarrow 2Li\left[(H_2O)_4\right]^+ + S^{2-} \quad (12.2)$$

$$S^{2-} + H_2O \rightarrow HS^- + OH^- \quad (12.3)$$

$$HS^- + H^+ \rightarrow H_2S \quad (12.4)$$

Hence, the formation hazardous gas such as H_2S can be avoided through introducing the basic solution for dissolution of metal powder. The dissolution reaction will occur through the below given reaction scheme (12.5).

$$Li_2S + 2NaOH \rightarrow Na_2S + 2LiOH \quad (12.5)$$

$$Li_2S + H_2O \rightarrow 2LiOH \quad (12.6)$$

$$Al_2O_3 + 2NaOH + 3H_2O \rightarrow 2NaAl(OH)_4 \quad (12.7)$$

Carbon will not get mix into basic solution, which will impact a bad effect on the recycling process. However, it can be addressed through performing the filtration process for the separation of carbon, which can be reused in the battery production or other potential applications, depending upon the purity and concentration. The resulting solution, having lithium, can be further sent for the purification through performing the precipitation or solvent extraction (Siczek 2019).

Conclusion

In the current chapter, recycling techniques, employed for the recycling of LiSBs, have been explained. The insight features of multiple steps including sorting, discharging, vacuum distillation, mechanical separation, and hydro-treatment along with their challenges and issues for the recycling of lithium sulfur battery systems have been critically analyzed. The lower intrinsic importance of lithium sulfur battery systems on product level than lithium ions battery can be addressed through introducing the organic solvents for gaining the better recycling efficiency. The designed recycling methods can give a recycling efficiency of more than 50% for the decent recovery of LiSBs fractions. In addition to that, suitable acidic and alkaline solution can be very useful in enhancing the yield and purity of lithium sulfur battery fractions, which can be further reused for the battery manufacturing or other potential applications.

References

Al Rashdan M, Al Zubi M, Al Okour M (2019) Effect of using new technology vehicles on the World's environment and petroleum resources. J Ecol Eng 20:16

Armand M, Axmann P, Bresser D, Copley M, Edström K, Ekberg C, Guyomard D, Lestriez B, Novák P, Petranikova M (2020) Lithium-ion batteries–current state of the art and anticipated developments. J Power Sources 479:228708

Becker J, Beverungen D, Winter M, Menne SJS (2019) Dienstleistungen und Entscheidungsunterstützung. In: Umwidmung und Weiterverwendung von Traktionsbatterien. Vieweg, Wiesbaden

Brode G, Mark HJW (1982) In Kirk RE, Othmer DF (eds) Encyclopedia of chemical technology, 17. Wiley, New York, pp 411–413

Bruce PG, Hardwick LJ, Abraham KM (2011) Lithium-air and lithium-sulfur batteries. MRS Bull 36:506–512

Bureau de Recherches Géologiques et Minières, Deloitte Sustainability (2017) Study on the review of the list of critical raw materials. British Geological Survey

Chagnes A, Swiatowska J (2015) Lithium process chemistry: resources, extraction, batteries, and recycling. Elsevier, Amsterdam

Chen Y, Liu N, Hu F, Ye L, Xi Y, Yang S (2018) Thermal treatment and ammoniacal leaching for the recovery of valuable metals from spent lithium-ion batteries. Waste Manag 75:469–476

Chen Y, Liu N, Jie Y, Hu F, Li Y, Wilson BP, Xi Y, Lai Y, Yang S (2019) Toxicity identification and evolution mechanism of thermolysis-driven gas emissions from cathodes of spent lithium-ion batteries. ACS Sustain Chem Eng 7:18228–18235

Commission Regulation (EU) No 493/2012 of 11 June 2012 Laying down, Pursuant to Directive 2006/66/EC of the European Parliament and of the Council, Detailed Rules Regarding the Calculation of Recycling Efficiencies of the Recycling Processes of Waste Batteries and Accumulators, in

Directive 2006/66/EC of the European Parliament and of the Council of 6 September 2006 on Batteries and Accumulators and Waste Batteries and Accumulators and Repealing Directive 91/157/EEC in

Dörfler S, Althues H, Härtel P, Abendroth T, Schumm B, Kaskel S (2020) Challenges and key parameters of lithium-sulfur batteries on pouch cell level. Joule 4:539–554

European Commission (2015) Horizon, R.a.I. Programme, HELIS Project Website HELIS. High energy lithium sulphur cells and batteries. https://cordis.europa.eu/project/id/814471/reporting
European Commission (2019) Horizon 2020, R.a.I.P. 2019, LISA Project Website LISA. Lithium sulphur for safe road electrification. https://cordis.europa.eu/project/id/814471/reporting
Ferreira DA, Prados LMZ, Majuste D, Mansur MB (2009) Hydrometallurgical separation of aluminium, cobalt, copper and lithium from spent Li-ion batteries. J Power Sources 187:238–246
Frey M (2017) Neue Energiespeichermaterialien für Lithium-Schwefel-Batterien
Harper G, Sommerville R, Kendrick E, Driscoll L, Slater P, Stolkin R, Walton A, Christensen P, Heidrich O, Lambert S (2019) Recycling lithium-ion batteries from electric vehicles. Nature 575:75–86
He Y, Zhang T, Wang F, Zhang G, Zhang W, Wang J (2017) Recovery of LiCoO2 and graphite from spent lithium-ion batteries by Fenton reagent-assisted flotation. J Clean Prod 143:319–325
He Y, Qiao Y, Chang Z, Cao X, Jia M, He P, Zhou H (2019) Developing a "polysulfide-phobic" strategy to restrain shuttle effect in lithium–sulfur batteries. Angew Chem 131:11900–11904
Heimes HH, Kampker A, Lienemann C, Locke M, Offermanns C, Michaelis S, Rahimzei E (2018), Produktionsprozess einer Lithium-Ionen-Batteriezelle/Autoren Dr.-Ing. Dipl.-Wirt.-Ing. Heiner Hans Heimes, Prof. Dr.-Ing. Achim Kampker, Christoph Lienemann, Marc Locke, Christian Offermanns [und weitere], PEM der RWTH Aachen
Helbig C, Bradshaw AM, Wietschel L, Thorenz A, Tuma A (2018) Supply risks associated with lithium-ion battery materials. J Clean Prod 172:274–286
Kawamoto R, Mochizuki H, Moriguchi Y, Nakano T, Motohashi M, Sakai Y, Inaba A (2019) Estimation of CO2 emissions of internal combustion engine vehicle and battery electric vehicle using LCA. Sustainability 11:2690
Kittner N, Tsiropoulos I, Tarvydas D, Schmidt O, Staffell I, Kammen DM (2020) Electric vehicles. In: Technological learning in the transition to a low-carbon energy system. Elsevier, London, pp 145–163
Korthauer R (2013) Handbuch lithium-ionen-batterien. Springer, Berlin
Kurzweil P, Dietlmeier OK (2018) Elektrochemische Speicher: Superkondensatoren, Batterien, Elektrolyse-Wasserstoff, Rechtliche Rahmenbedingungen. Springer-Verlag, Wiesbaden
Kwade A, Diekmann J (2018) Springer International Publishing AG. Recycling of lithium-ion batteries, Cham
Li T, Bai X, Gulzar U, Bai YJ, Capiglia C, Deng W, Zhou X, Liu Z, Feng Z, Proietti Zaccaria R (2019) A comprehensive understanding of lithium–sulfur battery technology. Adv Funct Mater 29:1901730
Liu J, Chen H, Chen W, Zhang Y, Zheng YJC (2019) New insight into the "shuttle mechanism" of rechargeable lithium-sulfur batteries. ChemElectroChem 6:2782–2787
Lombardo G, Ebin B, Foreman MRSJ, Steenari B-M, Petranikova M (2020) Incineration of EV Lithium-ion batteries as a pretreatment for recycling–determination of the potential formation of hazardous by-products and effects on metal compounds. J Hazard Mater 393:122372
Martens H, Goldmann D (2016) Recyclingtechnik: Fachbuch für Lehre und Praxis. Springer, Wiesbaden
Or T, Gourley SW, Kaliyappan K, Yu A, Chen Z (2020) Recycling of mixed cathode lithium-ion batteries for electric vehicles: current status and future outlook. Carbon Energy 2:6–43
Pinegar H, Smith YR (2019) Recycling of end-of-life lithium ion batteries, part I: commercial processes. J Sustain Metallurgy 5:402–416
Placke T, Kloepsch R, Dühnen S, Winter M (2017) Lithium ion, lithium metal, and alternative rechargeable battery technologies: the odyssey for high energy density. J Solid State Electrochem 21:1939–1964
Regett A, Fischhaber S (2017) Wien, Austria, Reduction of critical resources consumption through second life applications of lithium ion trection batteries, 15–17
Schwich L, Sabarny P, Friedrich BJM (2020) Recycling potential of lithium–sulfur batteries—a first concept using thermal and hydrometallurgical methods 10:1513

Siczek K (2019) Li-S battery recycling. Research in Logistics & Production 9:125–135

Thackeray MM, Wolverton C, Isaacs ED (2012) Electrical energy storage for transportation—approaching the limits of, and going beyond, lithium-ion batteries. Energy Environ Sci 5:7854–7863

Velázquez-Martínez O, Valio J, Santasalo-Aarnio A, Reuter M, Serna-Guerrero RJB (2019) A critical review of lithium-ion battery recycling processes from a circular economy perspective. Batteries 5:68

Werner D, Peuker UA, Mütze T (2020) Recycling chain for spent lithium-ion batteries. Metals 10:316

Xu J, Thomas HR, Francis RW, Lum KR, Wang J, Liang B (2008) A review of processes and technologies for the recycling of lithium-ion secondary batteries. J Power Sources 177:512–527

Yamaki J-I, Tobishima S-I, Sakurai Y, Saito K-I, Hayashi K (1998) Safety evaluation of rechargeable cells with lithium metal anodes and amorphous V2O5 cathodes. J Appl Electrochem 28:135–140

Yan J (2015) Handbook of clean energy systems, 6 volume set. Wiley

Yan J, Liu X, Li B (2016) Capacity fade analysis of sulfur cathodes in lithium–sulfur batteries. Adv Sci 3:1600101

Zhang Z, Wang J-N, Shao A, Xiong D-G, Liu J-W, Lao C-Y, Xi K, Lu S-Y, Jiang Q, Yu J (2020) Recyclable cobalt-molybdenum bimetallic carbide modified separator boosts the polysulfide adsorption-catalysis of lithium sulfur battery. Sci China Mater 63:2443–2455

Zheng X, Zhu Z, Lin X, Zhang Y, He Y, Cao H, Sun Z (2018) A mini-review on metal recycling from spent lithium ion batteries. Engineering 4:361–370

Zou H, Gratz E, Apelian D, Wang Y (2013) A novel method to recycle mixed cathode materials for lithium ion batteries. Green Chem 15:1183–1191

Chapter 13
Challenges and Future Perspectives of Li–S Batteries

Iqra Khalid, Muhammad Sagir, and M. B. Tahir

Abstract The growing acceptance of electric vehicles across the globe is the main factor driving the trend toward mobile electrification, which has resulted in an unparalleled increase in the demand for lithium-ion batteries. Consequently, it is now more important than ever to recycle batteries. The lithium-sulfur (Li-S) battery is being explored as a potential replacement for the present lithium-ion battery. It releases energy by combining lithium metal with highly plentiful sulfur. The Li-S battery can obtain a high theoretical specific capacity of 1675 mAh/g and specific energy of 2600 Wh/kg due to the sulfur cathode's lightweight and multi-electron reaction. The "shuttle effect" of lithium polysulfides, significant volume changes, weak conductivity of sulfur and its solid-state derivatives, and the self-discharge phenomenon, on the other hand, restrict its practical utilization. The "shuttle effect" is acknowledged as the most significant issue influencing electrochemical performance out of all of these. Some of these techniques have shown early promise, but more work is required to reach a high enough performance level so that lithium metal may be employed in commercial Li-S cells. In this chapter, we examine the advantages and disadvantages of numerous suggested approaches to these problems, keeping in mind the constant objective of outperforming traditional Li-ion energy densities.

Keywords Shuttle effect · Li-S batteries · Polysulfides · Conductivity · Sulfur cathodes

---

I. Khalid · M. B. Tahir (✉)
Institute of Physics, Khwaja Fareed University of Engineering and Information Technology, Rahim Yar Khan, Punjab, Pakistan
e-mail: m.bilaltahir@kfueit.edu.pk

M. Sagir
Institute of Chemical and Environmental Engineering, Khwaja Fareed University of Engineering and Information Technology, Rahim Yar Khan, Punjab, Pakistan
e-mail: dr.msagir@kfueit.edu.pk

M. S. Tahir et al. (eds.), *Lithium-Sulfur Batteries: Key Parameters, Recent Advances, Challenges and Applications*, Springer Tracts in Electrical and Electronics Engineering, https://doi.org/10.1007/978-981-99-2796-8_13

Lithium–sulfur (Li–S) cells are promising electronic systems thanks to their exceptional energy capacity, dynamic discharge capacity, cost-effectiveness, and ecofriendly nature (because the cathode's primary ingredient is sulfur) (Rana et al. 2019). Lithium-ion (Li^+ ion) batteries, which have the maximum power capacity among the affordable batteries, are being studied for use as rechargeable batteries in automobiles (Etacheri et al. 2011; Goodenough 2013). Conventional Li^+ ion batteries, however, encounter difficulties in trying to meet the ever-growing demands for electrical energy storage because of their finite capacity and the large price of the transition-metal oxide cathodes now in use (Yu and Manthiram 2017). Li–S batteries, on the other hand, use lithium and sulfur as the positive terminal and negative terminals and have large energy capacities (~2600 W·h/kg^{-1}), so they remain among the most highly valued small batteries (Li et al. 2021).

Still, even lithium–sulfur batteries face numerous difficulties that hamper their advertised effectiveness. For example, Li–S batteries are safety hazards, mainly when they come into contact with open flames, causing explosions. This remains a limitation of using Li–S batteries. This study aims to resolve such safety concerns by investigating what other scientists have learned about the safety of Li–S batteries. The underlying problem is the commercial membrane, which is made of combustible polyolefin and liquid electrolytes that burn rapidly (Peng et al. 2017). In applications, another difficult that Li–S batteries face is that they can slow down the execution of the machinery they power (Manthiram et al. 2015). First, the core problem here is the electronic and ionic insultation of the sulfur and lithium, which requires a large amount of lead in the cathode to reduce the amount of active substances in the reaction (Kang et al. 2016; Lin et al. 2017). Second, the sulfur needs to be removed, and the average amount of sulfur in Li_2S (Li_2S_8, Li_2S_6, and Li_2S_4) is removable with the application of carbon-based electrolytes, resulting in a redox shuttle, which equally reduces the number of anodes and cathodes (Zhang et al. 2018a).

Another problem is the type of the elongated sequence of polysulfides formed at the positive terminal throughout the series of battery processes (Park et al. 2017). The polysulfide class transfers from the sulfur-positive terminal to the lithium-negative terminal, producing knock-on reactions. Therefore, polysulfide shuttle occurrences reduce the coulombic proficiency and the level of charging, and any catalyst causes nonstop damage to the energetic cathode substance (Bucur et al. 2017). Following that, under favorable conditions, the reduced rate of presentations and the extensive polarization of another ejection plateau produce extra kinetic blocks in the sulfur oxidation-reduction reactions, which is where intense lithium polysulfide (LiPSs) is responsible for small ionic conduction (Zhao et al. 2020). The early precipitation of LiPSs taking place in the conductive background blocks the ionic transference path and the intensity of kinetically slow solid–solid alterations from having an effect. Previous studies have shown that the foregoing causes another discharge plateau, which develops enough solid precipitation to reach the voltage cut-off, reducing the battery's capacity (Fan et al. 2015; Fan and Chiang 2017). Battery problems result from cathode damage, which rapidly reduces the reactions in sack cells. These are the most common consequences of the following three features: (i) Advanced definite current concentration and cycling measurements are practically arranged in the Li metal anode, which consumed large loads of sulfur cathodes, increasing the lithium deposition and then anode transformation, and (ii) the excess LiPSs from improper LiPS absorption and an additional simple shuttle cause lithium erosion (Cheng et al. 2017; Kong et al. 2019).

A single substance, however, can remedy this problem: graphene, as a single sheet of sp^2 merged with carbon atoms (Novoselov et al. 2005). Until now, graphene has been largely dismissed in several fields even though it has shown exceptional activities (Song et al. 2017; Wang et al. 2017). Graphene and graphene oxide (GO) have specific characteristics that are highly useful for lithium–sulfur batteries. For example, graphene is useful thanks to its excellent conductivity, which facilitates the transfer of electrons when isolating lithium–sulfur materials (Balandin et al. 2008). Graphene's exceptional mechanical strength and elasticity protect against high-volume variation throughout the charge/discharge cycle and simplify the manufacture of flexible devices (Wang et al. 2018a). Its large surface area (2630 m^2/g) provides excellent permeability for sulfur oxidation-reduction reactions. GO layers are frequently involved in various functional groups that tend to chemically relate to polysulfides, conferring a powerful absorbent capacity (Haubner et al. 2010; Zhu et al. 2010).

Several battery features need to be studied—and are estimated to lead to considerable improvements over the upcoming year—to resolve the abovementioned new problems of using Li–S batteries.

1. **Next-generation sulfur cathodes**: Using complex carbon and sulfur cathodes has been an established strategy that enables batteries to increase the energy capacity of small Li–S batteries (Agostini et al. 2018; Chung et al. 2016; Fang et al. 2016). To control the significant sulfur loading and electrolytes, the execute carbon matrix is equipped with additional ways to handle new tasks.
2. **Kinetic promoters**: The sulfur oxidation-reduction reactions that happen at various segment joints and cause serious movements have always been disadvantaged by low oxidation-reduction kinetics. Numerous kinetic enhancements, including oxide (Li et al. 2013), sulfide (Yuan et al. 2016), nitride (Qi et al. 2018), polymers (Frischmann et al. 2016), and structural materials (Wang et al. 2016), have been shown to improve various battery features—according to rate performance assessments—such as extending battery life (Li et al. 2019a, b). Kinetic parameters can normally be separated into mixed electrocatalysts and regular redox intermediaries, both of which have been similarly examined for their performance values.
3. **Designing a specific ion–solvent complex for electrolytes**: Tetrahydrofuran and dimethyl sulfoxides dissolve in the LiPSs of Li–S batteries, but ionic liquid and fluorinated ether do not. The currently used dioxolane (DOL) and dimethoxyethane (DME) ether-based electrolytes, however, provide reasonable LiPS dissolvability (Park et al. 2013; Weller et al. 2019). The Li–S chemistry varies depending on the specific Li–S conversion pathway, which is determined by each electrolyte formula. With a specific number of dissolved LiPSs, the

currently used ether-based electrolyte enables the dissolution–precipitation pathway to carry out a uniform sulfur redox (Wang et al. 2018b). Furthermore, growing evidence shows that the existing ether-based electrolyte cannot adequately supply long-cycling Li–S batteries given their current capabilities and limitations (Zhang et al. 2015, 2018b).

For applied batteries, maintaining a large electrochemical balance is an additional condition of the efficacy of such an electrolyte. A significant change in the Li ion could lead to a lower voltage demand during the phase of the electrochemical balance window, which is approximately 2.5 V for an improved demand vs. 4 V for the current demand (Bresser et al. 2013). The development in an Li–S electrolyte is critical for resolving the new problems of using commercial Li–S batteries, and such a development could introduce or inspire new directions for future research.

References

Agostini M, Hwang JY, Kim HM, Bruni P, Brutti S, Croce F, . . . Sun YK (2018) Minimizing the electrolyte volume in Li–S batteries: a step forward to high gravimetric energy density. Adv Energy Mater 8(26):1801560

Balandin AA, Ghosh S, Bao W, Calizo I, Teweldebrhan D, Miao F, Lau CN (2008) Superior thermal conductivity of single-layer graphene. Nano Lett 8(3):902–907. https://doi.org/10.1021/nl0731872

Bresser D, Passerini S, Scrosati B (2013) Recent progress and remaining challenges in sulfur-based lithium secondary batteries–a review. Chem Commun 49(90):10545–10562

Bucur CB, Jones M, Kopylov M, Spear J, Muldoon J (2017) Inorganic–organic layer by layer hybrid membranes for lithium–sulfur batteries. Energy Environ Sci 10(4):905–911

Cheng X-B, Yan C, Huang J-Q, Li P., Zhu L, Zhao L, . . . Zhang Q (2017) The gap between long lifespan Li–S coin and pouch cells: the importance of lithium metal anode protection. Energy Storage Mater 6:18–25

Chung S-H, Chang C-H, Manthiram A (2016) A carbon-cotton cathode with ultrahigh-loading capability for statically and dynamically stable lithium–sulfur batteries. ACS Nano 10(11): 10462–10470

Etacheri V, Marom R, Elazari R, Salitra G, Aurbach D (2011) Challenges in the development of advanced Li-ion batteries: a review. Energy Environ Sci 4(9):3243–3262

Fan FY, Chiang Y-M (2017) Electrodeposition kinetics in Li–S batteries: effects of low electrolyte/sulfur ratios and deposition surface composition. J Electrochem Soc 164(4):A917

Fan FY, Carter WC, Chiang YM (2015) Mechanism and kinetics of Li2S precipitation in lithium–sulfur batteries. Adv Mater 27(35):5203–5209

Fan FY, Pan MS, Lau KC, Assary RS, Woodford WH, Curtiss LA, . . . Chiang Y-M (2016) Solvent effects on polysulfide redox kinetics and ionic conductivity in lithium-sulfur batteries. J Electrochem Soc 163(14):A3111

Fang R, Zhao S, Hou P, Cheng M, Wang S, Cheng HM, . . . Li F (2016) 3D interconnected electrode materials with ultrahigh areal sulfur loading for Li–S batteries. Adv Mater 28(17):3374–3382

Frischmann PD, Hwa Y, Cairns EJ, Helms BA (2016) Redox-active supramolecular polymer binders for lithium–sulfur batteries that adapt their transport properties in operando. Chem Mater 28(20):7414–7421

Goodenough JB (2013) Evolution of strategies for modern rechargeable batteries. Acc Chem Res 46(5):1053–1061

Haubner K, Murawski J, Olk P, Eng LM, Ziegler C, Adolphi B, Jaehne E (2010) The route to functional graphene oxide. ChemPhysChem 11(10):2131–2139

Kang W, Deng N, Ju J, Li Q, Wu D, Ma X, . . . Cheng B (2016) A review of recent developments in rechargeable lithium–sulfur batteries. Nanoscale 8(37):16541–16588

Kong L, Jin Q, Huang JQ, Zhao LD, Li P, Li BQ, . . . Zhang Q (2019) Nonuniform redistribution of sulfur and lithium upon cycling: probing the origin of capacity fading in lithium–sulfur pouch cells. Energ Technol 7(12):1900111

Li W, Cha J, Zheng G, Yang Y, McDowell M, Hsu P, Cui Y (2013) Sulphur-TiO2 yolk-shell nanoarchitecture with internal void space for long-cycle lithium-sulphur batteries. Nat Commun 4:1331–1331

Li BQ, Kong L, Zhao CX, Jin Q, Chen X, Peng HJ, . . . Zhang Q (2019a) Expediting redox kinetics of sulfur species by atomic-scale electrocatalysts in lithium–sulfur batteries. InfoMat 1(4): 533–541

Li B-Q, Peng H-J, Chen X, Zhang S-Y, Xie J, Zhao C-X, Zhang Q (2019b) Polysulfide electrocatalysis on framework porphyrin in high-capacity and high-stable lithium–sulfur batteries. CCS Chem 1(1):128–137

Li Y, Shapter JG, Cheng H, Xu G, Gao G (2021) Recent progress in sulfur cathodes for application to lithium–sulfur batteries. Particuology 58:1–15

Lin D, Liu Y, Cui Y (2017) Reviving the lithium metal anode for high-energy batteries. Nat Nanotechnol 12(3):194–206

Manthiram A, Chung SH, Zu C (2015) Lithium–sulfur batteries: progress and prospects. Adv Mater 27(12):1980–2006

Novoselov KS, Jiang D, Schedin F, Booth T, Khotkevich V, Morozov S, Geim AK (2005) Two-dimensional atomic crystals. Proc Natl Acad Sci 102(30):10451–10453

Park J-W, Ueno K, Tachikawa N, Dokko K, Watanabe M (2013) Ionic liquid electrolytes for lithium–sulfur batteries. J Phys Chem C 117(40):20531–20541

Park H, Lim H-D, Lim H-K, Seong WM, Moon S, Ko Y, . . . Kang K (2017) High-efficiency and high-power rechargeable lithium–sulfur dioxide batteries exploiting conventional carbonate-based electrolytes. Nat Commun 8(1):1–10

Peng H-J, Huang J-Q, Zhang Q (2017) A review of flexible lithium–sulfur and analogous alkali metal–chalcogen rechargeable batteries. Chem Soc Rev 46(17):5237–5288

Qi B, Zhao X, Wang S, Chen K, Wei Y, Chen G, . . . Li F (2018) Mesoporous TiN microspheres as an efficient polysulfide barrier for lithium–sulfur batteries. J Mater Chem A 6(29):14359–14366

Rana M, Li M, Huang X, Luo B, Gentle I, Knibbe R (2019) Recent advances in separators to mitigate technical challenges associated with re-chargeable lithium sulfur batteries. J Mater Chem A 7(12):6596–6615

Song B, Wang T, Sun H, Shao Q, Zhao J, Song K, . . . Guo Z (2017) Two-step hydrothermally synthesized carbon nanodots/WO 3 photocatalysts with enhanced photocatalytic performance. Dalton Trans 46(45):15769–15777

Wang H, Zhang W, Liu H, Guo Z (2016) A strategy for configuration of an integrated flexible sulfur cathode for high-performance lithium–sulfur batteries. Angew Chem Int Ed 55(12):3992–3996

Wang C, Zhao M, Li J, Yu J, Sun S, Ge S, . . . Wujcik EK (2017) Silver nanoparticles/graphene oxide decorated carbon fiber synergistic reinforcement in epoxy-based composites. Polymer 131:263–271

Wang X, Liu X, Yuan H, Liu H, Liu C, Li T, . . . Guo Z (2018a) Non-covalently functionalized graphene strengthened poly (vinyl alcohol). Mater Des 139:372–379

Wang L, Ye Y, Chen N, Huang Y, Li L, Wu F, Chen R (2018b) Development and challenges of functional electrolytes for high-performance lithium–sulfur batteries. Adv Funct Mater 28(38): 1800919

Weller C, Pampel J, Dörfler S, Althues H, Kaskel S (2019) Polysulfide shuttle suppression by electrolytes with low-density for high-energy lithium–sulfur batteries. Energ Technol 7(12): 1900625

Yu X, Manthiram A (2017) Electrode–electrolyte interfaces in lithium–sulfur batteries with liquid or inorganic solid electrolytes. Acc Chem Res 50(11):2653–2660

Yuan Z, Peng H-J, Hou T-Z, Huang J-Q, Chen C-M, Wang D-W, . . . Zhang, Q (2016) Powering lithium–sulfur battery performance by propelling polysulfide redox at sulfiphilic hosts. Nano Lett 16(1):519–527

Zhang S, Ueno K, Dokko K, Watanabe M (2015) Recent advances in electrolytes for lithium–sulfur batteries. Adv Energy Mater 5(16):1500117

Zhang Y, Gao Z, Song N, He J, Li X (2018a) Graphene and its derivatives in lithium–sulfur batteries. Mater Today Energy 9:319–335

Zhang G, Peng HJ, Zhao CZ, Chen X, Zhao LD, Li P, . . . Zhang Q (2018b) The radical pathway based on a lithium-metal-compatible high-dielectric electrolyte for lithium–sulfur batteries. Angew Chem 130(51):16974–16978

Zhao M, Li B-Q, Zhang X-Q, Huang J-Q, Zhang Q (2020) A perspective toward practical lithium–sulfur batteries. ACS Cent Sci 6(7):1095–1104

Zhu Y, Murali S, Cai W, Li X, Suk JW, Potts JR, Ruoff RS (2010) Graphene and graphene oxide: synthesis, properties, and applications. Adv Mater 22(35):3906–3924

SPRINGER NATURE

GPSR Compliance

The European Union's (EU) General Product Safety Regulation (GPSR) is a set of rules that requires consumer products to be safe and our obligations to ensure this.

If you have any concerns about our products, you can contact us on ProductSafety@springernature.com

In case Publisher is established outside the EU, the EU authorized representative is:

Springer Nature Customer Service Center GmbH
Europaplatz 3
69115 Heidelberg, Germany

The manufacturer's authorised representative in the EU is Springer Nature Customer Service Centre GmbH, Europaplatz 3, 69115 Heidelberg, Germany. If you have any concerns regarding our products, please contact ProductSafety@springernature.com

Printed and bound by CPI Group (UK) Ltd, Croydon, CR0 4YY
30/09/2026
02225318-0004